Two Manuals Combined into One Book

Military First Aid

and

Marine Operations on Urbanized Terrain

Department of Defense

Pentagon
Publishing

*FIELD MANUAL
NO. 4-25.11
NAVY TACTICAL
REFERENCE
PUBLICATION
NO. 4-02.1
AIR FORCE MANUAL
NO. 44-163(I)

HEADQUARTERS
DEPARTMENT OF THE ARMY,
THE NAVY, AND THE AIR FORCE
Washington, DC, 23 December 2002

FIRST AID

TABLE OF CONTENTS

*This publication supersedes FM 21-11, 27 October 1988

PREFACE

This manual meets the first aid training needs of individual service members. Because medical personnel will not always be readily available, the nonmedical service members must rely heavily on their own skills and knowledge of life-sustaining methods to survive on the integrated battlefield. This publication outlines both self-aid and aid to other service members (buddy aid). More importantly, it emphasizes prompt and effective action in sustaining life and preventing or minimizing further suffering and disability. First aid is the emergency care given to the sick, injured, or wounded before being treated by medical personnel. The term *first aid* can be defined as "urgent and immediate lifesaving and other measures, which can be performed for casualties by nonmedical personnel when medical personnel are not immediately available." Nonmedical service members have received basic first aid training and should remain skilled in the correct procedures for giving first aid. This manual is directed to *all* service members. The procedures discussed apply to all types of casualties and the measures described are for use by both male and female service members.

This publication is in consonance with the following North Atlantic Treaty Organization (NATO) International Standardization Agreements (STANAGs) and American, British. Canadian, and Australian Quadripartite Standardization Agreements (QSTAGs).

TITLE	STANAG	QSTAG
Medical Training in First Aid, Basic Hygiene and Emergency Care	2122	535
First Aid Kits and Emergency Medical Care Kits	2126	
Medical First Aid and Hygiene Training in NBC Operations	2358	
First Aid Material for Chemical Injuries	2871	

These agreements are available on request, using Department of Defense (DD) Form 1425 from the Standardization Documents Order Desk, 700 Robins Avenue, Building 4, Section D, Philadelphia, Pennsylvania 19111-5094.

Unless this publication states otherwise, masculine nouns and pronouns do not refer exclusively to men.

Use of trade or brand names in this publication is for illustrative purposes only and does not imply endorsement by the Department of Defense (DOD).

The proponent for this publication is the US Army Medical Department Center and School. Submit comments and recommendations for the improvement of this publication directly to the **Commander, US Army Medical Department Center and School, ATTN: MCCS-FCD-L, 1400 East Grayson Street, Fort Sam Houston, Texas 78234-5052**.

CHAPTER 1

FUNDAMENTAL CRITERIA FOR FIRST AID

"The fate of the wounded rests in the hands of the ones who apply the first dressing."

Nicholas Senn (1898) (49th President of the American Medical Association)

1-1. General

When a nonmedical service member comes upon an unconscious or injured service member, he must accurately evaluate the casualty to determine the first aid measures needed to prevent further injury or death. He should seek medical assistance as soon as possible, but he should not interrupt the performance of first aid measures. To interrupt the first aid measures may cause more harm than good to the casualty. Remember that in a chemical environment, the service member should not evaluate the casualty until the casualty has been masked. After performing first aid, the service member must proceed with the evaluation and continue to monitor the casualty for development of conditions which may require the performance of necessary basic lifesaving measures, such as clearing the airway, rescue breathing, preventing shock, and controlling bleeding. He should continue to monitor the casualty until relieved by medical personnel.

Service members may have to depend upon their first aid knowledge and skills to save themselves (self-aid) or other service members (buddy aid/ combat lifesaver). They may be able to save a life, prevent permanent disability, or reduce long periods of hospitalization by knowing **WHAT** to do, **WHAT NOT** to do, and **WHEN** to seek medical assistance.

NOTE

The prevalence of various body armor systems currently fielded to US service members, and those in development for future fielding, may present a temporary obstacle to effective evaluation of an injured service member. You may have to *carefully remove* the body armor from the injured service member to complete the evaluation or administer first aid. Begin by removing the outer-most hard or soft body armor components (open, unfasten or cut the closures, fasteners, or straps), then remove any successive layers in the same manner. Be sure to follow other notes, cautions and warnings regarding procedures in contaminated situations and when a broken back or neck is suspected. Continue to evaluate.

1-2. Terminology

To enhance the understanding of the material contained in this publication, the following terms are used—

- *Combat lifesaver.* This is a US Army program governed by Army Regulation (AR) 350-41. The combat lifesaver is a member of a nonmedical unit selected by the unit commander for additional training beyond basic first aid procedures (referred to as *enhanced first aid*). A minimum of one individual per squad, crew, team, or equivalent-sized unit should be trained. The primary duty of this individual does not change. The additional duty of combat lifesaver is to provide enhanced first aid for injuries based on his training before the trauma specialist (military occupational specialty [MOS] 91W) arrives. The combat lifesaver's training is normally provided by medical personnel assigned, attached, or in direct support (DS) of the unit. The senior medical person designated by the commander manages the training program.

- *Trauma Specialist (US Army) or Hospital Corpsman (HM).* A medical specialist trained in emergency medical treatment (EMT) procedures and assigned or attached in support of a combat or combat support unit or marine forces.

- *Casualty evacuation.* Casualty evacuation (CASEVAC) is a term used by nonmedical units to refer to the movement of casualties aboard nonmedical vehicles or aircraft. See also the term *transported* below. Refer to FM 8-10-6 for additional information.

CAUTION

Casualties transported in this manner do not receive en route medical care.

- *Enhanced first aid (US Army).* Enhanced first aid is administered by the combat lifesaver. It includes measures, which require an additional level of training above self-aid and buddy aid, such as the initiation of intravenous (IV) fluids.

- *Medical* evacuation. Medical evacuation is the timely, efficient movement of the wounded, injured, or ill service members from the battlefield and other locations to medical treatment facilities (MTFs). Medical personnel provide en route medical care during the evacuation. Once the casualty has entered the medical stream (trauma specialist, hospital corpsman, evacuation crew, or MTF), the role of first aid in the care of the casualty ceases and the casualty becomes the responsibility of the health service support (HSS) chain. Once he has entered the HSS chain he is referred to as a *patient*.

- *First aid measures*. Urgent and immediate lifesaving and other measures, which can be performed for casualties (or performed by the casualty himself) by nonmedical personnel when medical personnel are not immediately available.

- *Medical treatment*. Medical treatment is the care and management of wounded, injured, or ill service members by medically trained (MOS-trained) HM, and area of concentration (AOC) personnel. It may include EMT, advanced trauma management (ATM), and resuscitative and surgical intervention.

- *Medical treatment facility*. Any facility established for the purpose of providing medical treatment. This includes battalion aid stations, Level II facilities, dispensaries, clinics, and hospitals.

- *Self-aid/buddy aid*. Each individual service member is trained to be proficient in a variety of specific first aid procedures. This training enables the service member or a buddy to apply immediate first aid measures to alleviate a life-threatening situation.

- *Transported*. A casualty is moved to an MTF in a nonmedical vehicle without en route care provided by a medically-trained service member (such as a Trauma Specialist or HM). First aid measures should be continually performed while the casualty is being transported. If the casualty is acquired by a dedicated medical vehicle with a medically-trained crew, the role of first aid ceases and the casualty becomes the responsibility of the HSS chain, and is then referred to as a *patient*. This method of transporting a casualty is also referred to as *CASEVAC*.

1-3. Understanding Vital Body Functions for First Aid

In order for the service member to learn to perform first aid procedures, he must have a basic understanding of what the vital body functions are and what the result will be if they are damaged or not functioning.

a. Breathing Process. All humans must have oxygen to live. Through the breathing process, the lungs draw oxygen from the air and put it into the blood. The heart pumps the blood through the body to be used by the cells that require a constant supply of oxygen. Some cells are more dependent on a constant supply of oxygen than others. For example, cells of the brain may die within 4 to 6 minutes without oxygen. Once these cells die, they are lost forever since they do not regenerate. This could result in permanent brain damage, paralysis, or death.

b. Respiration. Respiration occurs when a person inhales (oxygen is taken into the body) and then exhales (carbon dioxide [CO_2] is expelled from the body). Respiration involves the—

- *Airway.* The airway consists of the nose, mouth, throat, voice box, and windpipe. It is the canal through which air passes to and from the lungs.

- *Lungs.* The lungs are two elastic organs made up of thousands of tiny air spaces and covered by an airtight membrane. The *bronchial tree* is a part of the lungs.

- *Rib cage.* The rib cage is formed by the muscle-connected ribs, which join the spine in back, and the breastbone in front. The top part of the rib cage is closed by the structure of the neck, and the bottom part is separated from the abdominal cavity by a large dome-shaped muscle called the *diaphragm* (Figure 1-1). The diaphragm and rib muscles, which are under the control of the respiratory center in the brain, automatically *contract* and *relax*. *Contraction* increases and *relaxation* decreases the size of the rib cage. When the rib cage increases and then decreases, the air pressure in the lungs is first less and then more than the atmospheric pressure, thus causing the air to rush into and out of the lungs to equalize the pressure. This cycle of inhaling and exhaling is repeated about 12 to 18 times per minute.

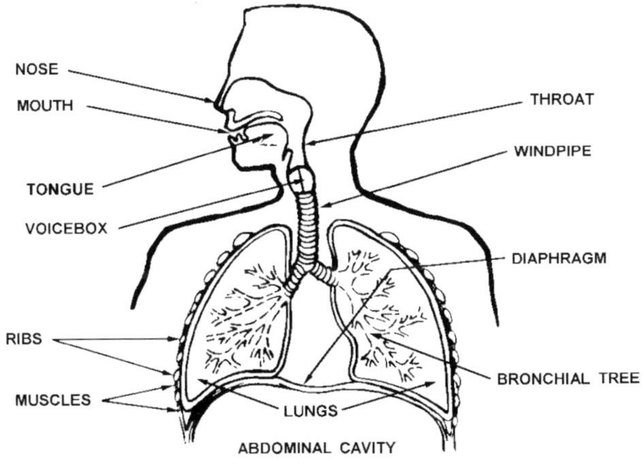

Figure 1-1. Airway, lungs, and rib cage.

c. *Blood Circulation.* The heart and the blood vessels (arteries, veins, and capillaries) circulate blood through the body tissues. The heart is divided into two separate halves, each acting as a pump. The left side pumps oxygenated blood (bright red) through the arteries into the capillaries; nutrients and oxygen pass from the blood through the walls of the capillaries into the cells. At the same time waste products and CO_2 enter the capillaries. From the capillaries the oxygen poor blood is carried through the veins to the right side of the heart and then into the lungs where it expels the CO_2 and picks up oxygen. Blood in the veins is dark red because of its low oxygen content. Blood does not flow through the veins in spurts as it does through the arteries. The entire system of the heart, blood vessels, and lymphatics is called the *circulatory system.*

(1) *Heartbeat.* The heart functions as a pump to circulate the blood continuously through the blood vessels to all parts of the body. It contracts, forcing the blood from its chambers; then it relaxes, permitting its chambers to refill with blood. The rhythmical cycle of contraction and relaxation is called the *heartbeat.* The normal heartbeat is from 60 to 80 beats per minute.

(2) *Pulse.* The heartbeat causes a rhythmical expansion and contraction of the arteries as it forces blood through them. This cycle of expansion and contraction can be felt (monitored) at various points in the body and is called the *pulse.* The common points for checking the pulse are at the—

- Side of the neck (*carotid*).
- Groin (*femoral*).
- Wrist (*radial*).
- Ankle (*posterior tibial*).

(*a*) *Carotid pulse.* To check the carotid pulse, feel for a pulse on the side of the casualty's neck closest to you. This is done by placing the tips of your first two fingers beside his Adam's apple (Figure 1-2).

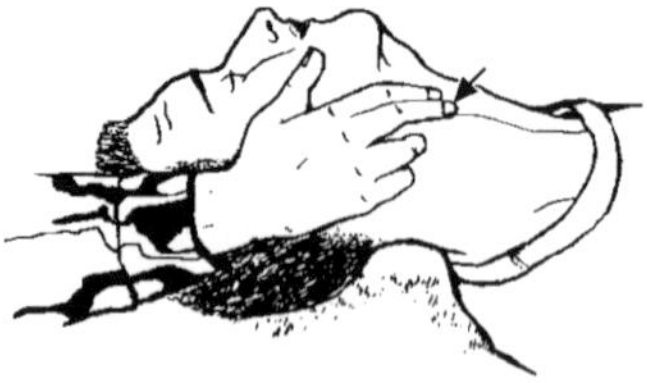

Figure 1-2. Carotid pulse.

(*b*) *Femoral pulse.* To check the femoral pulse, press the tips of your first two fingers into the middle of the groin (Figure 1-3).

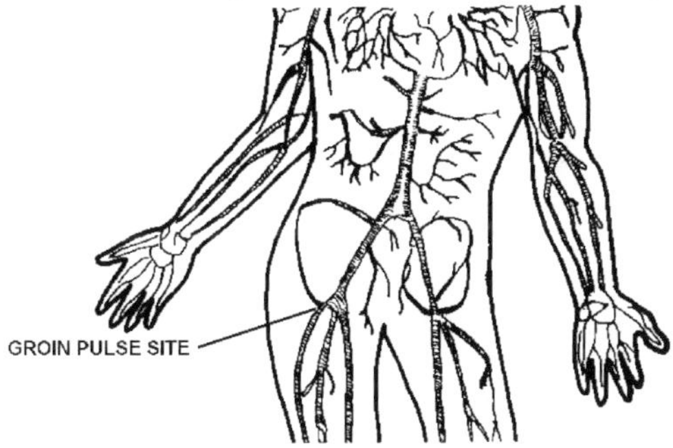

Figure 1-3. Femoral pulse.

(*c*) *Radial pulse.* To check the radial pulse, place your first two fingers on the thumb side of the casualty's wrist (Figure 1-4).

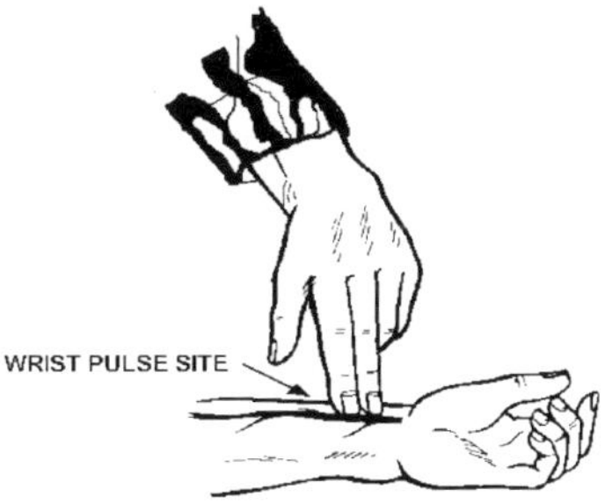

Figure 1-4. Radial pulse.

(*d*) *Posterior tibial pulse.* To check the posterior tibial pulse, place your first two fingers on the inside of the ankle (Figure 1-5).

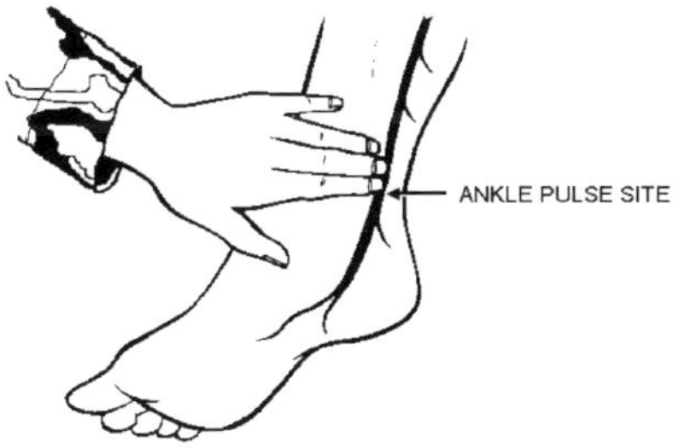

Figure 1-5. Posterior tibial pulse.

NOTE

DO NOT use your thumb to check a casualty's pulse because you may confuse the beat of your pulse with that of the casualty.

1-4. Adverse Conditions

a. Lack of Oxygen. Human life cannot exist without a continuous intake of oxygen. Lack of oxygen rapidly leads to death. First aid involves knowing how to open the airway and restore breathing.

b. Bleeding. Human life cannot continue without an adequate volume of blood circulating through the body to carry oxygen to the tissues. An important first aid measure is to stop the bleeding to prevent the loss of blood.

c. Shock. Shock means there is an inadequate blood flow to the vital tissues and organs. Shock that remains uncorrected may result in death even though the injury or condition causing the shock would not otherwise be fatal. Shock can result from many causes, such as loss of blood, loss of fluid from deep burns, pain, and reaction to the sight of a wound or blood. First aid includes preventing shock, since the casualty's chances of survival are much greater if he does not develop shock. Refer to paragraphs 2-21 through 2-24 for a further discussion of shock.

d. Infection. Recovery from a severe injury or a wound depends largely upon how well the injury or wound was initially protected. Infections result from the multiplication and growth (spread) of harmful microscopic organisms (sometimes referred to as germs). These harmful microscopic organisms are in the air, water, and soil, and on the skin and clothing. Some of these organisms will immediately invade (contaminate) a break in the skin or an open wound. The objective is to keep wounds clean and free of these organisms. A good working knowledge of basic first aid measures also includes knowing how to dress a wound to avoid infection or additional contamination.

1-5. Basics of First Aid

Most injured or ill service members are able to return to their units to fight or support primarily because they are given appropriate and timely first aid followed by the best medical care possible. Therefore, all service members must remember the basics.

- Check for **BREATHING**: Lack of oxygen intake (through a compromised airway or inadequate breathing) can lead to brain damage or death in very few minutes.

- Check for **BLEEDING**: Life cannot continue without an adequate volume of blood to carry oxygen to tissues.

- Check for **SHOCK**: Unless shock is prevented, first aid performed, and medical treatment provided, death may result even though the injury would not otherwise be fatal.

1-6. Evaluating a Casualty

a. The time may come when you must instantly apply your knowledge of first aid measures. This could occur during combat operations, in training situations, or while in a nonduty status. Any service member observing an unconscious and/or ill, injured, or wounded person must carefully and skillfully evaluate him to determine the first aid measures required to prevent further injury or death. He should seek help from medical personnel as soon as possible, but must not interrupt his evaluation of the casualty or fail to administer first aid measures. A second service member may be sent to find medical help. One of the cardinal principles for assisting a casualty is that you (the initial rescuer) must continue the evaluation and first aid measures, as the tactical situation permits, until another individual relieves you. If, during any part of the evaluation, the casualty exhibits the conditions (such as shock) for which the service member is checking, the service member must stop the evaluation and immediately administer first aid. In a chemical environment, the service member should not evaluate the casualty until both the individual and the casualty have been masked. If it is suspected that a nerve agent was used, administer the casualty's own nerve agent antidote autoinjector. After providing first aid, the service member must proceed with the evaluation and continue to monitor the casualty for further complications until relieved by medical personnel.

WARNING

Do not use your own nerve agent antidote autoinjector on the casualty.

NOTE

Remember, when evaluating and/or administering first aid to a casualty, you should seek medical aid as soon as possible. **DO NOT** stop first aid measures, but if the situation allows, send another service member to find medical aid.

b. To evaluate a casualty, perform the following steps:

(1) *Check the casualty for responsiveness.* This is done by gently shaking or tapping him while calmly asking, "Are you OK?" Watch for a response. If the casualty does not respond, go to step (2). If the casualty responds, continue with the evaluation.

(*a*) If the casualty is conscious, ask him where he feels different than usual or where it hurts. Ask him to identify the location of pain if he can, or to identify the area in which there is no feeling.

(*b*) If the casualty is conscious but is choking and cannot talk, stop the evaluation and begin first aid measures. Refer to paragraphs 2-10 and 2-11 for specific information on opening the airway.

WARNING

If a broken back or neck is suspected, do not move the casualty unless his life is in immediate danger (such as close to a burning vehicle). Movement may cause permanent paralysis or death.

(2) *Check for breathing.* (Refer to paragraph 2-6 for this procedure.)

(*a*) If the casualty is breathing, proceed to step (3).

(*b*) If the casualty is not breathing, stop the evaluation and begin first aid measures to attempt to ventilate the casualty. Attempt to open the airway, if an airway obstruction is apparent, clear the airway obstruction, then ventilate (see paragraphs 2-10 and 2-11).

(*c*) After successfully ventilating the casualty, proceed to step (3).

(3) *Check for pulse.* (Refer to paragraph 1-3*c*(2) for specific methods.) If a pulse is present and the casualty is breathing, proceed to step (4).

(*a*) If a pulse is present, but the casualty is still not breathing, start rescue breathing.

(*b*) If a pulse is not present, seek medical personnel for help.

(4) *Check for bleeding.* Look for spurts of blood or blood-soaked clothes. Also check for *both* entry and exit wounds. If the casualty is bleeding from an open wound, stop the evaluation and begin first aid procedures as follows for a—

(*a*) Wound of the arm or leg (refer to paragraphs 2-16 through 2-18 for information on putting on a field or pressure dressing).

(*b*) Partial or complete amputation, apply dressing (refer to paragraph 2-16 to 2-18) and then apply tourniquet if bleeding is not stopped (refer to paragraph 2-20 for information on putting on a tourniquet).

(*c*) Open head wound (refer to paragraph 3-10 for information on applying a dressing to an open head wound).

(*d*) Open chest wound (refer to paragraph 3-5 for information on applying a dressing to an open chest wound).

(*e*) Open abdominal wound (refer to paragraph 3-7 for information on applying a dressing to an open abdominal wound).

WARNING

In a chemically contaminated area, do not expose the wounds. Apply field dressing and then pressure dressing over wound area as needed.

(5) *Check for shock.* (Refer to paragraph 2-24 for first aid measures for shock.) If the signs and symptoms of shock are present, stop the evaluation, and begin first aid measures immediately. The following are the nine signs and symptoms of shock.

(*a*) Sweaty but cool skin (clammy skin).

(*b*) Paleness of skin. (In dark-skinned service members look for a grayish cast to the skin.)

(*c*) Restlessness or nervousness.

(*d*) Thirst.

(*e*) Loss of blood (bleeding).

(*f*) Confusion (does not seem aware of surroundings).

(*g*) Faster than normal breathing rate.

(*h*) Blotchy or bluish skin, especially around the mouth.

(*i*) Nausea or vomiting.

> **WARNING**
>
> **Leg fractures must be splinted before elevating the legs as a first aid measure for shock.**

(6) *Check for fractures.*

(*a*) Check for the following signs and symptoms of a back or neck injury and perform first aid procedures as necessary.

- Pain or tenderness of the back or neck area.
- Cuts or bruises on the back or neck area.
- Inability of a casualty to move or decreased sensation to extremities (paralysis or numbness).
  - Ask about ability to move (paralysis).
  - Touch the casualty's arms and legs and ask whether he can feel your hand (numbness).
- Unusual body or limb position.

(*b*) Immobilize any casualty suspected of having a back or neck injury by doing the following:

- Tell the casualty not to move.
- If a back injury is suspected, place padding (rolled or folded to conform to the shape of the arch) under the natural arch of the casualty's back. (For example, a blanket/poncho may be used as padding.)

WARNING

Do not move casualty to place padding.

- If a neck injury is suspected, immediately immobilize (manually) the head and neck. Place a roll of cloth under the casualty's neck, and put weighted boots (filled with dirt or sand) or rocks on both sides of his head.

(*c*) Check the casualty's arms and legs for open or closed fractures.

- Check for *open* fractures by looking for—
  - Bleeding.
  - Bones sticking through the skin.
  - Check for pulse.
- Check for *closed* fractures by looking for—
  - Swelling.
  - Discoloration.
  - Deformity.
  - Unusual body position.
  - Check for pulse.

(*d*) Stop the evaluation and begin first aid measures if a fracture to an arm or leg is suspected. Refer to Chapter 4 for information on splinting a suspected fracture.

(*e*) Check for signs/symptoms of fractures of other body areas (for example, shoulder or hip) and provide first aid as necessary.

(7) *Check for burns*. Look carefully for reddened, blistered, or charred skin; also check for singed clothing. If burns are found, stop the evaluation and begin first aid procedures. Refer to paragraph 3-9 for information on giving first aid for burns.

NOTE

Burns to the upper torso and face may cause respiratory complications. When evaluating the casualty, look for singed nose hair, soot around the nostrils, and listen for abnormal breath sounds or difficulty breathing.

(8) *Check for possible head injury.*

(*a*) Look for the following signs and symptoms:

- Unequal pupils.
- Fluid from the ear(s), nose, mouth, or injury site.
- Slurred speech.
- Confusion.
- Sleepiness.
- Loss of memory or consciousness.
- Staggering in walking.
- Headache.
- Dizziness.
- Nausea or vomiting.
- Paralysis.
- Convulsions or twitches.
- Bruising around the eyes and behind the ears.

(*b*) If a head injury is suspected, continue to watch for signs which would require performance of rescue breathing, first aid measures for shock, or control of bleeding; seek medical aid. Refer to paragraph 3-10 for information on first aid measures for head injuries.

CHAPTER 2

BASIC MEASURES FOR FIRST AID

2-1. General

Several conditions that require immediate attention are an inadequate airway, lack of breathing, and excessive loss of blood (circulation). A casualty without a clear airway or who is not breathing may die from lack of oxygen. Excessive loss of blood may lead to shock, and shock can lead to death; therefore, you must act immediately to control the loss of blood. All wounds are considered to be contaminated, since infection-producing organisms (germs) are always present on the skin and clothing, and in the soil, water, and air. Any missile or instrument (such as a bullet, shrapnel, knife, or bayonet) causing a wound pushes or carries the germs into that wound. Infection results as these organisms multiply. That a wound is contaminated does not lessen the importance of protecting it from further contamination. You must dress and bandage a wound as soon as possible to prevent further contamination.

NOTE

> It is also important that you attend to any airway, breathing, or bleeding problems **IMMEDIATELY** because these problems, if left unattended, may become life threatening.

Section I. OPEN THE AIRWAY AND RESTORE BREATHING

2-2. Breathing Process

All humans must have oxygen to live. Through the breathing process, the lungs draw oxygen from the air and put it into the blood. The heart pumps the blood through the body to be used by the cells that require a constant supply of oxygen. Some cells are more dependent on a constant supply of oxygen than others. For example, cells of the brain may die within 4 to 6 minutes without oxygen. Once these cells die, they are lost forever since they do not regenerate. This could result in permanent brain damage, paralysis, or death.

2-3. Assessment of and Positioning the Casualty

a. **CHECK** for responsiveness (Figure 2-1A)—establish whether the casualty is conscious by gently shaking him and asking, "Are you OK?"

b. **CALL** for help (Figure 2-1B).

c. **POSITION** the unconscious casualty so that he is lying on his back and on a firm surface (Figure 2-1C).

WARNING

If the casualty is lying on his chest (prone position), cautiously roll the casualty as a unit so that his body does not twist (which may further complicate a back, neck, or spinal injury).

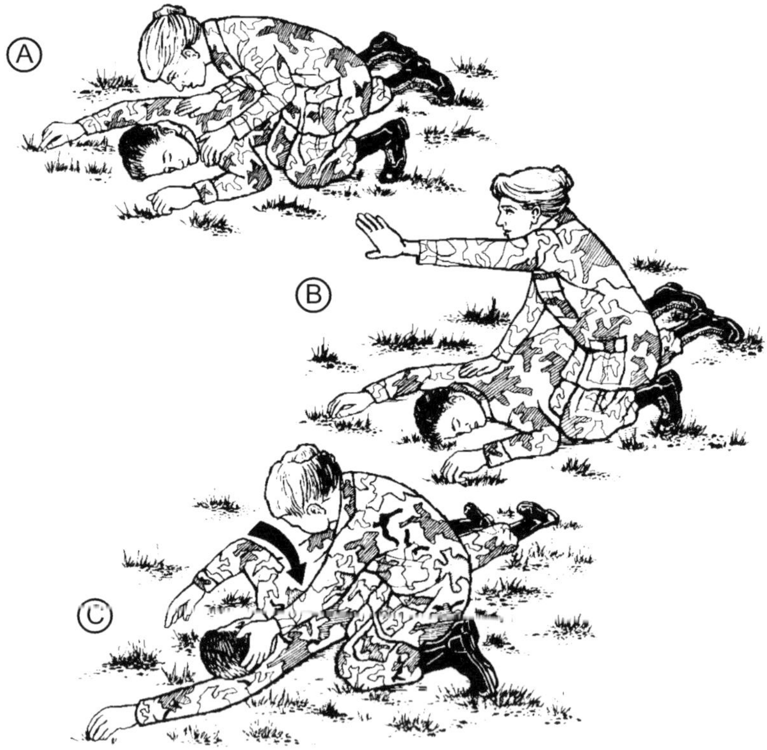

Figure 2-1. Assessment (Illustrated A—C).

(1) Straighten the casualty's legs. Take the casualty's arm that is nearest to you and move it so that it is straight and above his head. Repeat the procedure for the other arm.

(2) Kneel beside the casualty with your knees near his shoulders (leave space to roll his body) (Figure 2-1B). Place one hand behind his head and neck for support. With your other hand, grasp the casualty under his far arm (Figure 2-1C).

(3) Roll the casualty towards you using a steady, even pull. His head and neck should stay in line with his back.

(4) Return the casualty's arms to his side. Straighten his legs. Reposition yourself so that you are now kneeling at the level of the casualty's shoulders. However, if a neck injury is suspected and the jaw-thrust technique will be used, kneel at the casualty's head, looking towards his feet.

2-4. Opening the Airway of an Unconscious or Not Breathing Casualty

The tongue is the single most common cause of an airway obstruction (Figure 2-2). In most cases, simply using the head-tilt/chin-lift technique can clear the airway. This action pulls the tongue away from the air passage in the throat (Figure 2-3).

Figure 2-2. Airway blocked by tongue.

Figure 2-3. Airway opened by extending neck.

a. Call for help and then position the casualty. Move (roll) the casualty onto his back (Figure 2-1C). (Refer to paragraph 2-3*c* for information on positioning the casualty.)

NOTE

Perform finger sweep. If foreign material or vomitus is visible in the mouth, it should be removed, but do not spend an excessive amount of time doing so.

b. Open the airway using the jaw-thrust or head-tilt/chin-lift technique.

CAUTION

The head-tilt/chin-lift technique is an important procedure in opening the airway; however, use extreme care because excess force in performing this maneuver may cause further spinal injury. In a casualty with a suspected neck injury or severe head trauma, the safest approach to opening the airway is the jaw-thrust technique because in most cases it can be accomplished without extending the neck.

(1) *Perform the jaw-thrust technique.* The jaw-thrust may be accomplished by the rescuer grasping the angles of the casualty's lower jaw and lifting with both hands, one on each side, displacing the jaw forward and up (Figure 2-4). The rescuer's elbows should rest on the surface on which the casualty is lying. If the lips close, the lower lip can be retracted with the thumb. If mouth-to-mouth breathing is necessary, close the nostrils by placing your cheek tightly against them. The head should be carefully supported without tilting it backwards or turning it from side to side. If this is unsuccessful, the head should be tilted back very slightly. The jaw-thrust is the safest first approach to opening the airway of a casualty who has a suspected neck injury because in most cases it can be accomplished without extending the neck.

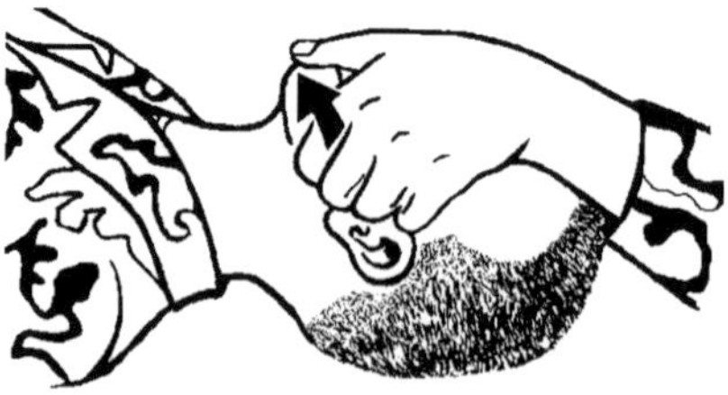

Figure 2-4. Jaw-thrust technique of opening airway.

(2) *Perform the head-tilt/chin-lift technique.* Place one hand on the casualty's forehead and apply firm, backward pressure with the palm to tilt the head back. Place the fingertips of the other hand under the bony part of the lower jaw and lift, bringing the chin forward. The thumb should not be used to lift the chin (Figure 2-5).

NOTE

The fingers should not press deeply into the soft tissue under the chin because the airway may be obstructed.

Figure 2-5. Head-tilt/chin-lift technique of opening airway.

(3) *Check for breathing (while maintaining an airway).* After establishing an open airway, it is important to maintain that airway in an open position. Often the act of just opening and maintaining the airway will allow the casualty to breathe properly. Once the rescuer uses one of the techniques to open the airway (jaw-thrust or head-tilt/chin-lift), he should maintain that head position to keep the airway open. Failure to maintain the open airway will prevent the casualty from receiving an adequate supply of oxygen. Therefore, while maintaining an open airway the rescuer should check for breathing by observing the casualty's chest and performing the following actions within 3 to 5 seconds:

(*a*) **LOOK** for the chest to rise and fall.

(*b*) **LISTEN** for air escaping during exhalation by placing your ear near the casualty's mouth.

(*c*) **FEEL** for the flow of air on your cheek (see Figure 2-6).

(*d*) **PERFORM** rescue breathing if the casualty does not resume breathing spontaneously.

NOTE

If the casualty resumes breathing, monitor and maintain the open airway. He should be transported to an MTF, as soon as practical.

2-5. Rescue Breathing (Artificial Respiration)

a. If the casualty does not promptly resume adequate spontaneous breathing after the airway is open, rescue breathing (artificial respiration) must be started. Be calm! Think and act quickly! The sooner you begin rescue breathing, the more likely you are to restore the casualty's breathing. If you are in doubt whether the casualty is breathing, give artificial respiration, since it can do no harm to a person who is breathing. If the casualty is breathing, you can feel and see his chest move. If the casualty is breathing, you can feel and hear air being expelled by putting your hand or ear close to his mouth and nose.

b. There are several methods of administering rescue breathing. The mouth-to-mouth method is preferred; however, it cannot be used in all situations. If the casualty has a severe jaw fracture or mouth wound or his jaws are tightly closed by spasms, use the mouth-to-nose method.

2-6. Preliminary Steps—All Rescue Breathing Methods

a. Establish unresponsiveness. Call for help. Turn or position the casualty.

b. Open the airway.

c. Check for breathing by placing your ear over the casualty's mouth and nose, and looking toward his chest.

(1) **LOOK** for rise and fall of the casualty's chest (Figure 2-6).

(2) **LISTEN** for sounds of breathing.

(3) **FEEL** for breath on the side of your face. If the chest does not rise and fall and no air is exhaled, then the casualty is not breathing.

(4) **PERFORM** rescue breathing if the casualty is not breathing.

NOTE

Although the rescuer may notice that the casualty is making respiratory efforts, the airway may still be obstructed and opening the airway may be all that is needed. If the casualty resumes breathing, the rescuer should continue to maintain an open airway.

Figure 2-6. Check for breathing.

2-7. Mouth-to-Mouth Method

In this method of rescue breathing, you inflate the casualty's lungs with air from your lungs. This can be accomplished by blowing air into the person's mouth. The mouth-to-mouth rescue breathing method is performed as follows:

a. If the casualty is not breathing, place your hand on his forehead, and pinch his nostrils together with the thumb and index finger of this hand. Let this same hand exert pressure on his forehead to maintain the backward head tilt and maintain an open airway. With your other hand, keep your fingertips on the bony part of the lower jaw near the chin and lift (Figure 2-7).

Figure 2-7. Head tilt/chin lift.

NOTE

If you suspect the casualty has a neck injury and you are using the jaw-thrust technique, close the nostrils by placing your cheek tightly against them.

b. Take a deep breath and place your mouth (in an airtight seal) around the casualty's mouth (Figure 2-8). (If the injured person is small, cover both his nose and mouth with your mouth, sealing your lips against the skin of his face.)

Figure 2-8. Rescue breathing.

c. Blow two full breaths into the casualty's mouth (1 to 1 1/2 seconds per breath), taking a breath of fresh air each time before you blow. Watch out of the corner of your eye for the casualty's chest to rise. If the chest rises, sufficient air is getting into the casualty's lungs. Therefore, proceed as described in step (1). If the chest does not rise, do the following (*a*, *b*, and *c* below) and then attempt to ventilate again.

(1) Take corrective action immediately by reestablishing the airway. Make sure that air is not leaking from around your mouth or out of the casualty's pinched nose.

(2) Reattempt to ventilate.

(3) If the chest still does not rise, take the necessary action to open an obstructed airway (paragraph 2-10).

NOTE

If the initial attempt to ventilate the casualty is unsuccessful, reposition the casualty's head and repeat rescue breathing. Improper chin and head positioning is the most common cause of difficulty with ventilation. If the casualty cannot be ventilated after repositioning the head, proceed with foreign-body airway obstruction maneuvers (see paragraph 2-10).

(4) After giving two slow breaths, which cause the chest to rise, attempt to locate a pulse on the casualty. Feel for a pulse on the side of the casualty's neck closest to you by placing the first two fingers (index and middle fingers) of your hand on the groove beside the casualty's Adam's apple (carotid pulse) (Figure 2-9). (Your thumb should not be used for pulse taking because you may confuse your pulse beat with that of the casualty.) Maintain the airway by keeping your other hand on the casualty's forehead. Allow 5 to 10 seconds to determine if there is a pulse.

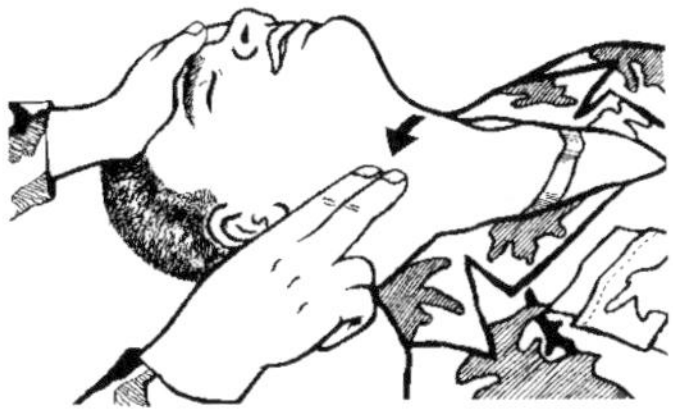

Figure 2-9. Placement of fingers to detect pulse.

(*a*) If signs of circulation are present and a pulse is found and the casualty is breathing—**STOP**; allow the casualty to breathe on his own. If possible, keep him warm and comfortable.

(*b*) If a pulse is found and the casualty is not breathing, continue rescue breathing.

(*c*) If a pulse is not found, seek medically trained personnel for help as soon as possible.

2-8. Mouth-to-Nose Method

Use this method if you cannot perform mouth-to-mouth rescue breathing because the casualty has a severe jaw fracture or mouth wound or his jaws are tightly closed by spasms. The mouth-to-nose method is performed in the same way as the mouth-to-mouth method except that you blow into the nose while you hold the lips closed with one hand at the chin. You then remove your mouth to allow the casualty to exhale passively. It may be necessary to separate the casualty's lips to allow the air to escape during exhalation.

2-9. Heartbeat

If a casualty's heart stops beating, you must immediately seek medical help. **SECONDS COUNT!** Stoppage of the heart is soon followed by cessation of respiration unless it has occurred first. Be calm! Think and act! When a casualty's heart has stopped, there is no pulse at all; the person is unconscious and limp, and the pupils of his eyes are open wide. When evaluating a casualty or when performing the preliminary steps of rescue breathing, feel for a pulse. If you DO NOT detect a pulse, seek medical help.

2-10. Airway Obstructions

In order for oxygen from the air to flow to and from the lungs, the upper airway must be unobstructed.

a. Upper airway obstructions often occur because—

(1) The casualty's tongue falls back into his throat while he is unconscious. The tongue *falls back* and *obstructs* the airway, it is not swallowed by the casualty.

NOTE

Ensure the correct positioning and maintenance of the open airway for an injured or unconscious casualty.

(2) Foreign bodies become lodged in the throat. These obstructions usually occur while eating. Choking on food (usually meat) is associated with—

- Attempting to swallow large pieces of poorly chewed food.
- Drinking alcohol.
- Slipping dentures.

(3) The contents of the stomach are regurgitated and may block the airway.

(4) Blood clots may form as a result of head and facial injuries.

b. Upper airway obstruction may cause either partial or complete airway blockage.

(1) *Partial airway obstruction.* The casualty may still have an air exchange. A good air exchange means that the casualty can cough

forcefully, though he may be wheezing between coughs. You, the rescuer, should not interfere, and should encourage the casualty to cough up the object obstructing his airway on his own. A poor air exchange may be indicated by weak coughing with a high pitched noise between coughs. Further, the casualty may show signs of shock (paragraph 1-6*b*[5]) indicating a need for oxygen. You should assist the casualty and treat him as though he had a complete obstruction.

(2) *Complete airway obstruction.* A complete obstruction (no air exchange) is indicated if the casualty cannot speak, breathe, or cough at all. He may be clutching his neck and moving erratically. In an unconscious casualty, a complete obstruction is also indicated if after opening his airway you cannot ventilate him.

2-11. Opening the Obstructed Airway—Conscious Casualty

Clearing a conscious casualty's airway obstruction can be performed with the casualty either standing or sitting and by following a relatively simple procedure.

WARNING

Once an obstructed airway occurs, the brain will develop an oxygen deficiency resulting in unconsciousness. Death will follow rapidly if breathing is not promptly restored.

a. Ask the casualty if he can speak or if he is choking. Check for the universal choking sign (Figure 2-10).

Figure 2-10. Universal sign of choking.

b. If the casualty can speak, encourage him to attempt to cough; the casualty still has a good air exchange. If he is able to speak or cough effectively, DO NOT interfere with his attempts to expel the obstruction.

c. Listen for high pitched sounds when the casualty breathes or coughs (poor air exchange). If there is poor air exchange or no breathing, CALL FOR HELP and immediately deliver manual thrusts (either an abdominal or chest thrust).

NOTE

The manual thrust with the hands centered between the waist and the rib cage is called an abdominal thrust (or Heimlich maneuver). The chest thrust (the hands are centered in the middle of the breastbone) is used only for an individual in the advanced stages of pregnancy, in the markedly obese casualty, or if there is a significant abdominal wound.

(1) Apply abdominal thrusts. This can be accomplished by using the following procedures:

(*a*) Stand behind the casualty and wrap your arms around his waist.

(*b*) Make a fist with one hand and grasp it with the other. The thumb side of your fist should be against the casualty's abdomen, in the midline and slightly above the casualty's navel, but well below the tip of the breastbone (Figure 2-11).

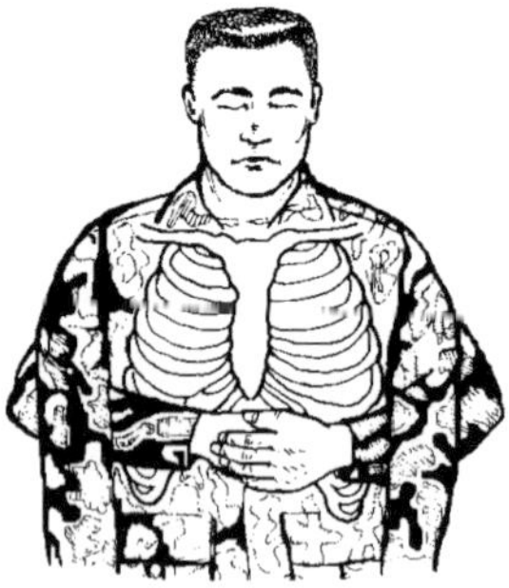

Figure 2-11. Anatomical view of abdominal thrust procedure.

(*c*) Press the fists into the abdomen with a quick backward and upward thrust (Figure 2-12).

Figure 2-12. Profile view of abdominal thrust.

(*d*) Each thrust should be a separate and distinct movement.

NOTE

Continue performing abdominal thrusts until the obstruction is expelled or the casualty becomes unresponsive.

(*e*) If the casualty becomes unresponsive, call for help as you proceed with steps to open the airway, and perform rescue breathing. (Refer to paragraph 2-7 for information on how to perform mouth-to-mouth resuscitation.)

(2) Apply chest thrusts. An alternate technique to the abdominal thrust is the chest thrust. This technique is useful when the casualty has an abdominal wound, when the casualty is pregnant, or when the casualty is so large that you cannot wrap your arms around the abdomen. To apply chest thrusts with casualty sitting or standing:

(*a*) Stand behind the casualty and wrap your arms around his chest with your arms under his armpits.

(*b*) Make a fist with one hand and place the thumb side of the fist in the middle of the breastbone (take care to avoid the tip of the breastbone and the margins of the ribs).

(*c*) Grasp the fist with the other hand and exert thrusts (Figure 2-13).

Figure 2-13. Profile view of chest thrust.

(*d*) Each thrust should be delivered slowly, distinctly, and with the intent of relieving the obstruction.

(*e*) Perform chest thrusts until the obstruction is expelled or the casualty becomes unresponsive.

(*f*) If the casualty becomes unresponsive, call for help as you proceed with steps to open the airway and perform rescue breathing.

2-12. Opening the Obstructed Airway—Casualty Lying Down or Unresponsive

The following procedures are used to expel an airway obstruction in a casualty who is lying down, who becomes unconscious, or who is found unconscious (the cause unknown):

- If a conscious casualty who is choking becomes unresponsive, call for help, open the airway, perform a finger sweep, and attempt rescue breathing (paragraphs 2-4 through 2-8). If you still cannot administer rescue breathing due to an airway blockage, then remove the airway obstruction using the procedures as in *b* below.

- If a casualty is unresponsive when you find him (the cause unknown), assess or evaluate the situation, call for help, position the casualty on his back, open the airway, establish breathlessness, and attempt to perform rescue breathing (paragraphs 2-4 through 2-8).

a. Open the airway and attempt rescue breathing (refer to paragraph 2-7 for information on how to perform mouth-to-mouth resuscitation).

b. If still unable to ventilate the casualty, perform 6 to 10 manual (abdominal or chest) thrusts.

(1) To perform the abdominal thrusts:

(*a*) Kneel astride the casualty's thighs (Figure 2-14).

Figure 2-14. Abdominal thrust on unresponsive casualty.

(*b*) Place the heel of one hand against the casualty's abdomen (in the midline slightly above the navel but well below the tip of the breastbone). Place your other hand on top of the first one. Point your fingers toward the casualty's head.

(*c*) Press into the casualty's abdomen with a quick, forward and upward thrust. You can use your body weight to perform the maneuver. Deliver each thrust quickly and distinctly.

(*d*) Repeat the sequence of abdominal thrusts, finger sweep, and rescue breathing (attempt to ventilate) as long as necessary to remove the object from the obstructed airway.

(*e*) If the casualty's chest rises, proceed to feeling for pulse.

(2) To perform chest thrusts:

(*a*) Place the unresponsive casualty on his back, face up, and open his mouth. Kneel close to the side of the casualty's body.

1. Locate the lower edge of the casualty's ribs with your fingers. Run the fingers up along the rib cage to the notch (Figure 2-15A).

2. Place the middle finger on the notch and the index finger next to the middle finger on the lower edge of the breastbone. Place the heel of the other hand on the lower half of the breastbone next to the two fingers (Figure 2-15B).

3. Remove the fingers from the notch and place that hand on top of the positioned hand on the breastbone, extending or interlocking the fingers (Figure 2-15C).

4. Straighten and lock your elbows with your shoulders directly above your hands without bending the elbows, rocking, or allowing the shoulders to sag. Apply enough pressure to depress the breastbone 1 1/2 to 2 inches, then release the pressure completely (Figure 2-15D). Do this 6 to 10 times. Each thrust should be delivered quickly and distinctly. See Figure 2-16 for another view of the breastbone being depressed.

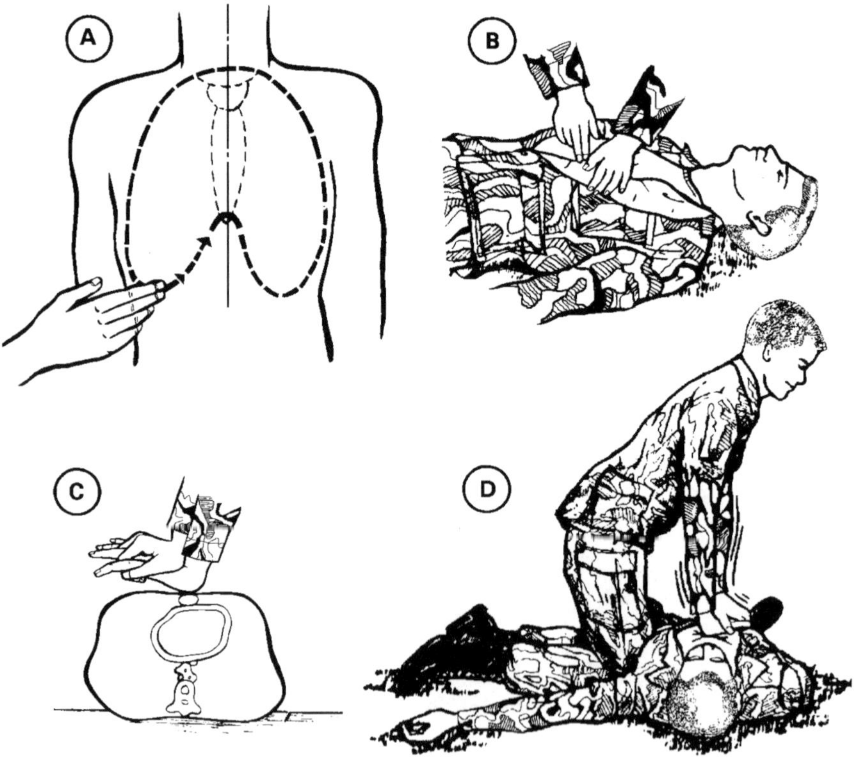

Figure 2-15. Hand placement for chest thrust (Illustrated A-D).

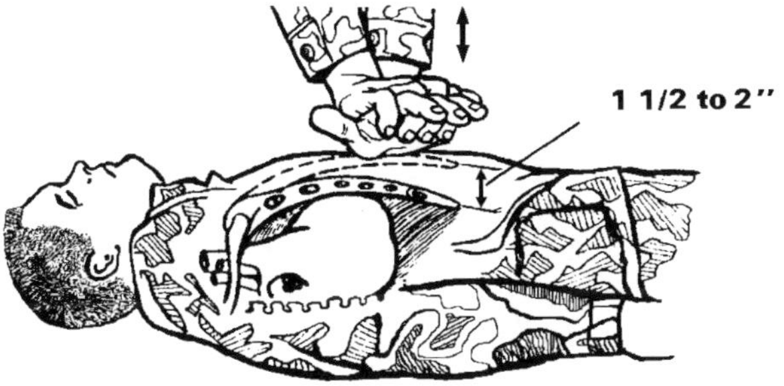

Figure 2-16. Breastbone depressed 1 1/2 to 2 inches.

(*b*) Repeat the sequence of chest thrust, finger sweep, and rescue breathing as long as necessary to clear the object from the obstructed airway. See paragraph (3) below.

(*c*) If the casualty's chest rises, proceed to feeling for his pulse.

(3) If you still cannot administer rescue breathing due to an airway obstruction, then remove the airway obstruction using the procedures in steps (*a*) and (*b*) below.

(*a*) Place the casualty on his back, face up, turn the unresponsive casualty as a unit, and call out for help.

(*b*) Perform finger sweep, keep casualty face up, use tongue-jaw lift to open mouth.

1. Open the casualty's mouth by grasping both his tongue and lower jaw between your thumb and fingers and lifting (tongue-jaw lift) (Figure 2-17). If you are unable to open his mouth, cross your fingers and thumb (crossed-finger method) and push his teeth apart (Figure 2-18) by pressing your thumb against his upper teeth and pressing your finger against his lower teeth.

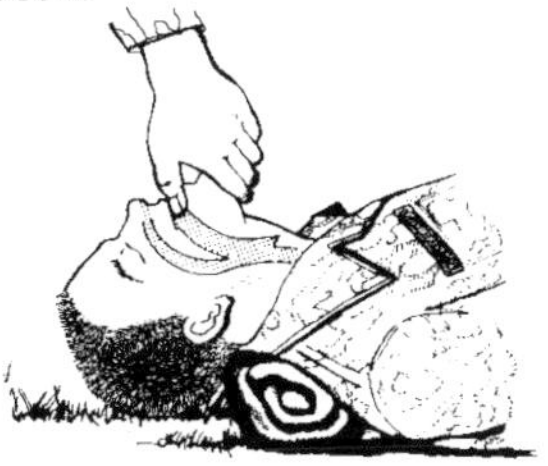

Figure 2-17. Opening casualty's mouth (tongue-jaw lift).

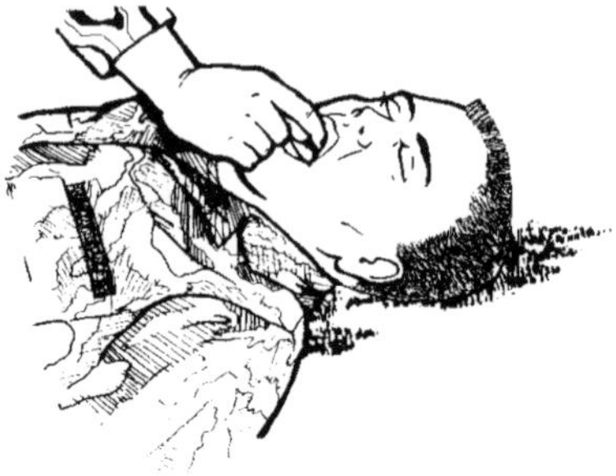

Figure 2-18. Opening casualty's mouth (crossed-finger method).

2. Insert the index finger of the other hand down along the inside of his cheek to the base of the tongue. Use a hooking motion from the side of the mouth toward the center to dislodge the foreign body (Figure 2-19).

Figure 2-19. Using finger to dislodge a foreign body.

WARNING

Take care not to force the object deeper into the airway by pushing it with the finger.

Section II. STOP THE BLEEDING AND PROTECT THE WOUND

2-13. General

The longer a service member bleeds from a major wound, the less likely he will be able to survive his injuries. It is, therefore, important that the first aid provider promptly stop the external bleeding.

2-14. Clothing

In evaluating the casualty for location, type, and size of the wound or injury, cut or tear his clothing and carefully expose the entire area of the wound. This procedure is necessary to properly visualize injury and avoid further contamination. Clothing stuck to the wound should be left in place to avoid further injury. DO NOT touch the wound; keep it as clean as possible.

> **WARNING**
>
> **DO NOT REMOVE protective clothing in a chemical environment. Apply dressings over the protective clothing.**

2-15. Entrance and Exit Wounds

Before applying the dressing, carefully examine the casualty to determine if there is more than one wound. A missile may have entered at one point and exited at another point. The *EXIT* wound is usually *LARGER* than the entrance wound.

> **WARNING**
>
> **The casualty should be continually monitored for development of conditions which may require the performance of necessary basic lifesaving measures, such as clearing the airway and mouth-to-mouth resuscitation. All open (or penetrating) wounds should be checked for a point of entry and exit and first aid measures applied accordingly.**

> **WARNING**
>
> **If the missile lodges in the body (fails to exit), DO NOT attempt to remove it or probe the wound. Apply a dressing. If there is an object extending from (impaled in) the wound, DO NOT remove the object. Apply a dressing around the object and use additional improvised bulky materials/dressings (use the cleanest material available) to build up the area around the object to stabilize the object and prevent further injury. Apply a supporting bandage over the bulky materials to hold them in place.**

2-16. Field Dressing

a. Use the casualty's field dressing; remove it from the wrapper and grasp the tails of the dressing with both hands (Figure 2-20).

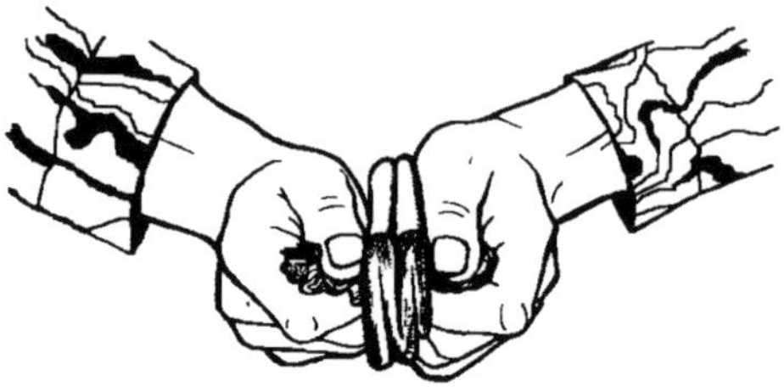

Figure 2-20. Grasping tails of dressing with both hands.

> **WARNING**
>
> **DO NOT touch the white (sterile) side of the dressing, and DO NOT allow it to come in contact with any surface other than the wound.**

b. Hold the dressing directly over the wound with the white side down. Pull the dressing open (Figure 2-21) and place it directly over the wound (Figure 2-22).

Figure 2-21. Pulling dressing open.

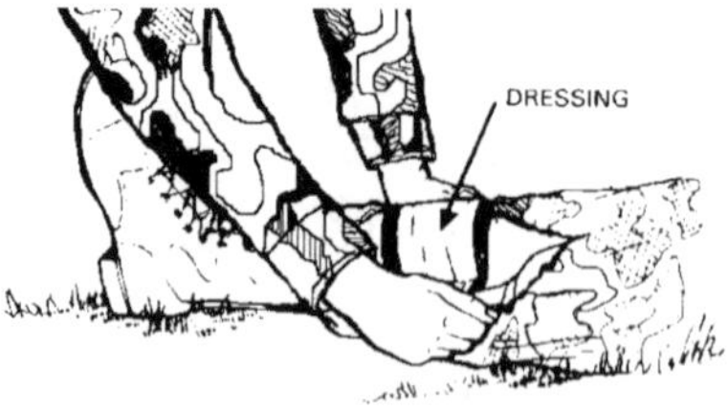

Figure 2-22. Placing dressing directly on wound.

c. Hold the dressing in place with one hand. Use the other hand to wrap one of the tails around the injured part, covering about one-half of the dressing (Figure 2-23). Leave enough of the tail for a knot. If the casualty is able, he may assist by holding the dressing in place.

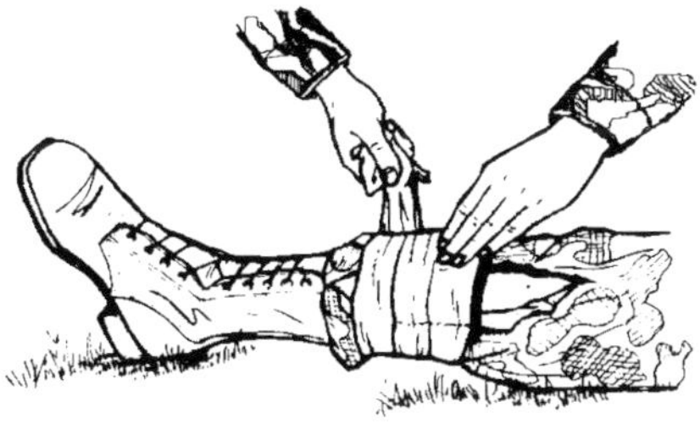

Figure 2-23. Wrapping tail of dressing around injured part.

d. Wrap the other tail in the opposite direction until the remainder of the dressing is covered. The tails should seal the sides of the dressing to keep foreign material from getting under it.

e. Tie the tails into a nonslip knot over the outer edge of the dressing (Figure 2-24). **DO NOT TIE THE KNOT OVER THE WOUND**. In order to allow blood to flow to the rest of an injured limb, tie the dressing firmly enough to prevent it from slipping but without causing a tourniquet-like effect; that is, the skin beyond the injury should not becomes cool, blue, or numb.

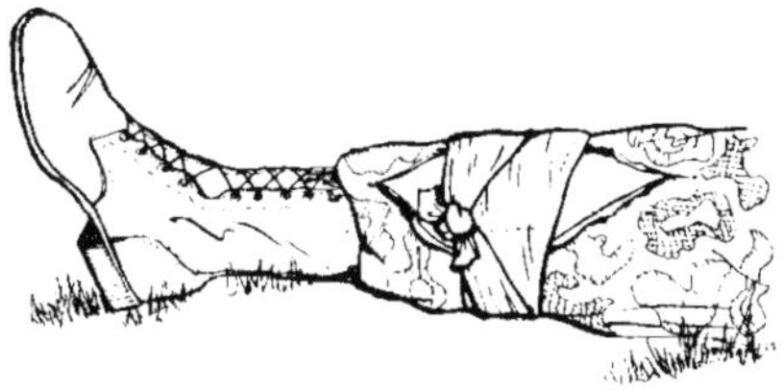

Figure 2-24. Tails tied into nonslip knot.

2-17. Manual Pressure

a. If bleeding continues after applying the sterile field dressing, direct manual pressure may be used to help control bleeding. Apply such pressure by placing a hand on the dressing and exerting firm pressure for 5 to 10 minutes (Figure 2-25). The casualty may be asked to do this himself if he is conscious and can follow instructions.

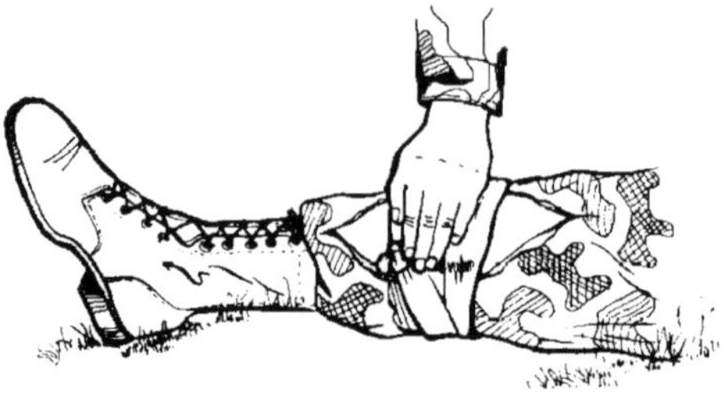

Figure 2-25. Direct manual pressure applied.

b. Elevate an injured limb slightly above the level of the heart to reduce the bleeding (Figure 2-26).

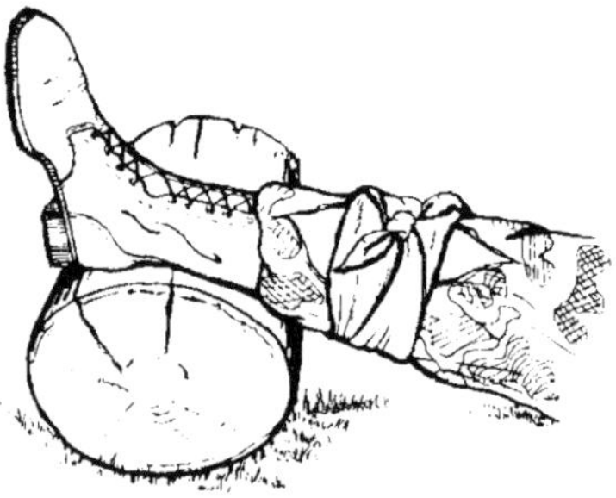

Figure 2-26. Injured limb elevated.

WARNING

DO NOT elevate a suspected fractured limb unless it has been properly splinted.

c. If the bleeding stops, check shock; administer first aid for shock as necessary. If the bleeding continues, apply a pressure dressing.

2-18. Pressure Dressing

Pressure dressings aid in blood clotting and compress the open blood vessel. If bleeding continues after the application of a field dressing, manual pressure, and elevation, then a pressure dressing must be applied as follows:

a. Place a wad of padding on top of the field dressing, directly over the wound (Figure 2-27). Keep the injured extremity elevated.

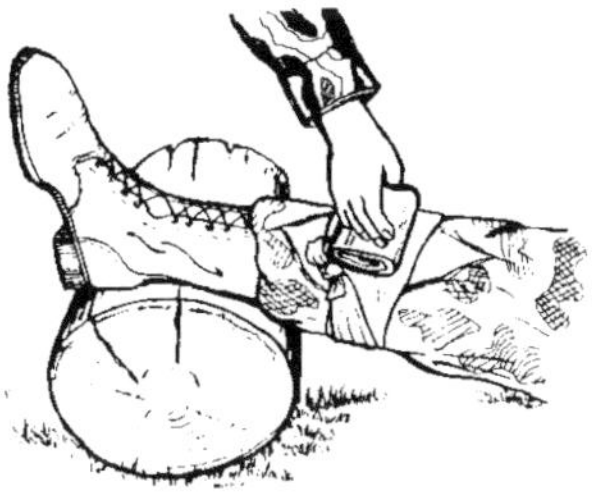

Figure 2-27. Wad of padding on top of field dressing.

NOTE

Improvised bandages may be made from strips of cloth. These strips may be made from T-shirts, socks, or other garments.

b. Place an improvised dressing (or cravat, if available) over the wad of padding (Figure 2-28). Wrap the ends tightly around the injured limb, covering the previously placed field dressing (Figure 2-29).

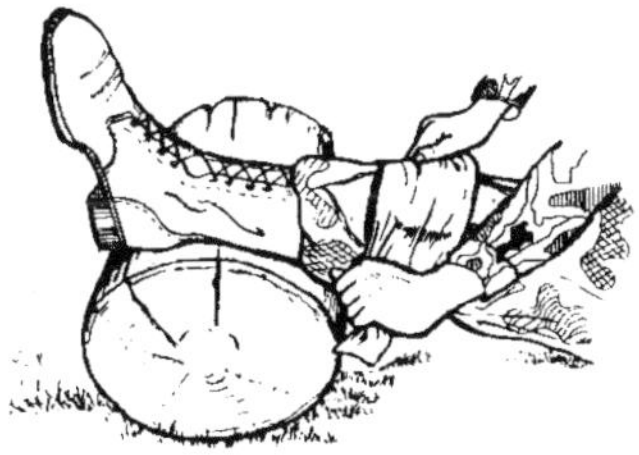

Figure 2-28. Improvised dressing over wad of padding

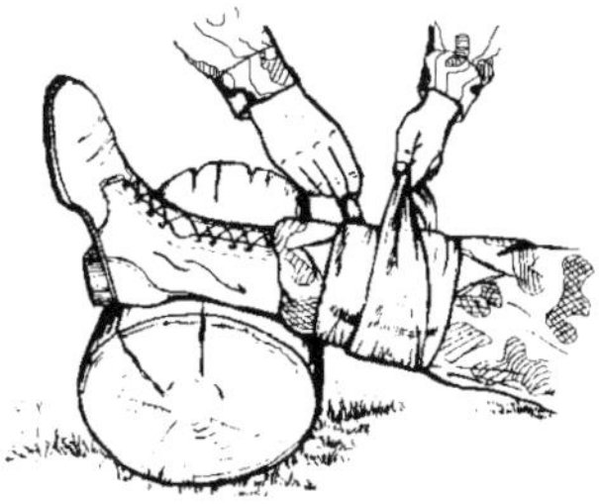

Figure 2-29. Ends of improvised dressing wrapped tightly around limb.

c. Tie the ends together in a nonslip knot, directly over the wound site (Figure 2-30). DO NOT tie so tightly that it has a tourniquet-like effect. If bleeding continues and all other measures have failed, or if the limb is severed, then apply a tourniquet. Use the tourniquet as a **LAST RESORT**. When the bleeding stops, check for shock; administer first aid for shock as necessary.

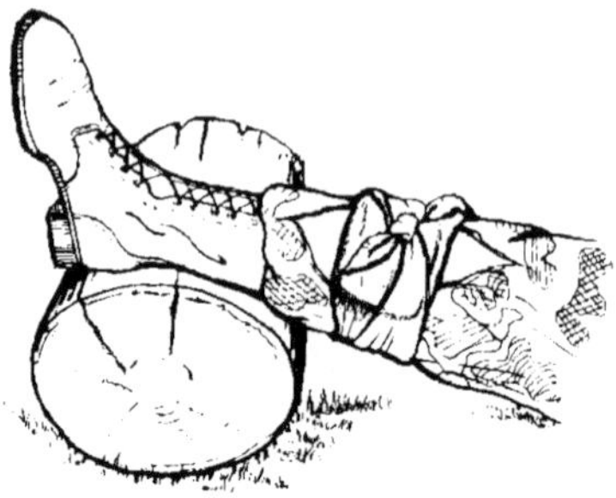

Figure 2-30. Ends of improvised dressing tied together in nonslip knot.

NOTE

Distal end of wounded extremities (fingers and toes) should be checked periodically for adequate circulation. The dressing must be loosened if the extremity becomes cool, blue, or numb.

NOTE

If bleeding continues and all other measures have failed (dressings and covering wound, applying direct manual pressure, elevating the limb above the heart level, and applying a pressure dressing while maintaining limb elevation) *then apply digital pressure* (see paragraph 2-19).

2-19. Digital Pressure

Digital pressure (often called "pressure points") is an alternative method to control bleeding. This method uses pressure from the fingers, thumbs, or hands to press at the site or point where a main artery supplying the wounded area lies near the skin surface or over bone (Figure 2-31). This pressure may help shut off or slow down the flow of blood from the heart to the wound and is used in combination with direct pressure and elevation. It may help in instances where bleeding is not easily controlled, where a pressure dressing has not yet been applied, or where pressure dressings are not readily available.

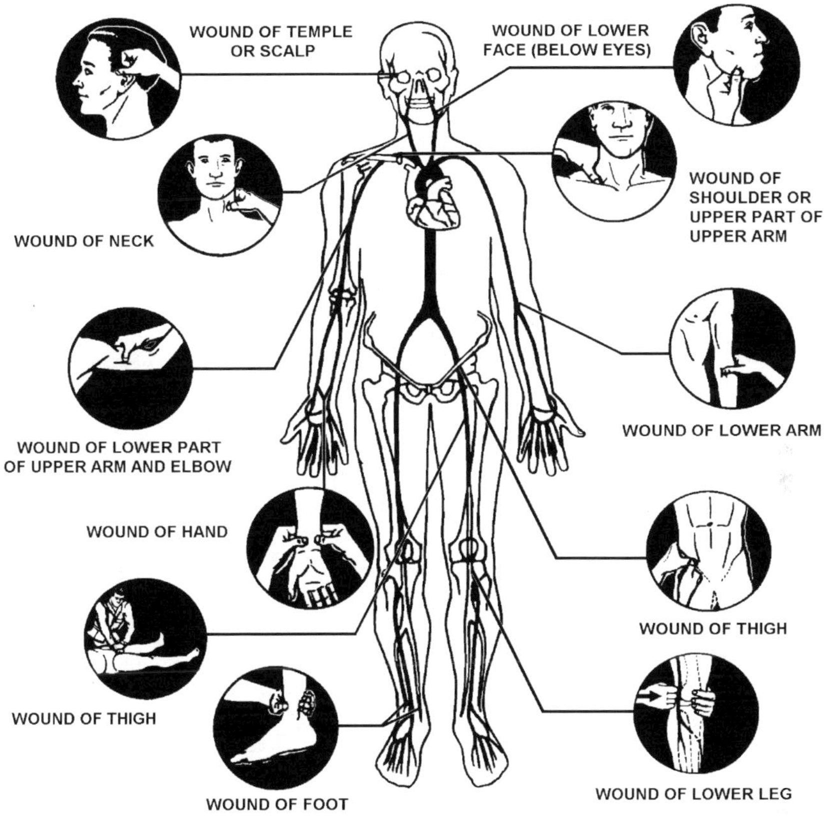

Figure 2-31. Digital pressure (pressure with fingers, thumbs or hands).

2-20. Tourniquet

DANGER

A tourniquet is only used on an arm or leg where there is a danger of the casualty losing his life (bleeding to death).

A tourniquet is a constricting band placed around an arm or leg to control bleeding. A service member whose arm or leg has been completely amputated may not be bleeding when first discovered, but a tourniquet should be applied anyway. This absence of bleeding is due to the body's normal defenses (contraction or clotting of blood vessels) as a result of the amputation, but after a period of time bleeding will start as the blood vessels relax or the clot may be knocked loose by moving the casualty. Bleeding from a major artery of the thigh, lower leg, or arm and bleeding from multiple arteries (which occurs in a traumatic amputation) may prove to be beyond control by manual pressure. If the pressure dressing (see paragraph 2-18, above) under firm hand pressure becomes soaked with blood and the wound continues to bleed, apply a tourniquet.

WARNING

Casualty should be continually monitored for development of conditions which may require the performance of necessary basic lifesaving measures, such as: clearing the airway, performing mouth-to-mouth resuscitation, preventing shock, and/or bleeding control. All open (or penetrating) wounds should be checked for a point of entry or exit and treated accordingly.

The tourniquet should not be used unless a pressure dressing has failed to stop the bleeding or an arm or leg has been cut off. On occasion, tourniquets have injured blood vessels and nerves. If left in place too long, a tourniquet can cause loss of an arm or leg. Once applied, it must stay in place, and the casualty must be taken to the nearest MTF as soon as possible. *DO NOT loosen or release a tourniquet after it has been applied as release could precipitate bleeding and potentially lead to shock.*

a. Improvising a Tourniquet. In the absence of a specially designed tourniquet, a tourniquet may be made from a strong, pliable material, such as gauze or muslin bandages, clothing, or cravats. An improvised tourniquet is used with a rigid stick-like object. To minimize skin damage, ensure that the improvised tourniquet is at least 2 inches wide.

WARNING

The tourniquet must be easily identified or easily seen.

WARNING

DO NOT use wire or shoestring for a tourniquet band.

b. Placing the Improvised Tourniquet.

(1) Place the tourniquet around the limb, between the wound and the body trunk (or between the wound and the heart). Never place it directly over a wound, a fracture, or joint. Tourniquets, for maximum effectiveness, should be placed on the upper arm or above the knee on the thigh (Figure 2-32).

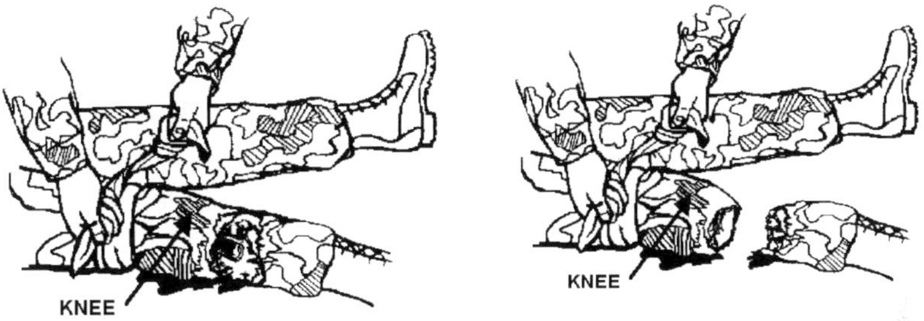

Figure 2-32. Tourniquet above knee.

(2) The tourniquet should be well-padded. If possible, place the tourniquet over the smoothed sleeve or trouser leg to prevent the skin from being pinched or twisted. If the tourniquet is long enough, wrap it around the limb several times, keeping the material as flat as possible. Damaging the skin may deprive the surgeon of skin required to cover an amputation. Protection of the skin also reduces pain.

c. Applying the Tourniquet.

(1) Tie a half-knot. (A half-knot is the same as the first part of tying a shoe lace.)

(2) Place a stick (or similar rigid object) on top of the half-knot (Figure 2-33).

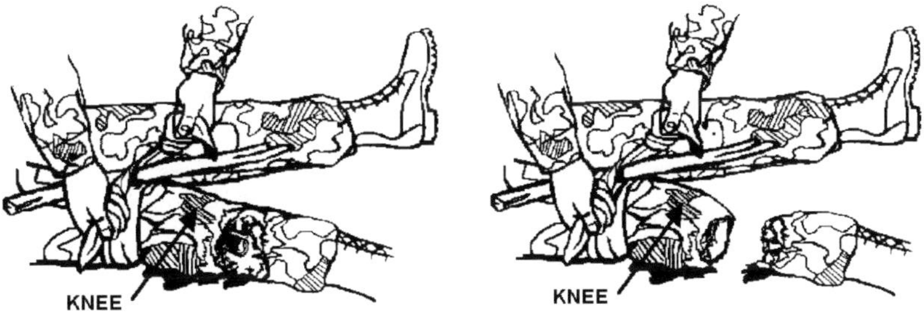

Figure 2-33. Rigid object on top of half-knot.

(3) Tie a full knot over the stick (Figure 2-34).

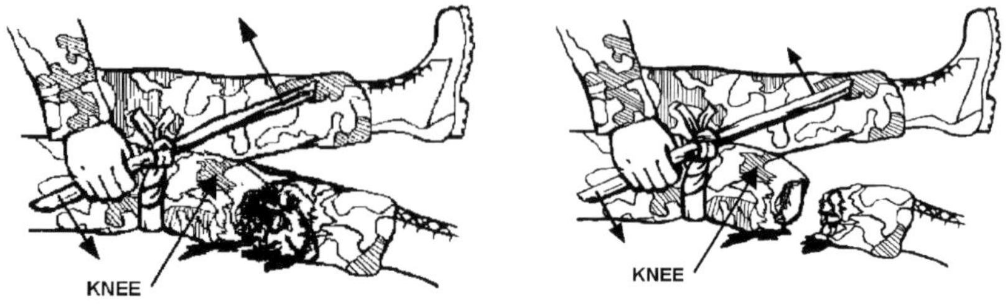

Figure 2-34. Full knot over rigid object.

(4) Twist the stick (Figure 2-35) until the tourniquet is tight around the limb and/or the bright red bleeding has stopped. In the case of amputation, dark oozing blood may continue for a short time. This is the blood trapped in the area between the wound and tourniquet.

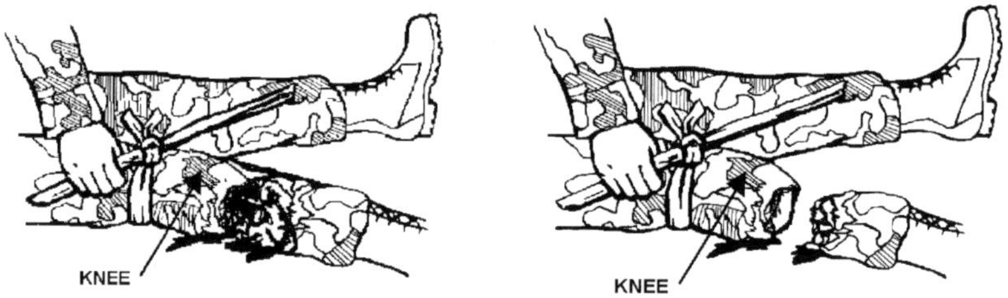

Figure 2-35. Stick twisted.

(5) Fasten the tourniquet to the limb by looping the free ends of the tourniquet over the ends of the stick. Then bring the ends around the limb to prevent the stick from loosening. Tie them together on the side of the limb (Figure 2-36).

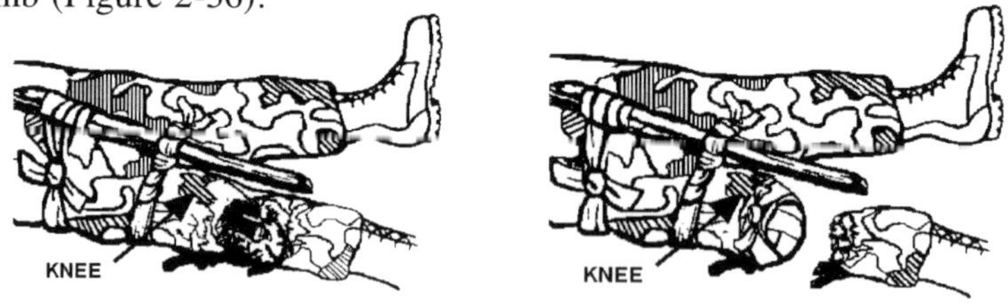

Figure 2-36. Tie free ends on side of limb.

NOTE

Other methods of securing the stick may be used as long as the stick does not unwind and no further injury results.

NOTE

If possible, save and transport any severed (amputated) limbs or body parts with (but out of sight of) the casualty.

(6) DO NOT cover the tourniquet—you should leave it in full view. If the limb is missing (total amputation), apply a dressing to the stump. All wounds should have a dressing to protect the wound from contamination.

(7) Mark the casualty's forehead with a "T" and the time to indicate a tourniquet has been applied. If necessary, use the casualty's blood to make this mark.

(8) Check and treat for shock.

(9) Seek medical aid.

CAUTION

Only appropriately skilled medical personnel may adjust or otherwise remove/release the tourniquet in the appropriate setting.

Section III. CHECK FOR SHOCK AND ADMINISTER FIRST AID MEASURES

2-21. General

The term *shock* has a variety of meanings. In medicine, it refers to a collapse of the body's cardiovascular system which includes an inadequate supply of blood to the body's tissues. Shock stuns and weakens the body. When the normal blood flow in the body is upset, death can result. Early recognition and proper first aid may save the casualty's life.

2-22. Causes and Effects

a. There are three basic mechanisms associated with shock. These are—

- The heart is damaged and fails to work as a pump.
- Blood loss (heavy bleeding) causes the volume of fluid within the vascular system to be insufficient.
- The blood vessels dilate (open wider) so that the blood within the system (even though it is a normal volume [the casualty is not bleeding or dehydrated]) is insufficient to provide adequate circulation within the body.

b. Shock may be the result of a number of conditions. These include—

- Dehydration.
- Allergic reaction to foods, drugs, insect stings, and snakebites.
- Significant loss of blood.
- Reaction to the sight of a wound, blood, or other traumatic scene.
- Traumatic injuries, such as—
  - Burns.
  - Gunshot or shrapnel wounds.
  - Crush injuries.
  - Blows to the body (which can cause broken bones or damage to internal organs).
  - Head injuries.
  - Penetrating wounds (such as from a knife, bayonet, or missile).

2-23. Signs and Symptoms of Shock

Examine the casualty to see if he has any of the following signs and symptoms:

- Sweaty but cool skin (clammy skin).

- Weak and rapid pulse.
- Paleness of skin (in dark-skinned individuals they may have a grayish look to their skin).
- Restlessness, nervousness.
- Thirst.
- Loss of blood (bleeding).
- Confusion (or loss of awareness).
- Faster-than-normal breathing rate.
- Blotchy or bluish skin (especially around the mouth and lips).
- Nausea and/or vomiting.

2-24. First Aid Measures for Shock

In the field, the first aid procedures administered for shock are identical to procedures that would be performed *to prevent shock*. When treating a casualty, assume that shock is present or will occur shortly. By waiting until actual signs and symptoms of shock are noticeable, the rescuer may jeopardize the casualty's life.

a. Position the Casualty. (DO NOT move the casualty or his limbs if suspected fractures have not been splinted. See Chapter 4 for details.)

(1) Move the casualty to cover, if cover is available and the situation permits.

(2) Lay the casualty on his back.

NOTE

A casualty in shock from a chest wound or one who is experiencing breathing difficulty, may breathe easier in a sitting position. If this is the case, allow him to sit upright, but monitor carefully in case his condition worsens.

(3) Elevate the casualty's feet higher than the level of his heart. Use a stable object (field pack or rolled up clothing) so that his feet will not slip off (Figure 2-37).

WARNING

DO NOT elevate legs if the casualty has an unsplinted broken leg, head injury, or abdominal injury.

Figure 2-37. Clothing loosened and feet elevated.

WARNING

Check casualty for leg fracture(s) and splint, if necessary, before elevating his feet. For a casualty with an abdominal wound, place his knees in an upright (flexed) position.

(4) Loosen clothing at the neck, waist, or wherever it may be binding.

CAUTION

DO NOT loosen or remove protective clothing in a chemical environment.

(5) Prevent chilling or overheating. The key is to maintain body temperature. In cold weather, place a blanket or other like item over him to keep him warm and under him to prevent chilling (Figure 2-38). However, if a tourniquet has been applied, leave it exposed (if possible). In hot weather, place the casualty in the shade and protect him from becoming chilled; however, avoid the excessive use of blankets or other coverings.

Figure 2-38. Body temperature maintained.

(6) Calm the casualty. Throughout the entire procedure of providing first aid for a casualty, the rescuer should reassure the casualty and keep him calm. This can be done by being authoritative (taking charge) and by showing self-confidence. Assure the casualty that you are there to help him.

(7) Seek medical aid.

b. *Food and/or Drink.* When providing first aid for shock, DO NOT give the casualty any food or drink. If you must leave the casualty or if he is unconscious, turn his head to the side to prevent him from choking if he vomits (Figure 2-39).

Figure 2-39. Casualty's head turned to side.

c. *Evaluate Casualty.* Continue to evaluate the casualty until medical personnel arrives or the casualty is transported to an MTF.

CHAPTER 3

FIRST AID FOR SPECIFIC INJURIES

3-1. General

Basic lifesaving steps are discussed in Chapters 1 and 2; they apply to first aid measures for all injuries. Some wounds and burns will require special precautions and procedures when applying these measures. This chapter discusses specific first aid procedures for wounds of the head, face, and neck; chest and stomach wounds; and burns. It also discusses the techniques for applying dressings and bandages to specific parts of the body.

3-2. Head, Neck, and Facial Injuries

a. Head Injuries.

(1) Head injuries range from minor abrasions or cuts on the scalp to severe brain injuries that may result in unconsciousness and sometimes death. Head injuries are classified as open or closed wounds. An open wound is one that is visible, has a break in the skin, and usually has evidence of bleeding. A closed wound may be visible (such as a depression in the skull) or the first aid provider may not be able to see any apparent injury (such as internal bleeding). Some head injuries result in unconsciousness; however, a service member may have a serious head wound and still be conscious. Casualties with head and neck injuries should be treated as though they also have a spinal injury. The casualty should not be moved until the head and neck is stabilized unless he is in immediate danger (such as close to a burning vehicle).

(2) Prompt first aid measures should be initiated for casualties with suspected head and neck injuries. The conscious casualty may be able to provide information on the extent of his injuries. However, as a result of the head injury, he may be confused and unable to provide accurate information. The signs and symptoms a first aid provider might observe are—

- Nausea and vomiting.
- Convulsions or twitches.
- Slurred speech.
- Confusion and loss of memory. (Does he know who he is? Does he know where he is? Does he know what day it is?)
- Recent unconsciousness.

- Dizziness.
- Drowsiness.
- Blurred vision, unequal pupils, or bruising (black eyes).
- Paralysis (partial or full).
- Complaint of headache.
- Bleeding or other fluid discharge from the scalp, nose, or ears.
- Deformity of the head (depression or swelling).
- Staggering while walking.

b. Neck Injuries. Neck injuries may result in heavy bleeding. Apply pressure above and below the injury, *but do not interfere with the breathing process*, and attempt to control the bleeding. Apply a dressing. Always evaluate the casualty for a possible neck fracture/spinal cord injury; if suspected, seek medical treatment immediately.

NOTE

Establish and maintain the airway in cases of facial or neck injuries. If a neck fracture or spinal cord injury is suspected, immobilize the injury and, if necessary, perform basic life support measures.

c. Facial Injuries. Soft tissue injuries of the face and scalp are common. Abrasions (scrapes) of the skin cause no serious problems. Contusions (injury without a break in the skin) usually cause swelling. A contusion of the scalp looks and feels like a lump. Laceration (cut) and avulsion (torn away tissue) injuries are also common. Avulsions are frequently caused when a sharp blow separates the scalp from the skull beneath it. Because the face and scalp are richly supplied with blood vessels (arteries and veins), wounds of these areas usually bleed heavily.

3-3. General First Aid Measures

a. General Considerations. The casualty with a head injury (or suspected head injury) should be continually monitored for the development of conditions that *may require* basic lifesaving measures. After initiating first aid measures, request medical assistance and evacuation. If dedicated medical evacuation assets are not available, transport the casualty to an MTF as soon as the situation permits. The first aid provider should not attempt to remove a protruding object from the head or give the casualty anything to eat or drink. Further, the first aid provider should be prepared to—

- Clear the airway.
- Control bleeding (external).
- Administer first aid measures for shock.
- Keep the casualty warm.
- Protect the wound.

b. *Unconscious Casualty*. An unconscious casualty does not have control of all of his body's functions and may choke on his tongue, blood, vomitus, or other substances. (Refer to Figure 2-39.)

(1) *Breathing*. The brain requires a constant supply of oxygen. A bluish (or in an individual with dark skin—grayish) color of skin around the lips and nail beds indicates that the casualty is not receiving enough oxygen. Immediate action must be taken to clear the airway, to position the casualty on his side, or to initiate rescue breathing.

(2) *Bleeding*. Bleeding from a head injury usually comes from blood vessels within the scalp. Bleeding can also develop inside the skull or within the brain. In most instances visible bleeding from the head can be controlled by application of the field first aid dressing.

CAUTION

DO NOT attempt to put unnecessary pressure on the wound or attempt to push any brain matter back into the head (skull). **DO NOT** apply a pressure dressing.

c. *Concussion*. If an individual receives a heavy blow to the head or face, he may suffer a brain concussion (an injury to the brain that involves a temporary loss of some or all of the brain's ability to function). For example, the casualty may not breathe properly for a short period of time, or he may become confused and stagger when he attempts to walk. Symptoms of a concussion may only last for a short period of time. However, if a casualty is suspected of having suffered a concussion, he should be transported to an MTF as soon as conditions permit.

d. Convulsions. Convulsions (seizures/involuntary jerking) may occur even after a mild head injury. When a casualty is convulsing, protect him from hurting himself. Take the following measures:

(1) Ease him to the ground if he is standing or sitting.

(2) Support his head and neck.

(3) Maintain his airway.

(4) Protect him from further injury (such as hitting close-by objects).

NOTE

DO NOT forcefully hold the arms and legs if they are jerking because this can lead to broken bones. **DO NOT** force anything between the casualty's teeth—especially if they are tightly clenched because this may obstruct the casualty's airway. Maintain the casualty's airway if necessary.

e. Brain Damage. In *severe* head injuries where brain tissue is protruding, *leave the wound alone*; carefully place a loose moistened dressing (moistened with sterile normal saline if available) and also a first aid dressing over the tissue to protect it from further contamination. ***DO NOT** remove or disturb any foreign matter that may be in the wound.* Position the casualty so that his head is higher than his body. Keep him warm and *seek medical assistance immediately.*

NOTE

If there is an object extending from the wound, **DO NOT** remove the object. Improvise bulky dressings from the cleanest material available and place this material around the protruding object for support, then apply the field dressing.

3-4. Chest Wounds

Blunt trauma, bullet or missile wounds, stab wounds, or falls may cause chest injuries. These injuries can be serious and may cause death quickly if first aid is not administered in a timely manner. A casualty with a chest injury may complain of pain in the chest or shoulder area; he may have difficulty breathing. His chest may not rise normally when he breathes. The injury may cause the casualty to cough up blood and to have a rapid or a weak heartbeat. A casualty with an open chest wound has a punctured chest wall. The sucking sound heard when he breathes is caused by air leaking into his chest cavity. This particular type of wound is dangerous and will collapse the injured lung (Figure 3-1). Breathing becomes difficult for the casualty because the wound is open. The service members life may depend upon how quickly you apply an occlusive dressing over the wound (refer to paragraph 3-5).

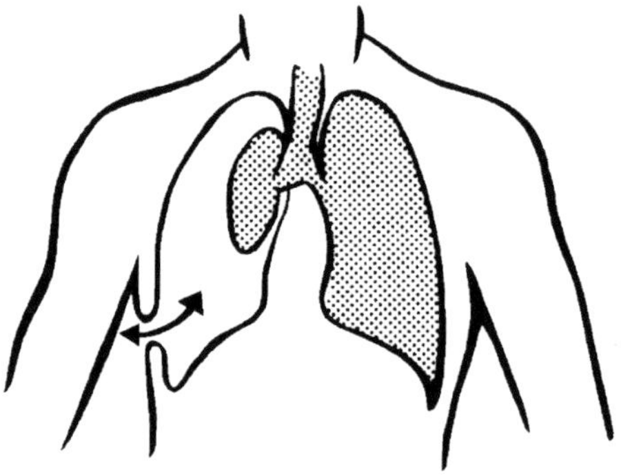

Figure 3-1. Collapsed lung.

3-5. First Aid for Chest Wounds

a. Evaluate the Casualty. Be prepared to perform first aid measures. These measures may include clearing the airway, rescue breathing, treatment for shock, and/or bleeding control.

b. Expose the Wound. If appropriate, cut or remove the casualty's clothing to expose the wound. Remember, **DO NOT** remove clothing that is stuck to the wound because additional injury may result. **DO NOT** attempt to clean the wound.

NOTE

Examine the casualty to see if there is an entry and exit wound. If there are two wounds (entry, exit), perform the same procedure for both wounds. Treat the more serious (heavier bleeding, larger) wound first. It may be necessary to improvise a dressing for the second wound by using strips of cloth, such as a torn T-shirt, or whatever material is available. Also, listen for sucking sounds to determine if the chest wall is punctured.

CAUTION

If there is an object impaled in the wound, **DO NOT** remove it. Apply a dressing around the object and use additional improvised bulky materials/dressings (use the cleanest materials available) to build up the area around the object. Apply a supporting bandage over the bulky materials to hold them in place.

CAUTION

DO NOT REMOVE protective clothing in a chemical environment. Apply dressings *over* the protective clothing.

c. Open the Casualty's Field Dressing Plastic Wrapper. In cases where there is a sucking chest wound, the plastic wrapper is used with the field dressing to create an occlusive dressing. If a plastic wrapper is not available, or if an additional wound needs to be treated; cellophane, foil, the casualty's poncho, or similar material may be used. The covering should be wide enough to extend 2 inches or more beyond the edges of the wound in all directions.

(1) Tear open one end of the casualty's plastic wrapper covering the field dressing. Be careful not to destroy the wrapper and **DO NOT** touch the inside of the wrapper.

(2) Remove the inner packet (field dressing).

(3) Complete tearing open the empty plastic wrapper using as much of the wrapper as possible to create a flat surface.

d. Place the Wrapper Over the Wound. Place the inside surface of the plastic wrapper directly over the wound *when the casualty exhales* and hold it in place (Figure 3-2). The casualty may hold the plastic wrapper in place if he is able.

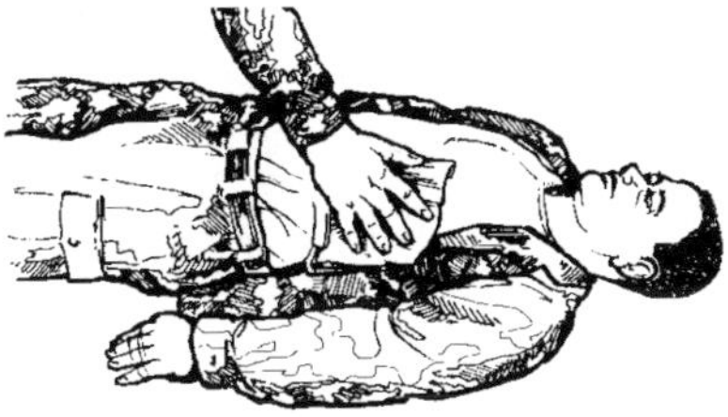

Figure 3-2. Open chest wound sealed with an occlusive dressing.

e. Apply the Dressing to the Wound.

(1) Use your free hand and shake open the field dressing (Figure 3-3).

Figure 3-3. Shaking open the field dressing.

(2) Place the white side of the dressing on the plastic wrapper covering the wound (Figure 3-4).

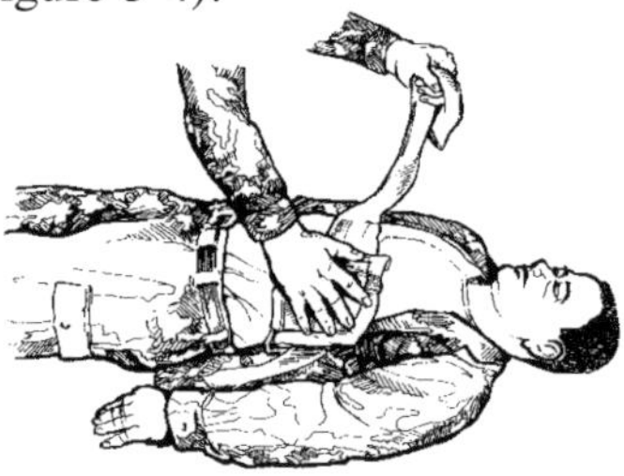

Figure 3-4. Field dressing placed on plastic wrapper.

NOTE

Use the casualty's field dressing, not your own.

(3) Have the casualty breathe normally.

(4) While maintaining pressure on the dressing, grasp one tail of the field dressing with the other hand and wrap it around the casualty's back. If tape is available, tape three sides of the plastic wrapper to the chest wall to provide occlusive type dressing. Leave one side untapped to provide emergency escape for air that may build up in the chest. If tape is not available, secure wrapper on three sides with field dressing leaving the fourth side as a flap.

(5) Wrap the other tail in the opposite direction, bringing both tails over the dressing (Figure 3-5).

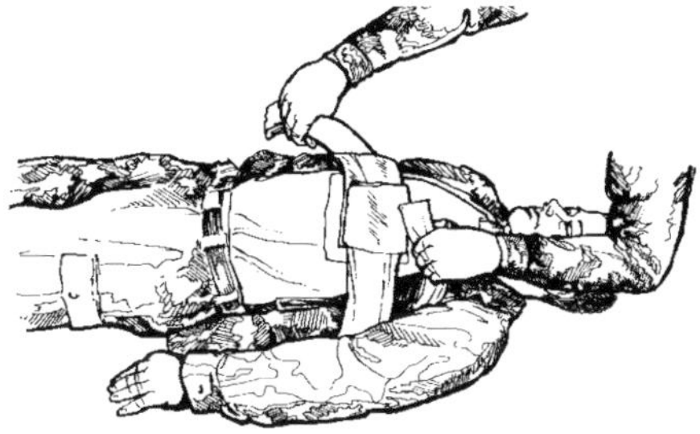

Figure 3-5. Tails of field dressing wrapped around casualty in opposite direction.

(6) Tie the tails into a square knot in the center of the dressing *after* the casualty exhales and *before* he inhales. This will aid in maintaining pressure on the bandage after it has been tied (Figure 3-6). Tie the dressing firmly enough to secure the dressing without interfering with the casualty's breathing.

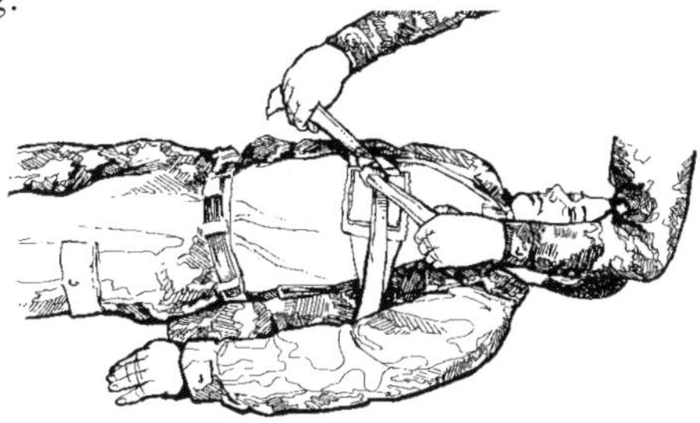

Figure 3-6. Tails of dressing tied into square knot over center of dressing.

NOTE

When practical, apply direct manual pressure over the dressing for 5 to 10 minutes to help control the bleeding.

f. Position the Casualty. Position the casualty on his injured side or in a sitting position, whichever makes breathing easier (Figure 3-7).

Figure 3-7. Casualty positioned (lying) on injured side.

g. Seek Medical Assistance. Contact medical personnel.

WARNING

If an occlusive dressing has been improperly placed, air may enter the chest cavity with no means of escape. This causes a life-threatening condition called *tension pneumothorax*. If the casualty's condition (for example, difficulty breathing, shortness of breath, restlessness, or blueness/grayness of the skin) worsens after placing the dressing, quickly lift or remove, and then replace the occlusive dressing.

3-6. Abdominal Wounds

The most serious abdominal wound is one in which an object penetrates the abdominal wall and pierces internal organs or large blood vessels. In these instances, bleeding may be severe and death can occur rapidly.

3-7. First Aid for Abdominal Wounds

a. Evaluate the Casualty. Be prepared to perform basic first aid measures. Always check for both entry and exit wounds. If there are two wounds (entry and exit), treat the wound that appears more serious first (for example, the heavier bleeding, protruding organs, larger wound, and so forth). It may be necessary to improvise dressings for the second wound by using strips of cloth, a T-shirt, or the cleanest material available.

b. Position the Casualty. Place and maintain the casualty on his back with his knees in an upright (flexed) position (Figure 3-8). The knees-up position helps relieve pain, assists in the treatment of shock, prevents further exposure of the bowel (intestines) or abdominal organs, and helps relieve abdominal pressure by allowing the abdominal muscles to relax.

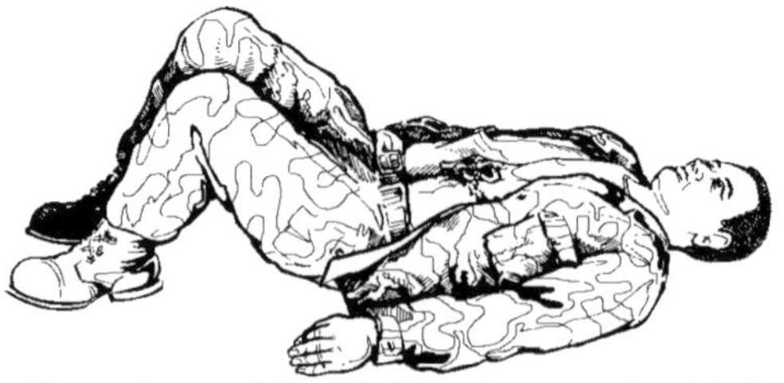

Figure 3-8. Casualty positioned (lying) on back with knees (flexed) up.

c. Expose the Wound.

(1) Remove the casualty's loose clothing to expose the wound. However, **DO NOT** attempt to remove clothing that is stuck to the wound; removing it may cause further injury.

CAUTION

DO NOT REMOVE protective clothing in a chemical environment. Apply dressings *over* the protective clothing.

(2) Gently pick up any organs that may be on the ground. Do this with a clean, dry dressing or with the cleanest available material. Place the organs on top of the casualty's abdomen (Figure 3-9).

Figure 3-9. Protruding organs placed near wound.

NOTE

DO NOT probe, clean, or try to remove any foreign object from the abdomen. **DO NOT** touch with bare hands any exposed organs. **DO NOT** push organs back inside the body.

d. Apply the Field Dressing. Use the casualty's field dressing, not your own. If the field dressing is not large enough to cover the entire wound, the plastic wrapper from the dressing may be used to cover the wound first (placing the field dressing on top). Open the plastic wrapper carefully without touching the inner surface. If necessary, other improvised dressings may be made from clothing, blankets, or the cleanest materials available.

WARNING

If there is an object extending from the wound, DO NOT remove it. Place as much of the wrapper over the wound as possible without dislodging or moving the object. DO NOT place the wrapper over the object.

(1) Grasp the tails in both hands.

(2) Hold the dressing with the white side down directly over the wound. **DO NOT** touch the white (sterile) side of the dressing or allow anything except the wound to come in contact with it.

(3) Pull the dressing open and place it directly over the wound (Figure 3-10). If the casualty is able, he may hold the dressing in place.

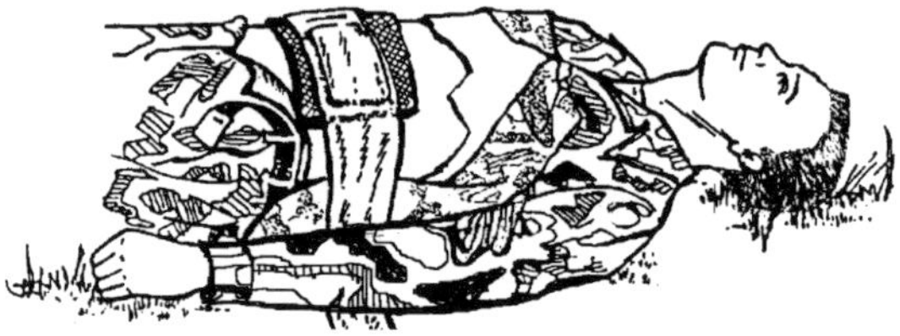

Figure 3-10. Dressing placed directly over the wound.

(4) Hold the dressing in place with one hand and use the other hand to wrap one of the tails around the body.

(5) Wrap the other tail in the opposite direction until the dressing is completely covered. Leave enough of the tail for a knot.

(6) Loosely tie the tails with a square knot at the casualty's side (Figure 3-11).

Figure 3-11. Dressing applied and tails tied with a square knot.

WARNING

When the dressing is applied, DO NOT put pressure on the wound or exposed internal parts, because pressure could cause further injury (vomiting, ruptured intestines, and so forth). Therefore, tie the dressing ties (tails) loosely at casualty's side, not directly over the dressing.

(7) Tie the dressing firmly enough to prevent slipping without applying pressure to the wound site (Figure 3-12).

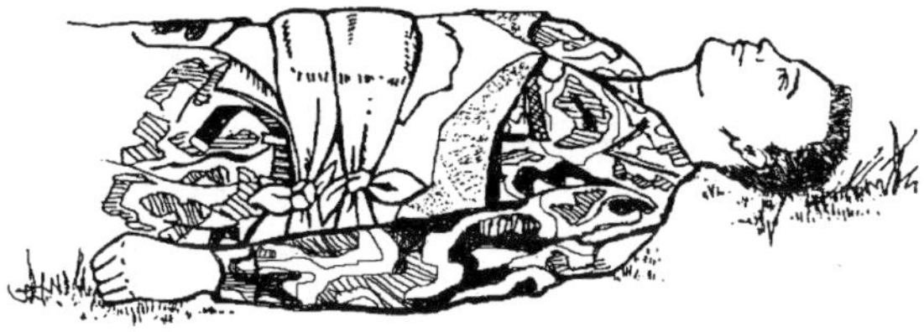

Figure 3-12. Field dressing covered with improvised material and loosely tied.

Field dressings can be covered with improvised reinforcement material (cravats, strips of torn T-shirt, or other cloth) for additional support and protection. Tie improvised bandage on the opposite side of the dressing ties firmly enough to prevent slipping but without applying additional pressure to the wound.

> **CAUTION**
>
> **DO NOT** give casualties with abdominal wounds food or water (moistening the lips is allowed).

e. Seek Medical Assistance. Notify medical personnel.

3-8. Burn Injuries

Burns often cause extreme pain, scarring, or even death. Before administering first aid, you must be able to recognize the type of burn. There are four types of burns:

- Thermal burns caused by fire, hot objects, hot liquids, and gases; or by nuclear blast or fireball.
- Electrical burns caused by electrical wires, current, or lightning.
- Chemical burns caused by contact with wet or dry chemicals or white phosphorus (WP)—from marking rounds and grenades.
- Laser burns (eye [ocular] injury).

3-9. First Aid for Burns

a. Eliminate the Source of the Burn. The source of the burn must be eliminated before any evaluation of the casualty can occur and first aid administered.

(1) Quickly remove the casualty from danger and cover the *thermal burn* with any large nonsynthetic material, such as a field jacket. If the casualty's clothing is still on fire, roll the casualty on the ground to smother (put out) the flames (Figure 3-13).

Figure 3-13. Casualty covered and rolled on ground.

CAUTION

Synthetic materials, such as nylon, may melt and cause further injury.

(2) Remove the *electrical burn* casualty from the electrical source by turning off the electrical current. **DO NOT** attempt to turn off the electricity if the source is not close by. Speed is critical, so **DO NOT** waste unnecessary time. If the electricity cannot be turned off, wrap any *nonconductive* material (*dry* rope, clothing, wood, and so forth) around the casualty's back and shoulders and drag the casualty away from the electrical source (Figure 3-14). **DO NOT** make body-to-body contact with the casualty or touch any wires because you could also become an electrical burn casualty.

Figure 3-14. Casualty removed from electrical source (using nonconductive material).

> **WARNING**
>
> **High voltage electrical burns may cause temporary unconsciousness, difficulties in breathing, or difficulties with the heart (heartbeat).**

(3) Remove the *chemical* from the *burned casualty*. Remove *liquid* chemicals by flushing with as much water as possible. Remove *dry* chemicals by brushing off loose particles (**DO NOT** use the bare surface of your hand because you could become a chemical burn casualty) and then flush with large amounts of water, if available. If large amounts of water are not available, then **NO** water should be applied because small amounts of water applied to a dry chemical burn may cause a chemical reaction. When WP strikes the skin, smother with a wet cloth or mud. Keep WP covered with a wet material to exclude air; this should help prevent the particles from burning.

(4) Remove the *laser burn* casualty from the source. When removing the casualty from the laser beam source, be careful not to enter the

beam or you may become a casualty. Never look directly at the beam source and if possible, wear appropriate eye protection.

NOTE

After the casualty is removed from the source of the burn, he should be evaluated for conditions requiring basic first aid measures.

b. Expose the Burn. Cut and gently lift away any clothing covering the burned area, without pulling clothing over the burns. Leave in place any clothing that is stuck to the burn. If the casualty's hands or wrists have been burned, remove jewelry if possible without causing further injury (rings, watches, and so forth) and place in his pockets. This prevents the necessity to cut off jewelry since swelling usually occurs as a result of a burn.

CAUTION

DO NOT lift or cut away clothing if in a chemical environment. Apply the dressing directly over the casualty's protective clothing. **DO NOT** attempt to decontaminate skin where blisters have formed.

c. Apply a Field Dressing to the Burn.

(1) Grasp the tails of the casualty's dressing in both hands.

(2) Hold the dressing directly over the wound with the white side down, pull the dressing open, and place it directly over the wound. **DO NOT** touch the white (sterile) side of the dressing or allow anything except the wound to come in contact with it. If the casualty is able, he may hold the dressing in place.

(3) Hold the dressing in place with one hand and use the other hand to wrap one of the tails around the limbs or the body.

(4) Wrap the other tail in the opposite direction until the dressing is completely covered.

(5) Tie the tails into a square knot over the outer edge of the dressing. The dressing should be applied lightly over the burn. Ensure that dressing is applied firmly enough to prevent it from slipping.

NOTE

Use the cleanest improvised dressing material available if a field dressing is not available or if it is not large enough for the entire wound.

d. *Take the Following Precautions*:

- **DO NOT** place the dressing over the face or genital area.
- **DO NOT** break the blisters.
- **DO NOT** apply grease or ointments to the burns.
- For electrical burns, check for both an entry and exit burn from the passage of electricity through the body. Exit burns may appear on any area of the body despite location of entry burn.
- For burns caused by wet or dry chemicals, flush the burns with large amounts of water and cover with a dry dressing.
- For burns caused by WP, flush the area with water, then cover with a wet material, dressing, or mud to exclude the air and keep the WP particles from burning.
- For laser burns, apply a field dressing.
- If the casualty is conscious and not nauseated, give him small amounts of water.

e. *Seek Medical Assistance.* Notify medical personnel.

3-10. Dressings and Bandages

a. *Head Wounds.*

(1) *Position the casualty.*

WARNING

DO NOT move the casualty if you suspect he has sustained a neck, spine, or head injury (which produces any signs or symptoms other than minor bleeding).

- If the casualty has a minor (superficial) scalp wound and is conscious:

  - Have the casualty sit up (unless other injuries prohibit or he is unable to).

  - If the casualty is lying down and is not accumulating fluids or drainage in his throat, elevate his head slightly.

  - If the casualty is bleeding from or into his mouth or throat, turn his head to the side or position him on his side so that the airway will be clear. Avoid putting pressure on the wound and place him on his uninjured side (Figure 3-15).

Figure 3-15. Casualty lying on side opposite injury.

- If the casualty is unconscious or has a severe head injury, then suspect and treat him as having a potential neck or spinal injury, *immobilize and **DO NOT** move the casualty.*

NOTE

If the casualty is choking or vomiting or is bleeding from or into his mouth (thus compromising his airway), position him on his uninjured side to allow for drainage and to help keep his airway clear.

WARNING

If it is necessary to turn a casualty with a suspected neck/spine injury; roll the casualty gently onto his side, keeping the head, neck, and body aligned while providing support for the head and neck. DO NOT roll the casualty by yourself but seek assistance. *Move him only if absolutely necessary*, otherwise keep the casualty immobilized to prevent further damage to the neck/spine.

(2) *Expose the wound.* Remove the casualty's helmet (if necessary). In a nuclear, biological, and chemical (NBC) environment, the first aid provider must leave the casualty as much protection (such as protective mask, mission-oriented protective posture [MOPP] overgarments) as possible. What items of protective equipment can be removed is dependent upon the casualty's injuries (where on the body and what type), the MOPP level, integrity of protective equipment (such as tears in the garment or mask seal), availability of chemical protective shelters, and the tactical situation.

WARNING

DO NOT attempt to clean the wound or remove a protruding object.

NOTE

Always use the casualty's field dressing, not your own.

(3) *Apply a dressing to a wound of the forehead or back of head.* To apply a dressing to a wound of the forehead or back of the head—

(*a*) Remove the dressing from the wrapper.

(*b*) Grasp the tails of the dressing in both hands.

(*c*) Hold the dressing (white side down) directly over the wound. **DO NOT** touch the white (sterile) side of the dressing or allow anything except the wound to come in contact with it.

(*d*) Place it directly over the wound.

(*e*) Hold it in place with one hand. If the casualty is able, he may assist.

(*f*) Wrap the first tail horizontally around the head; ensure the tail covers the dressing (Figure 3-16).

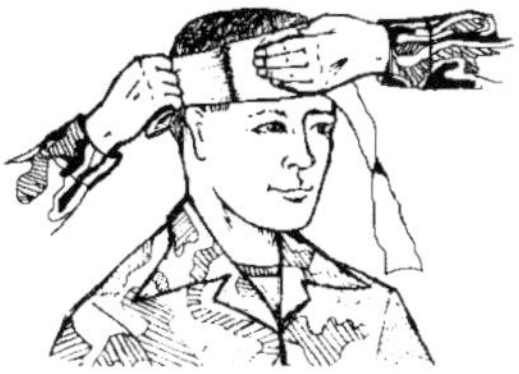

Figure 3-16. First tail of dressing wrapped horizontally around head.

(*g*) Hold the first tail in place and wrap the second tail in the opposite direction, covering the dressing (Figure 3-17).

Figure 3-17. Second tail wrapped in opposite direction.

(*h*) Tie a square knot and secure the tails at the side of the head, making sure they **DO NOT** cover the eyes or ears (Figure 3-18).

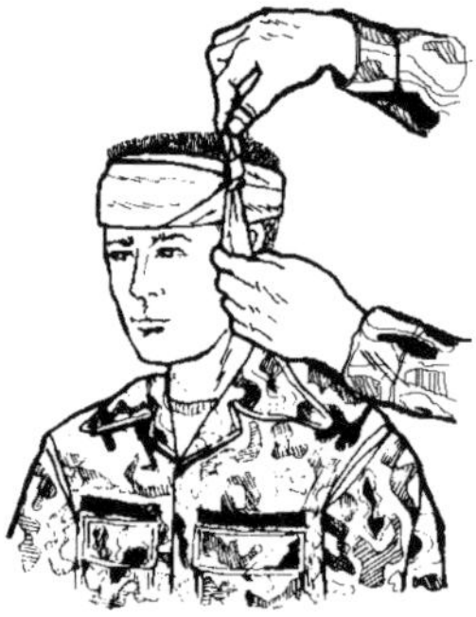

Figure 3-18. Tails tied in square knot at side of head.

(4) *Apply a dressing to a wound on top of the head.* To apply a dressing to a wound on top of the head—

(*a*) Remove the dressing from the wrapper.

(*b*) Grasp the tails of the dressing in both hands.

(*c*) Hold it (white side down) directly over the wound. **DO NOT** touch the white (sterile) side of the dressing or allow anything except the wound to come in contact with it.

(*d*) Place it over the wound (Figure 3-19).

Figure 3-19. Dressing placed over wound.

(*e*) Hold it in place with one hand. If the casualty is able, he may assist.

(*f*) Wrap one tail down under the chin (Figure 3-20), up in front of the ear, over the dressing, and in front of the other ear.

Figure 3-20. One tail of dressing wrapped under chin.

WARNING

Ensure the tails remain wide and close to the front of the chin to avoid choking the casualty.

(*g*) Wrap the remaining tail under the chin in the opposite direction and up the side of the face to meet the first tail (Figure 3-21).

Figure 3-21. Remaining tail wrapped under chin in opposite direction.

(*h*) Cross the tails (Figure 3-22), bringing one around the forehead (above the eyebrows) and the other around the back of the head (at the base of the skull) to a point just above and in front of the opposite ear, and tie them using a square knot (Figure 3-23).

Figure 3-22. Tails of dressing crossed with one around forehead.

Figure 3-23. Tails tied in square knot (in front of and above ear).

(5) *Apply a triangular bandage to the head.* To apply a triangular bandage to the head—

(*a*) Turn the base (longest side) of the bandage up and center its base on the center of the forehead, letting the point (apex) fall on the back of the neck (Figure 3-24A).

(*b*) Take the ends behind the head and cross the ends over the apex.

(*c*) Take them over the forehead and tie them (Figure 3-24B).

(*d*) Tuck the apex behind the crossed part of the bandage or secure it with a safety pin, if available (Figure 3-24C).

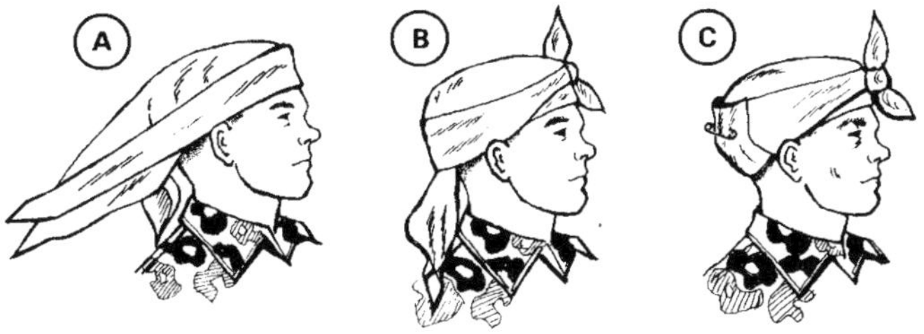

Figure 3-24. Triangular bandage applied to head (Illustrated A—C)

(6) *Apply a cravat bandage to the head.* To apply a cravat bandage to the head—

(*a*) Place the middle of the bandage over the dressing (Figure 3-25A).

(*b*) Cross the two ends of the bandage in opposite directions completely around the head (Figure 3-25B).

(*c*) Tie the ends over the dressing (Figure 3-25C).

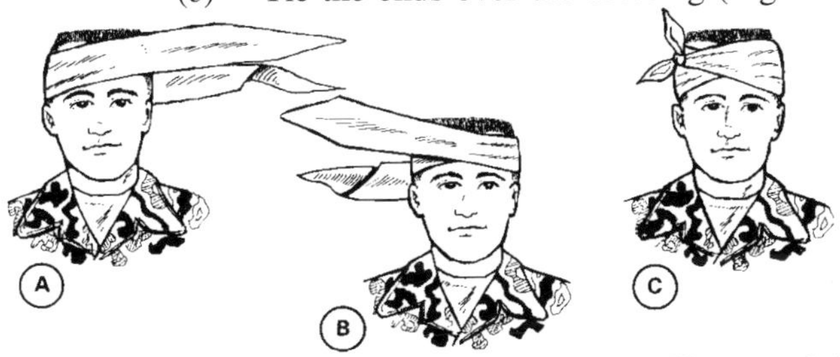

Figure 3-25. Cravat bandage applied to head (Illustrated A—C).

b. Eye Injuries. The eye is a vital sensory organ, and blindness is a severe physical handicap. Timely first aid of the eye may relieve pain and may also help to prevent shock, permanent eye injury, and possible loss of vision. Because the eye is very sensitive, any injury can be easily aggravated if it is improperly handled. Injuries of the eye may be quite severe. Cuts of the eyelids can appear to be very serious, but if the eyeball is not involved, a person's vision usually will not be damaged. However, lacerations (cuts) of the eyeball can cause permanent damage or loss of sight.

(1) *Lacerated/torn eyelids*. Lacerated eyelids may bleed heavily, but bleeding usually stops quickly. Cover the injured eye with a sterile dressing. **DO NOT** put pressure on the wound because you may injure the eyeball. Handle torn eyelids very carefully to prevent further injury. Place any detached pieces of the eyelid on a clean bandage or dressing and immediately send them with the casualty to the medical facility.

(2) *Lacerated eyeball (injury to the globe)*. Lacerations or cuts to the eyeball may cause serious and permanent eye damage. Cover the injury with a loose sterile dressing. **DO NOT** put pressure on the eyeball because additional damage may occur. An important point to remember is that when one eyeball is injured, you should immobilize both eyes. This is done by applying a bandage to both eyes. Because the eyes move together, covering both will lessen the chances of further damage to the injured eye. (However, in hazardous surroundings, leave uninjured eye uncovered to enable casualty to see.)

CAUTION

DO NOT apply pressure when there is a possible laceration of the eyeball. The eyeball contains fluid. Pressure applied over the eye will force the fluid out, resulting in permanent injury. **APPLY PROTECTIVE DRESSING WITHOUT ADDED PRESSURE**.

(3) *Extruded eyeballs*. Service members may encounter casualties with severe eye injuries that include an extruded eyeball (eyeball out-of-socket). In such instances you should gently cover the extruded eye with a loose moistened dressing and also cover the unaffected eye. **DO NOT** bind or exert pressure on the injured eye while applying the dressing. Keep the casualty quiet, place him on his back, treat for shock, and evacuate him immediately.

(4) *Burns of the eyes*. Chemical burns, thermal (heat) burns, and light burns can affect the eyes.

(*a*) *Chemical burns.* Injuries from chemical burns require immediate first aid. Mainly acids or alkalies cause chemical burns. The first aid measures consist of flushing the eyes immediately with large amounts of water for at least 5 to 20 minutes, or as long as necessary to flush out the chemical and, once flushed, bandaging the eyes. If the burn is an acid burn, you should flush the eye for at least 5 to 10 minutes. If the burn is an alkali burn, you should flush the eye for at least 20 minutes. After the eye has been flushed evacuate the casualty immediately.

(*b*) *Thermal burns.* When an individual suffers burns of the face from a fire, the eyes will close quickly due to extreme heat. This reaction is a natural reflex to protect the eyeballs; however, the eyelids remain exposed and are frequently burned. If a casualty receives burns of the eyelids or face—

- **DO NOT** apply a dressing.
- **DO NOT** touch.
- **SEEK** medical assistance immediately.

(*c*) *Light burns.* Exposure to intense light can burn an individual. Infrared rays, eclipse light (if the casualty has looked directly at the sun), or laser burns cause injuries of the exposed eyeball. Ultraviolet rays from arc welding can cause a superficial burn to the surface of the eye. These injuries are generally not painful but may cause permanent damage to the eyes. Immediate first aid is usually not required. Loosely bandaging the eyes may make the casualty more comfortable and protect his eyes from further injury caused by exposure to other bright lights or sunlight.

CAUTION

With impaled objects or significant sized foreign bodies, both eyes are usually bandaged to help secure the foreign body in the injured eye. In a battlefield environment, leave the uninjured eye uncovered so that the casualty can see.

c. *Side-of-Head or Cheek Wound.* Facial injuries to the side of the head or the cheek may bleed profusely (Figure 3-26). Prompt action is necessary to ensure that the airway remains open and also to control the bleeding. It may be necessary to apply a dressing. To apply a dressing—

(1) Remove the dressing from its wrapper.

(2) Grasp the tails in both hands.

(3) Hold the dressing directly over the wound with the white side down and place it directly on the wound (Figure 3-27A). **DO NOT** touch the white (sterile) side of the dressing or allow anything except the wound to come in contact with it.

(4) Hold the dressing in place with one hand (the casualty may assist if able). Wrap the top tail over the top of the head and bring it down in front of the ear (on the side opposite the wound), under the chin (Figure 3-27B) and up over the dressing to a point just above the ear (on the wound side).

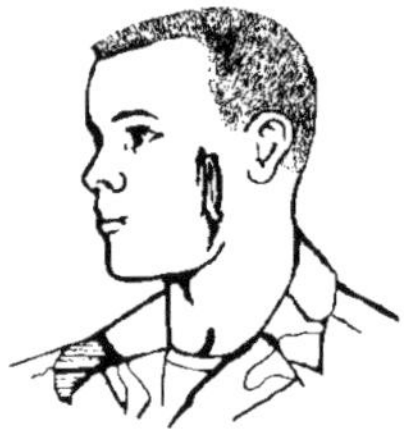

Figure 3-26. Side of head or cheek wound.

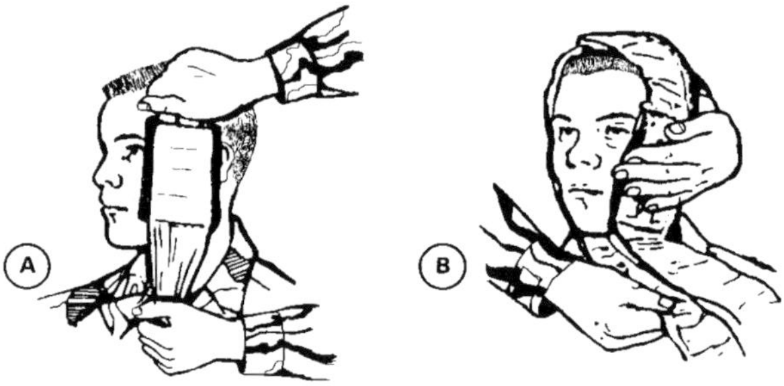

Figure 3-27. Dressing placed directly on wound. Top tail wrapped over top of head, down in front of ear, and under chin (Illustrated A—B).

NOTE

When possible, avoid covering the casualty's ear with the dressing, as this will decrease his ability to hear.

(5) Bring the second tail under the chin, up in front of the ear (on the side opposite the wound), and over the head to meet the other tail (on the wounded side) (Figure 3-28).

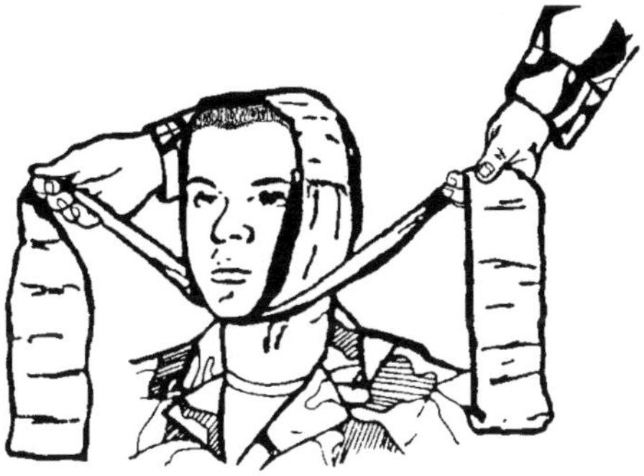

Figure 3-28. Bringing second tail under the chin.

(6) Cross the two tails (on the wound side) (Figure 3-29) and bring one end across the forehead (above the eyebrows) to a point just in front of the opposite ear (on the uninjured side).

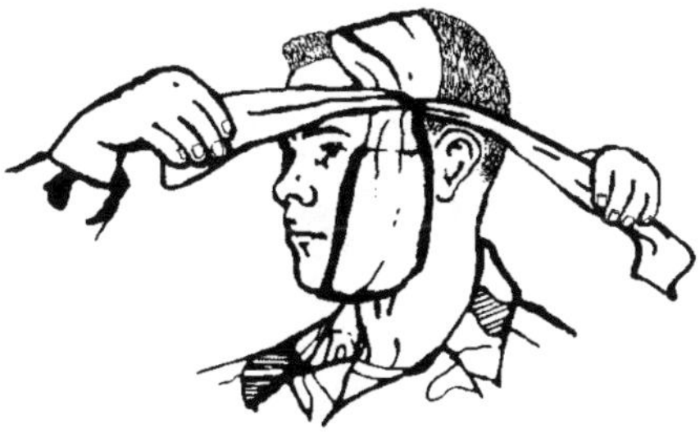

Figure 3-29. Crossing the tails on the side of the wound.

(7) Wrap the other tail around the back of the head (at the base of the skull), and tie the two ends just in front of the ear on the uninjured side with a square knot (Figure 3-30).

Figure 3-30. Tying the tails of the dressing in a square knot.

d. *Ear Injuries.* Lacerated (cut) or avulsed (torn) ear tissue may not, in itself, be a serious injury. Bleeding, or the drainage of fluids from the ear canal, however, may be a sign of a head injury, such as a skull fracture. **DO NOT** attempt to stop the flow from the inner ear canal nor put anything into the ear canal to block it. Instead, you should cover the ear lightly with a dressing. For minor cuts or wounds to the external ear, apply a cravat bandage as follows:

(1) Place the middle of the bandage over the ear (Figure 3-31A).

(2) Cross the ends, wrap them in opposite directions around the head, and tie them (Figures 3-31B and 3-31C).

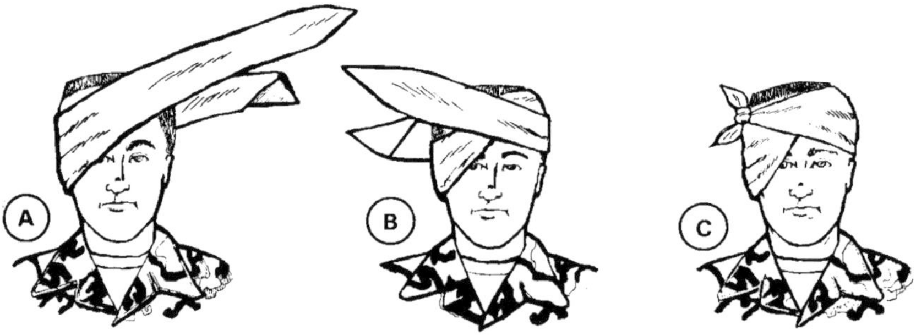

Figure 3-31. Applying cravat bandage to ear (Illustrated A—C).

(3) If possible, place some dressing material between the back of the ear and the side of the head to avoid crushing the ear against the head with the bandage.

e. *Nose Injuries.* Nose injuries generally produce bleeding. The bleeding may be controlled by placing an ice pack (if available) over the nose, or pinching the nostrils together. The bleeding may also be controlled by placing torn gauze (rolled) between the upper teeth and the lip.

CAUTION
DO NOT attempt to remove objects inhaled into the nose. An untrained person who removes such an object could worsen the casualty's condition and cause permanent injury.

f. *Jaw Injuries.* Before applying a bandage to a casualty's jaw, remove all loose or free-floating foreign material from the casualty's mouth.

If the casualty is unconscious, check for obstructions in the airway and remove if possible. If there is profuse bleeding in the oral cavity, the cavity may require loose packing with soft bandaging material (for example: Kerlix™ gauze) prior to applying a bandage. Care should be taken to avoid occluding the airway. When applying the bandage, allow the jaw enough freedom to permit passage of air and drainage from the mouth.

(1) *Apply bandages attached to field first aid dressing to the jaw.* After dressing the wound, apply the bandages using the same technique illustrated in Figure 3-32A—C.

NOTE

The dressing and bandaging procedure outlined for the jaw serves a twofold purpose. In addition to stopping the bleeding and protecting the wound, it also immobilizes a fractured jaw.

(2) *Apply a cravat bandage to the jaw.*

(*a*) Place the bandage under the chin and pull its ends upward. Adjust the bandage to make one end longer than the other (Figure 3-32A).

(*b*) Take the longer end over the top of the head to meet the short end at the temple and cross the ends over (Figure 3-32B).

(*c*) Take the ends in opposite directions to the other side of the head and tie them over the part of the bandage that was applied first (Figure 3-32C).

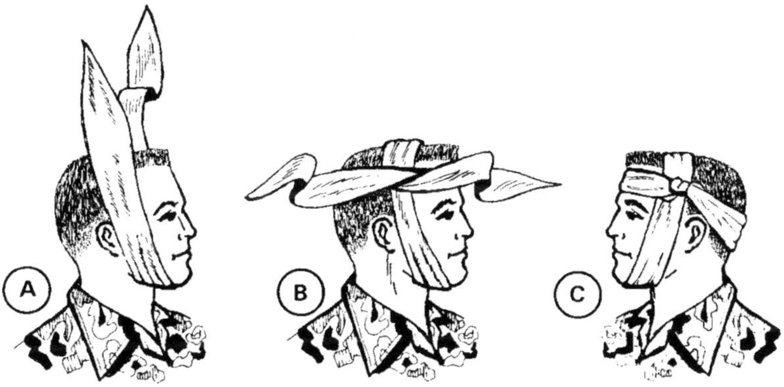

Figure 3-32. Applying a cravat bandage to jaw (Illustrated A—C).

NOTE

The cravat bandage technique is used to immobilize a fractured jaw or to maintain a sterile dressing that does not have tail bandages attached.

3-11. Shoulder Bandage

a. To apply bandages attached to the field first aid dressing—

(1) Take one bandage across the chest and the other across the back and under the arm opposite the injured shoulder.

(2) Tie the ends with a square knot (Figure 3-33).

Figure 3-33. Shoulder bandage.

b. To apply a cravat bandage to the shoulder or armpit—

(1) Make an extended cravat bandage by using two triangular bandages (Figure 3-34A); place the end of the first triangular bandage along the base of the second one (Figure 3-34B).

(2) Fold the two bandages into a single extended bandage (Figure 3-34C).

(3) Fold the extended bandage into a single cravat bandage (Figure 3-34D). After folding, secure the thicker part (overlap) with two or more safety pins (Figure 3-34E).

(4) Place the middle of the cravat bandage under the armpit so that the front end is longer than the back end and safety pins are on the outside (Figure 3-34F).

(5) Cross the ends on top of the shoulder (Figure 3-34G).

(6) Take one of the bandage ends across the back and under the arm on the opposite side and the other end across the chest. Tie the ends (Figure 3-34H).

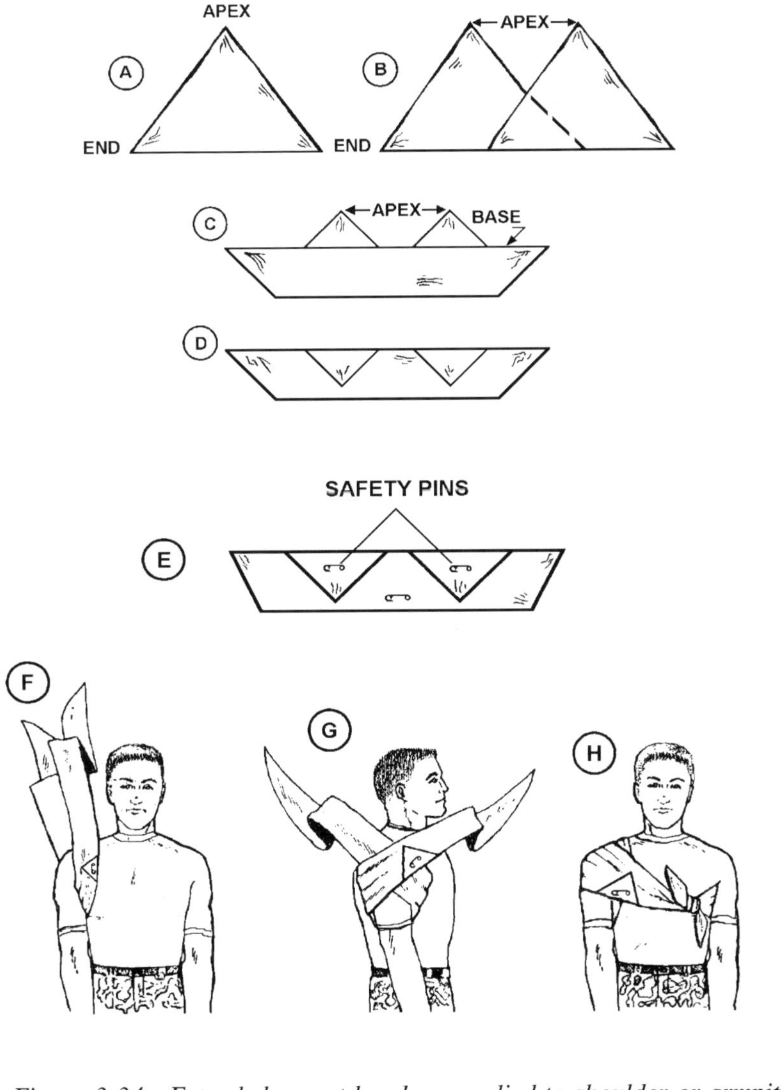

Figure 3-34. Extended cravat bandage applied to shoulder or armpit (Illustrated A—H).

Be sure to place sufficient wadding in the armpit. **DO NOT** tie the cravat bandage too tightly. Avoid compressing the major blood vessels in the armpit.

3-12. Elbow Bandage

To apply a cravat bandage to the elbow—

a. Bend the arm at the elbow and place the middle of the cravat at the point of the elbow bringing the ends upward (Figure 3-35A).

b. Bring the ends across, extending both downward (Figure 3-35B).

c. Take both ends around the arm and tie them with a square knot at the front of the elbow (Figure 3-35C).

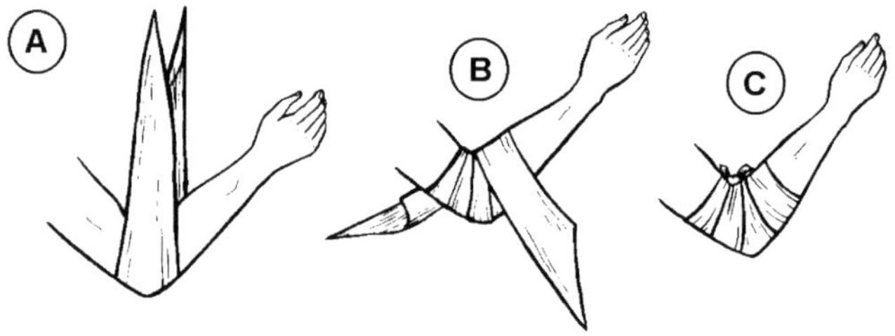

Figure 3-35. Elbow bandage (Illustrated A—C).

> **CAUTION**
>
> If an elbow fracture is suspected, **DO NOT** bend the elbow; bandage it in the position found.

3-13. Hand Bandage

a. To apply a triangular bandage to the hand—

(1) Place the hand in the middle of the triangular bandage with the wrist at the base of the bandage (Figure 3-36A). Ensure that the fingers are separated with absorbent material to prevent chafing and irritation of the skin.

(2) Place the apex over the fingers and tuck any excess material into the pleats on each side of the hand (Figure 3-36B).

(3) Cross the ends on top of the hand, take them around the wrist, and tie them (Figures 3-36C—E) with a square knot.

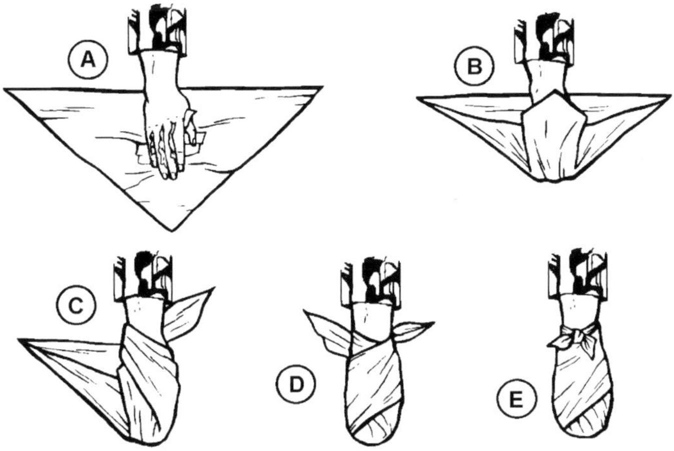

Figure 3-36. Triangular bandage applied to hand (Illustrated A—E).

b. To apply a cravat bandage to the palm of the hand—

(1) Lay the middle of the cravat over the palm of the hand with the ends hanging down on each side (Figure 3-37A).

(2) Take the end of the cravat at the little finger across the back of the hand, extending it upward over the base of the thumb; then bring it downward across the palm (Figure 3-37B).

(3) Take the thumb end across the back of the hand, over the palm, and through the hollow between the thumb and palm (Figure 3-37C).

(4) Take the ends to the back of the hand and cross them; then bring them up over the wrist and cross them again (Figure 3-37D).

(5) Bring both ends down and tie them with a square knot on top of the wrist (Figure 3-37E—F).

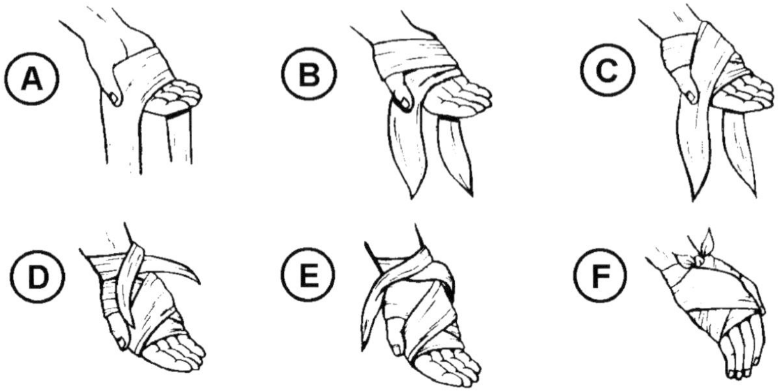

Figure 3-37. Cravat bandage applied to palm of hand (Illustrated A—F).

3-14. Leg (Upper and Lower) Bandage

To apply a cravat bandage to the leg—

a. Place the center of the cravat over the dressing (Figure 3-38A).

b. Take one end around and up the leg in a spiral motion and the other end around and down the leg in a spiral motion, overlapping part of each preceding turn (Figure 3-38B).

c. Bring both ends together and tie them (Figure 3-38C) with a square knot.

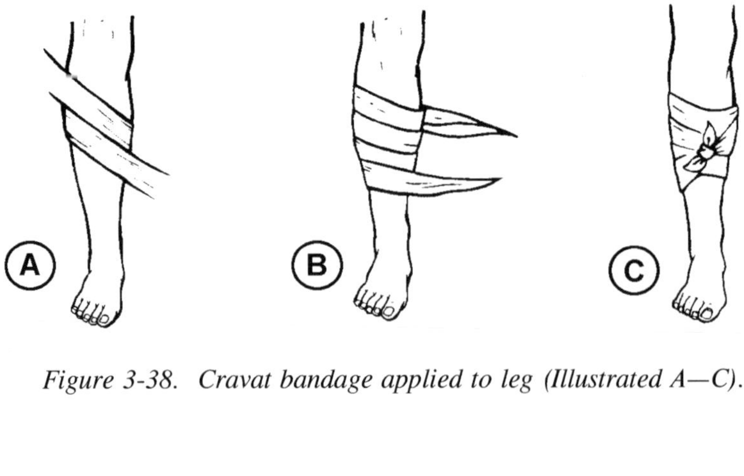

Figure 3-38. Cravat bandage applied to leg (Illustrated A—C).

3-15. Knee Bandage

To apply a cravat bandage to the knee as illustrated in Figure 3-39, use the same technique applied in bandaging the elbow.

> **CAUTION**
>
> If a fracture of the kneecap is suspected, **DO NOT** bend the knee; bandage it in the position found.

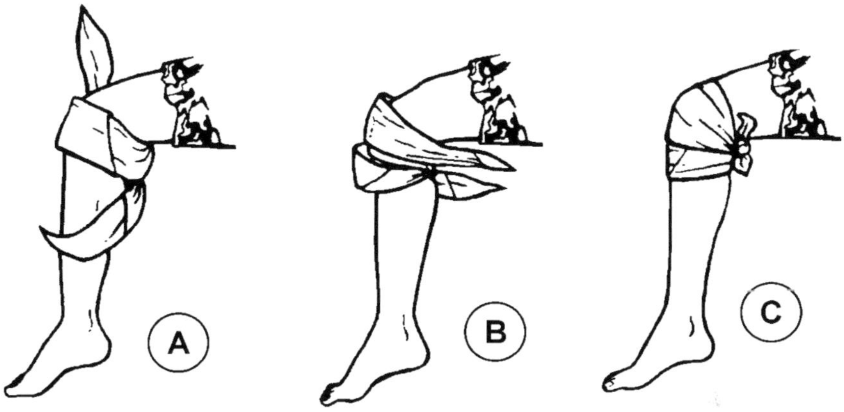

Figure 3-39. Cravat bandage applied to knee (Illustrated A—C).

3-16. Foot Bandage

To apply a triangular bandage to the foot—

a. Place the foot in the middle of the triangular bandage with the heel well forward of the base (Figure 3-40A). Ensure that the toes are separated by absorbent material to prevent chafing and irritation of the skin.

b. Place the apex over the top of the foot and tuck any excess material into the pleats on each side of the foot (Figure 3-40B).

c. Cross the ends on top of the foot, take them around the ankle, and tie them at the front of the ankle (Figure 3-40C—E).

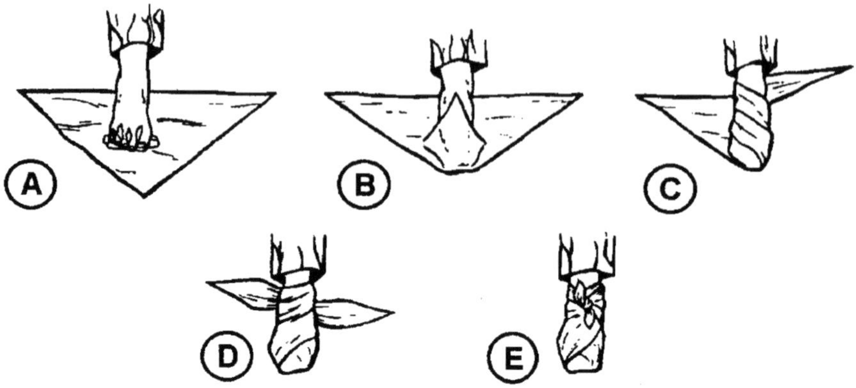

Figure 3-40. Triangular bandage applied to foot (Illustrated A—E).

CHAPTER 4

FIRST AID FOR FRACTURES

4-1. General

A fracture is any break in the continuity of a bone. Fractures can cause total disability or in some cases death by severing vital organs and/or arteries. On the other hand, they can most often be treated so there is a complete recovery. The potential for recovery depends greatly upon the first aid the individual receives before he is moved. First aid includes immobilizing the fractured part in addition to applying lifesaving measures when necessary. The basic splinting principle is to immobilize the joints above and below the fracture.

4-2. Kinds of Fractures

Figure 4-1 depicts types of fractures.

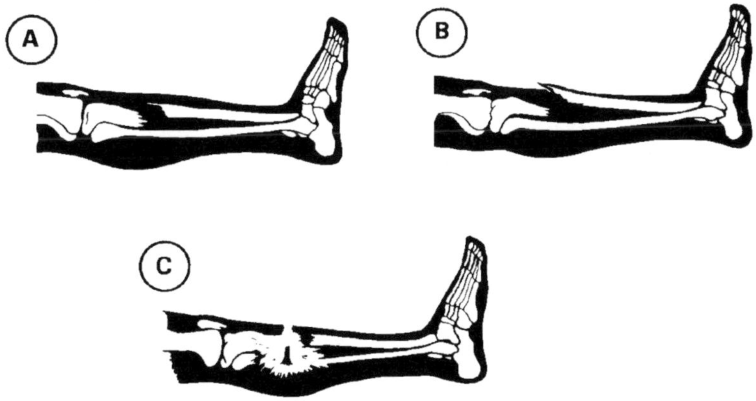

Figure 4-1. Types of fractures (Illustrated A—C).

a. Closed Fracture (Figure 4-1A). A closed fracture is a broken bone that does not break the overlying skin. The tissue beneath the skin may be damaged. A *dislocation* is when a joint, such as a knee, ankle, or shoulder, is not in the proper position. A *sprain* is when the connecting tissues of the joints have been torn. Dislocations and sprains (swelling, possible deformity, and discoloration) should be treated as closed fractures.

b. Open Fracture (Figure 4-1B and 4-1C). An open fracture is a broken bone that breaks (pierces) the overlying skin. The broken bone may come through the skin or a missile such as a bullet or shell fragment may go through the flesh and break the bone.

NOTE

An open fracture is contaminated and subject to infection.

4-3. Signs and Symptoms of Fractures

Indications of a fracture are deformity, tenderness, swelling, pain, inability to move the injured part, protruding bone, bleeding, or discolored skin at the injury site. A sharp pain when the service member attempts to move the part is also a sign of a fracture.

WARNING

DO NOT encourage the casualty to move the injured part in order to identify a fracture since such movement could cause further damage to surrounding tissues and promote shock. If you are not sure whether a bone is fractured, care for the injury as a fracture. At the site of the fracture, the bone ends are sharp and could cause vessel (artery and/or vein) damage.

4-4. Purposes of Immobilizing Fractures

A fracture is immobilized to prevent the sharp edges of the bone from moving and cutting tissue, muscle, blood vessels, and nerves. This reduces pain and helps prevent or control shock. In a closed fracture, immobilization keeps bone fragments from causing an open wound, which can become contaminated and subject to infection.

4-5. Splints, Padding, Bandages, Slings, and Swathes

a. Splints. Splints may be improvised from such items as boards, poles, sticks, tree limbs, or cardboard. If nothing is available for a splint, the chest wall can be used to immobilize a fractured arm and the uninjured leg can be used to immobilize (to some extent) the fractured leg.

b. Padding. Padding may be improvised from such items as a jacket, blanket, poncho, shelter half, or leafy vegetation.

c. Bandages. Bandages may be improvised from belts, rifle slings, kerchiefs, or strips torn from clothing or blankets. Narrow materials such as wire or cord should not be used to secure a splint in place. The application of wire and/or narrow material to an extremity could cause tissue damage and a tourniquet effect.

d. Slings. A sling is a bandage suspended from the neck to support an upper extremity. If a bandage is not available, a sling can be improvised by using the tail of a coat or shirt or pieces of cloth torn from such items as clothing and blankets. The triangular bandage is ideal for this purpose. Remember that the casualty's hand should be higher than his elbow, and the fingers should be showing at all times. The sling should be applied so that the supporting pressure is on the uninjured side.

e. Swathes. Swathes are any bands (pieces of cloth or load bearing equipment [LBE]) that are used to further immobilize a splinted fracture. Triangular and cravat bandages are often used and are called *swathe bandages*. The purpose of the swathe is to immobilize; therefore, the swathe bandage is placed above and/or below the fracture—not over it.

4-6. Procedures for Splinting Suspected Fractures

Before beginning first aid procedures for a fracture, gather whatever splinting materials are available. Ensure that splints are long enough to immobilize the joint above and below the suspected fracture. If possible, use at least four ties (two above and two below the fracture) to secure the splints. The ties should be square knots and should be tied away from the body on the splint. Distal pulses of the affected extremity should be checked before and after the application of the splint.

a. Evaluate the Casualty. Be prepared to perform any necessary lifesaving measures. Monitor the casualty for development of conditions that may require you to perform necessary lifesaving measures.

WARNING

Unless there is immediate life-threatening danger, such as a fire or an explosion, DO NOT move the casualty with a suspected back or neck injury. Improper movement may cause permanent paralysis or death.

WARNING

In a chemical environment, DO NOT remove any protective clothing. Apply the dressings and splints over the garments.

b. *Locate the Site of the Suspected Fracture.*

(1) Ask the casualty for the location of the injury.

- Does he have any pain?
- Where is it tender?
- Can he move the extremity?

NOTE

With the presence of an obvious deformity, do not make the casualty move extremity.

(2) Look for an unnatural position of the extremity.

(3) Look for a bone sticking out (protruding).

c. *Prepare the Casualty for Splinting the Suspected Fracture.*

(1) Reassure the casualty. Tell him that you will be providing first aid for him and that medical help is on the way.

(2) Loosen any tight or binding clothing.

(3) Remove all jewelry from the injured part and place it in the casualty's pocket. Tell the casualty you are doing this because if the jewelry is not removed and swelling occurs later, he may not be able to get it off and further bodily injury could result.

(4) Boots should not be removed from the casualty unless they are needed to stabilize a neck injury or there is actual bleeding from the foot.

d. *Gather Splinting Materials.* If standard splinting materials (splints, padding, and cravats) are not available, gather improvised materials. If splinting material is not available and the suspected fracture **CANNOT** be splinted, then swathes, or a combination of swathes and slings can be used to immobilize the extremity.

e. Pad the Splints. Pad the splints where they touch any bony part of the body, such as the elbow, wrist, knee, ankle, crotch, or armpit areas. Padding prevents excessive pressure on the area, which could lead to circulation problems.

f. Check the Circulation Below the Site of the Injury.

(1) Note any pale, white, or bluish-gray color of the skin, which may indicate impaired circulation. Circulation can also be checked by depressing the toe or fingernail beds and observing how quickly the color returns. A slower return of color to the injured side when compared with the uninjured side indicates a problem with circulation. The fingernail bed is the method to use to check the circulation in a dark-skinned casualty.

(2) Check the temperature of the injured extremity. Use your hand to compare the temperature of the injured side with the uninjured side. The body area below the injury may be colder to the touch indicating poor circulation.

(3) Question the casualty about the presence of numbness, tightness, cold, or tingling sensations.

WARNING

Casualties with fractures of the extremities may show impaired circulation, such as numbness, tingling, cold or pale to bluish skin tone. These casualties should be evacuated by medical personnel and treated as soon as possible. Prompt medical treatment may prevent possible loss of the limb.

WARNING

If it is an open fracture and the bone is protruding from the skin, *DO NOT ATTEMPT TO PUSH THE BONE BACK UNDER THE SKIN.* Apply a field dressing over the wound to protect the area.

g. Apply the Splint in Place.

(1) Splint the fracture in the position found. **DO NOT** attempt to reposition or straighten the injury. If it is an open fracture, stop the bleeding and protect the wound. Cover all wounds with field dressings before applying a splint. Remember to use the casualty's field dressing, not your own.

(2) Place one splint on each side of the fracture. Make sure that the splints reach, if possible, beyond the joints above and below the fracture.

(3) Tie the splints. Secure each splint in place above and below the fracture site with improvised (or actual) cravats. Improvised cravats, such as strips of cloth, belts, or whatever else you have, may be used. With minimal motion to the injured areas, place and tie the splints with the bandages. Push cravats through and under the natural body curvatures, and then gently position improvised cravats and tie in place. Use square knots. Tie all knots on the splint away from the casualty (Figure 4-2). **DO NOT** tie cravats directly over the suspected fracture site.

Figure 4-2. Square knots tied away from casualty.

h. Check the Splint for Tightness.

(1) **CHECK** to be sure that bandages are tight enough to securely hold splinting materials in place, but not so tight that circulation is impaired.

(2) **RECHECK** the circulation after application of the splint. Check the skin color and temperature. This is to ensure that the bandages holding the splint in place have not been tied too tightly. A fingertip check can be made by inserting the tip of the finger between the bandaged knot and the skin.

(3) **MAKE** any necessary adjustment without allowing the splint to become ineffective.

i. Apply a Sling. An improvised sling may be made from any available nonstretching piece of cloth, such as a battle dress uniform (BDU) shirt or trousers, poncho, or shelter half. Slings may also be improvised using the tail of a coat, belt, or a piece of cloth. Figure 4-3 depicts a shirttail used for support. A trousers belt or LBE may also be used for support (Figure 4-4). A sling should place the supporting pressure on the casualty's uninjured side. The supported arm should have the hand positioned slightly higher than the elbow showing the fingers.

Figure 4-3. Shirttail used for support.

Figure 4-4. Belt used for support.

(1) Insert the splinted arm in the center of the sling (Figure 4-5).

Figure 4-5. Arm inserted in center of improvised sling.

(2) Bring the ends of the sling up and tie them at the side (or hollow) of the neck on the uninjured side (Figure 4-6).

Figure 4-6. Ends of improvised sling tied to side of neck.

(3) Twist and tuck the corner of the sling at the elbow (Figure 4-7).

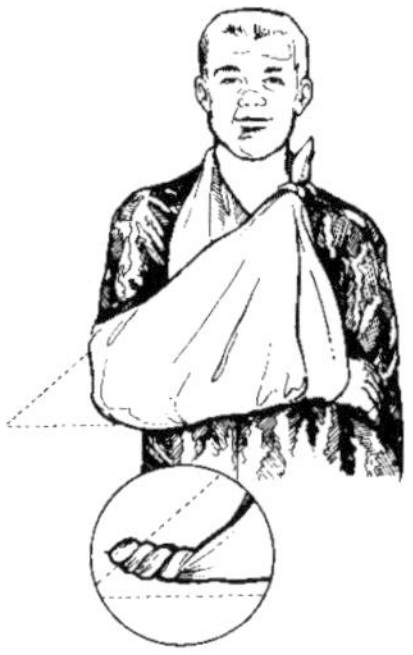

Figure 4-7. Corner of sling twisted and tucked at elbow.

j. Apply a Swathe. You may use any large piece of cloth, service member's belt, or pistol belt, to improvise a swathe.

WARNING

The swathe should not be placed directly on top of the injury, but positioned either above or below the fracture site.

(1) Apply swathes to the injured arm by wrapping the swathe over the injured arm, around the casualty's back, and under the arm on the uninjured side. Tie the ends on the uninjured side (Figure 4-8).

Figure 4-8. Arm immobilized with strip of clothing.

(2) A swathe is applied to an injured leg by wrapping the swathe around both legs and securing it on the uninjured side.

k. Seek Medical Assistance. Notify medical personnel, watch closely for development of life-threatening conditions and/or impaired circulation to the injured extremity. (Refer to Chapter 1 for additional information on life-threatening conditions.)

4-7. Upper Extremity Fractures

Figures 4-9 through 4-17 show how to apply slings, splints, and cravats (swathes) to immobilize and support fractures of the upper extremities. *Although the padding is not visible in some of the illustrations, it is always preferable to apply padding along the injured part for the length of the splint and especially where it touches any bony parts of the body.*

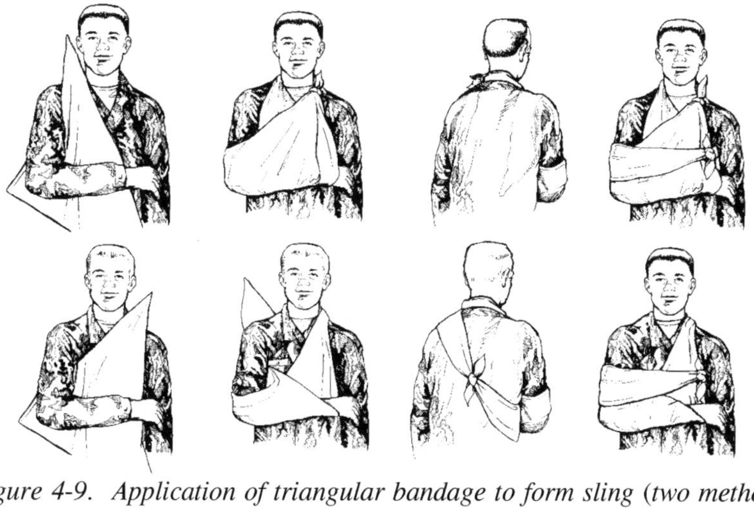

Figure 4-9. Application of triangular bandage to form sling (two methods).

Figure 4-10. Completing sling sequence by twisting and tucking the corner of the sling at the elbow.

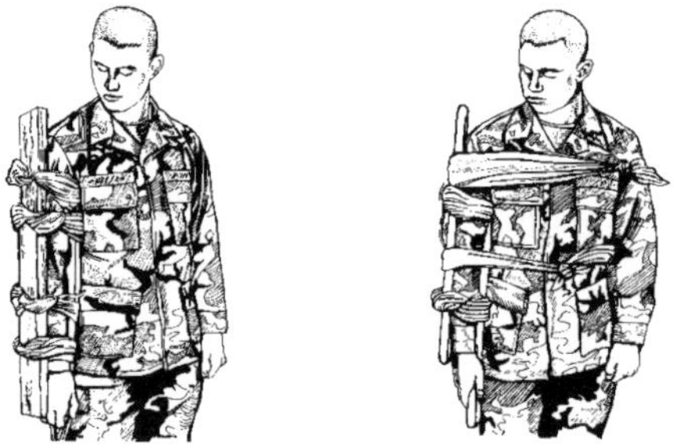

Figure 4-11. Board splints applied to fractured elbow when elbow is not bent (two methods).

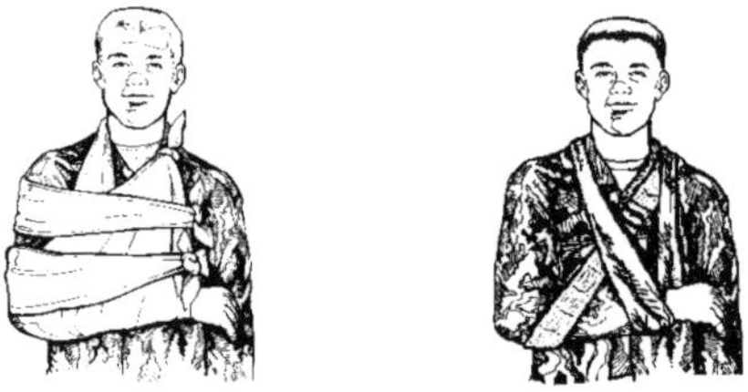

Figure 4-12. Chest wall used as splint for upper arm fracture when no splint is available.

Figure 4-13. Chest wall, sling, and cravat used to immobilize fractured elbow when elbow is bent.

Figure 4-14. Board splint applied to fractured forearm.

Figure 4-15. Fractured forearm or wrist splinted with sticks and supported with tail of shirt and strips of material.

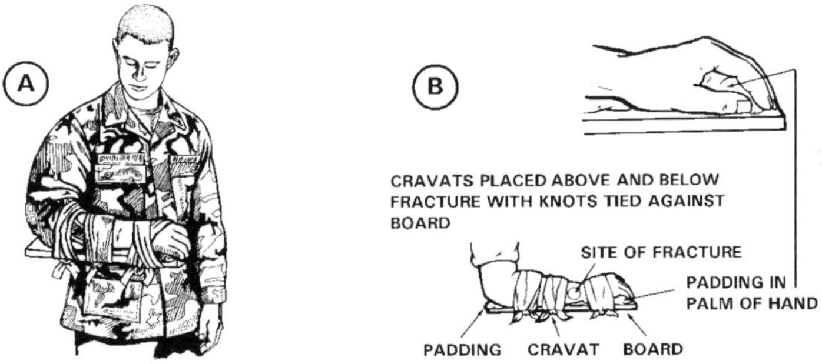

Figure 4-16. Board splint applied to fractured wrist and hand (Illustrated A—B).

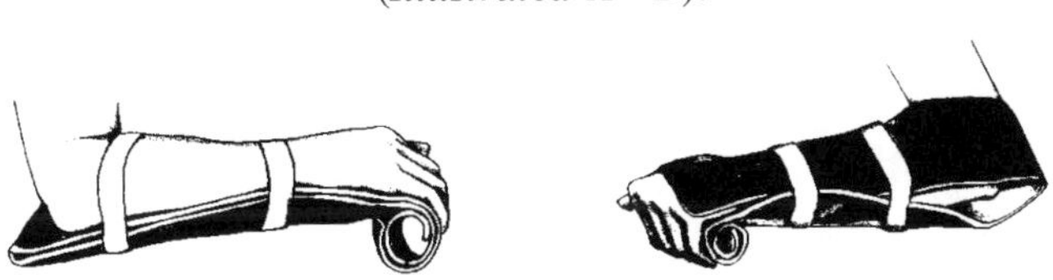

Figure 4-17. SAM® splint applied to fractured wrist or forearm.

4-8. Lower Extremity Fractures

Figures 4-18 through 4-24 show how to apply splints to immobilize fractures of the lower extremities. *Although padding is not visible in some of the illustrations, it is always preferable to apply padding along the injured part for the length of the splint and especially where it touches any bony parts of the body.*

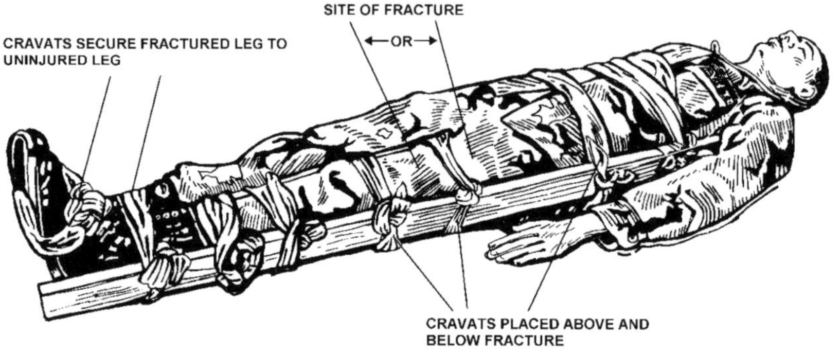

Figure 4-18. Board splints applied to fractured hip or thigh.

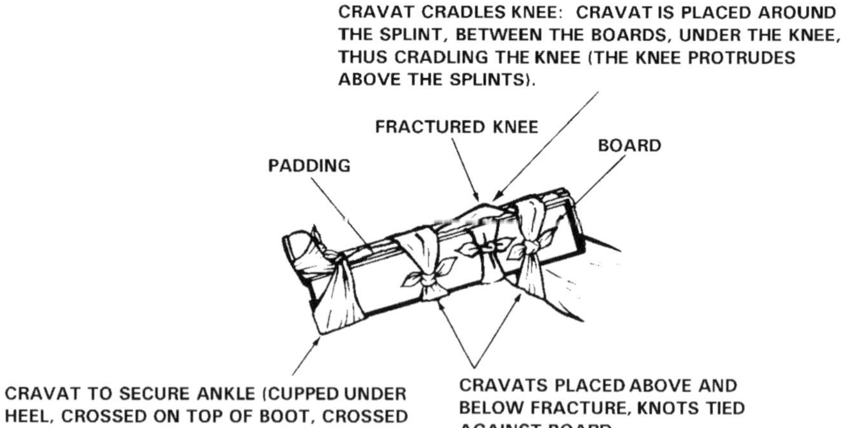

Figure 4-19. Board splint applied to fractured or dislocated knee.

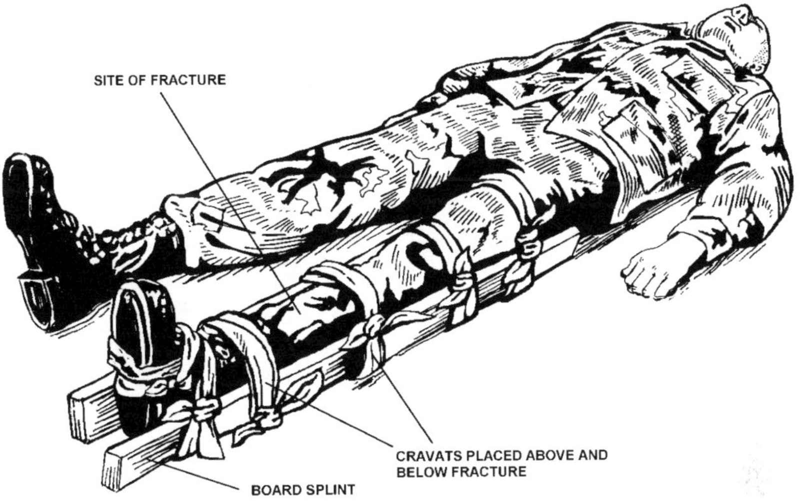

Figure 4-20. Board splints applied to fractured lower leg or ankle.

Figure 4-21. SAM® splint applied to fractured lower leg or ankle.

Figure 4-22. Improvised splints applied to fractured lower leg or ankle.

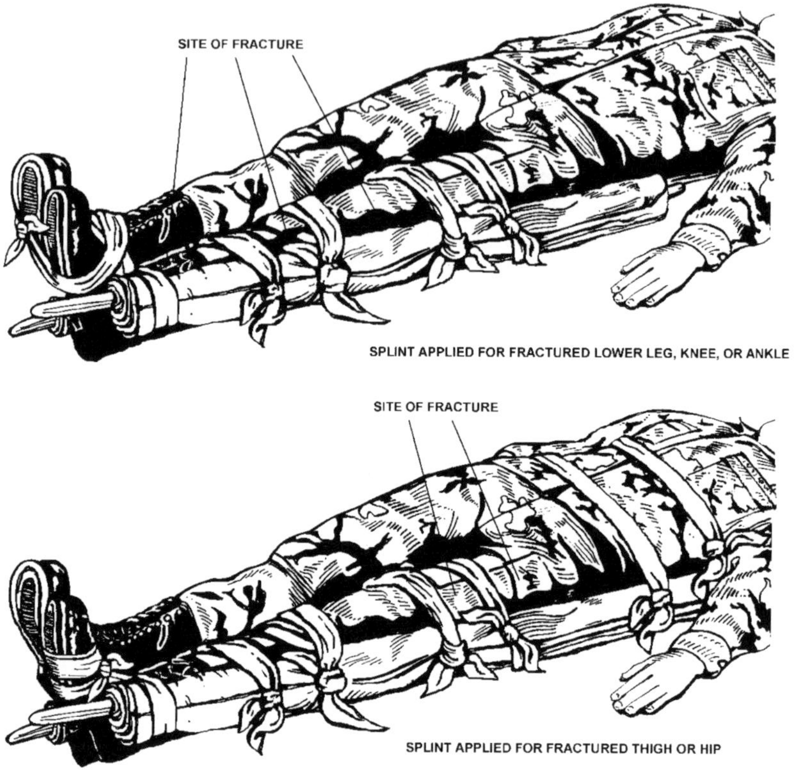

Figure 4-23. Poles rolled in a blanket and used as splints applied to fractured lower extremity.

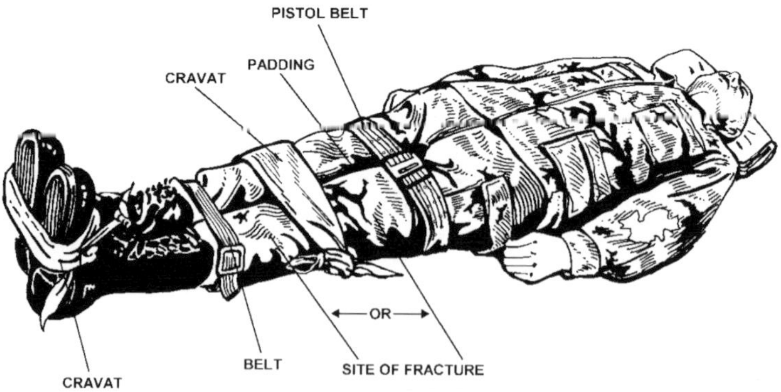

Figure 4-24. Uninjured leg used as splint for fractured leg (anatomical splint).

4-9. Jaw, Collarbone, and Shoulder Fractures

a. Apply a cravat to immobilize a fractured jaw as illustrated in Figure 4-25. Direct all bandaging support to the top of the casualty's head, not to the back of his neck. If incorrectly placed, the bandage will pull the casualty's jaw back and interfere with his breathing.

Figure 4-25. Fractured jaw immobilized.

> **WARNING**
>
> **Casualties with lower jaw (mandible) fractures cannot be laid flat on their backs because facial muscles will relax and may cause an airway obstruction.**

b. Apply two belts, a sling, and a cravat to immobilize a fractured collarbone, as illustrated in Figure 4-26.

Figure 4-26. Application of belts, sling, and cravat to immobilize a fractured collarbone.

c. Apply a sling and a cravat to immobilize a fractured or dislocated shoulder, using the technique illustrated in Figure 4-27.

Figure 4-27. Application of sling and cravat to immobilize a fractured or dislocated shoulder.

4-10. Spinal Column Fractures

It is often impossible to be sure a casualty has a fractured spinal column. Be suspicious of any back injury, especially if the casualty has fallen or if his back has been sharply struck or bent. If a casualty has received such an injury and does not have feeling in his legs or cannot move them, you can be reasonably sure that he has a severe back injury, which should be managed as a fracture. Remember, that the possibility of a neck fracture or injury to the back should always be suspected, and it is often impossible to be sure if a casualty has a fractured spinal column. If the spine is fractured, bending it can cause the sharp bone fragments to bruise or cut the spinal cord and result in permanent paralysis or death (Figure 4-28A). The spinal column must maintain normal spinal position at the lower back (lumbar region) to help remove pressure from the spinal cord.

a. If the casualty is not to be transported until medical personnel arrive—

- Caution him not to move. Ask him if he is in pain or if he is unable to move any part of his body.

- Leave him in the position in which he is found. **DO NOT** move any part of his body, unless he is in imminent danger.

- If the casualty is lying face up, slip a blanket or other supporting material under the arch of his lower back to help support the spine in a normal position (Figure 4-28B). Take care not to place so much bulky padding as to cause potential damage by causing undo pressure on the spine. If he is lying face down, **DO NOT** put anything under any part of his body.

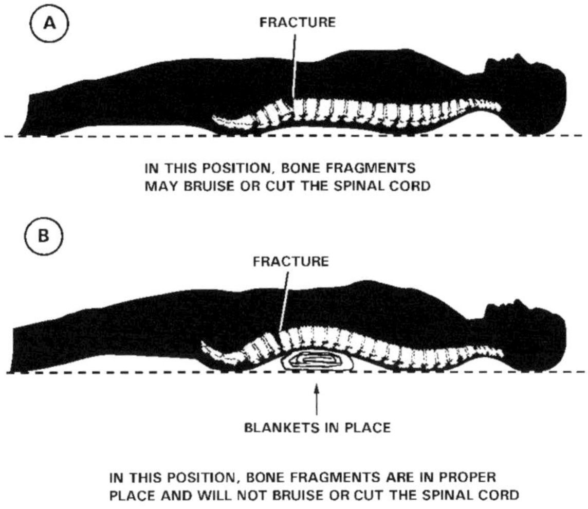

Figure 4-28. Spinal column must maintain a normal spine position.

b. If the casualty must be transported to a safe location before medical personnel arrive and if the casualty is in a—

- Face-up position, transport him by litter or use a firm substitute, such as a wide board or a door longer than his height. Loosely tie the casualty's wrists together over his waistline, using a cravat or a strip of cloth. Tie his feet together to prevent the accidental dropping or shifting of his legs. Lay a folded blanket across the litter where the arch of his back is to be placed. Using a four-man team (Figure 4-29), place the casualty on the litter without bending his spinal column or his neck.

Figure 4-29. Placing face-up casualty with fractured back onto litter.

- The number two man positions himself at the casualty's head. His responsibility is to provide manual in-line (neutral) stabilization of the head and neck. The number three, and four men position themselves on one side of the casualty; all kneel on one knee along the side of the casualty. The number one man positions himself to the opposite side of the casualty (or can be on the same side of number three and four). The numbers two, three, and four men gently place their hands under the casualty. The number one man on the opposite side places his hands under the injured part to assist.

- When all four men are in position to lift, the number two man commands, "**PREPARE TO LIFT**" and then, "**LIFT**." All men, in unison, gently lift the casualty about 8 inches. Once the casualty is lifted, the number one man recovers and slides the litter under the casualty, ensuring that the blanket is in proper position. The number one man then returns to his original lift position (Figure 4-29).

- When the number two man commands, "**LOWER CASUALTY**," all men, in unison, gently lower the casualty onto the litter.

- Facedown position, he must be transported in this same position. The four-man team lifts him onto a regular or improvised litter, keeping the spinal column in a normal spinal position. If a regular litter is used, first place a folded blanket on the litter at the point where the chest will be placed.

4-11. Neck Fractures

A fractured neck is extremely dangerous. Bone fragments may bruise or cut the spinal cord just as they might in a fractured back.

a. If the casualty is not to be transported until medical personnel arrive—

- Caution him not to move. Moving may cause permanent injury or death.

- Leave the casualty in the position in which he is found. If his neck and head (cervical spine) are in an abnormal position, immediately immobilize the neck and head.

- Keep his head still, if the casualty is lying face up, raise his shoulders slightly, and slip a roll of cloth that has the bulk of a bath towel under his neck (Figure 4-31). The roll should be thick enough to arch his neck only slightly, leaving the back of his head on the ground. **DO NOT** bend his neck or head forward. **DO NOT** raise or twist his head. Immobilize the casualty's head (Figure 4-32). Do this by padding heavy objects (such as rocks or his boots filled with dirt, sand, gravel, or rock) and placing them on each side of his head. If it is necessary to use boots, after filling them, tie the top tightly or stuff with pieces of cloth to secure the contents.)

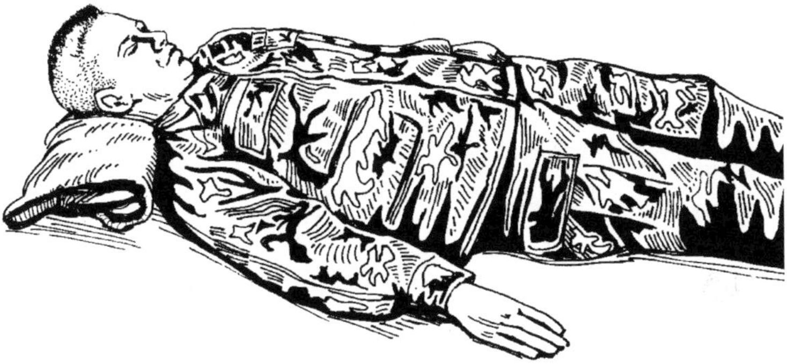

Figure 4-30. Casualty with roll of cloth (bulk) under neck.

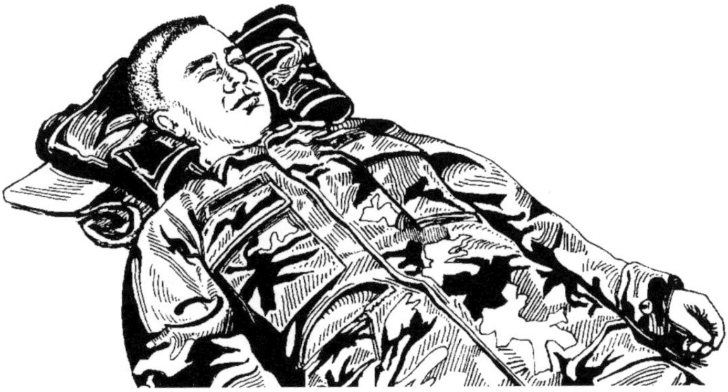

Figure 4-31. Immobilization of fractured neck.

- **DO NOT** move him if the casualty is lying face down. Immobilize the head and neck by padding heavy objects and placing them on each side of his head. DO NOT put a roll of cloth under the neck. **DO NOT** bend the neck or head, nor roll the casualty onto his back.

b. If the casualty must be prepared for transportation before medical personnel arrive—

- If the casualty has a fractured neck, at least two persons are needed because the casualty's head and trunk must be moved in unison. The two persons must work in close coordination (Figure 4-32) to avoid bending of the neck.

- A wide board is placed lengthwise beside the casualty. It should extend at least 4 inches beyond the casualty's head and feet (Figure 4-32A).

- If the casualty is lying face up, the number one man steadies the casualty's head and neck between his hands. At the same time, the number two man positions one foot and one knee against the board to prevent it from slipping. He then grasps the casualty underneath his shoulder and hip and gently slides him onto the board (Figure 4-32B).

- If the casualty is lying face down, the number one man steadies the casualty's head and neck between his hands, while the number two man gently rolls the casualty over onto the board (Figure 4-32C).

- The number one man continues to steady the casualty's head and neck. The number two man simultaneously raises the casualty's shoulders slightly, places padding under his neck, and immobilizes the casualty's head (Figures 4-32D—E).

- Any improvised supports are secured in position with a cravat or strip of cloth extended across the casualty's forehead and under the board (Figure 4-32D).

- The board is lifted onto a litter or blanket in order to transport the casualty (Figure 4-32E).

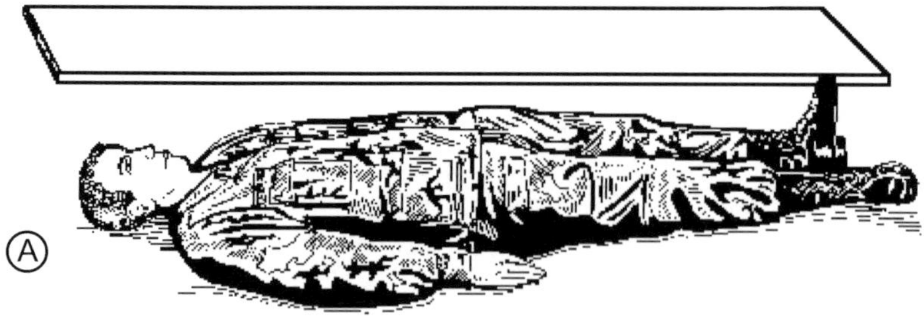

Figure 4-32. Preparing casualty with fractured neck for transportation (Illustrated A—E).

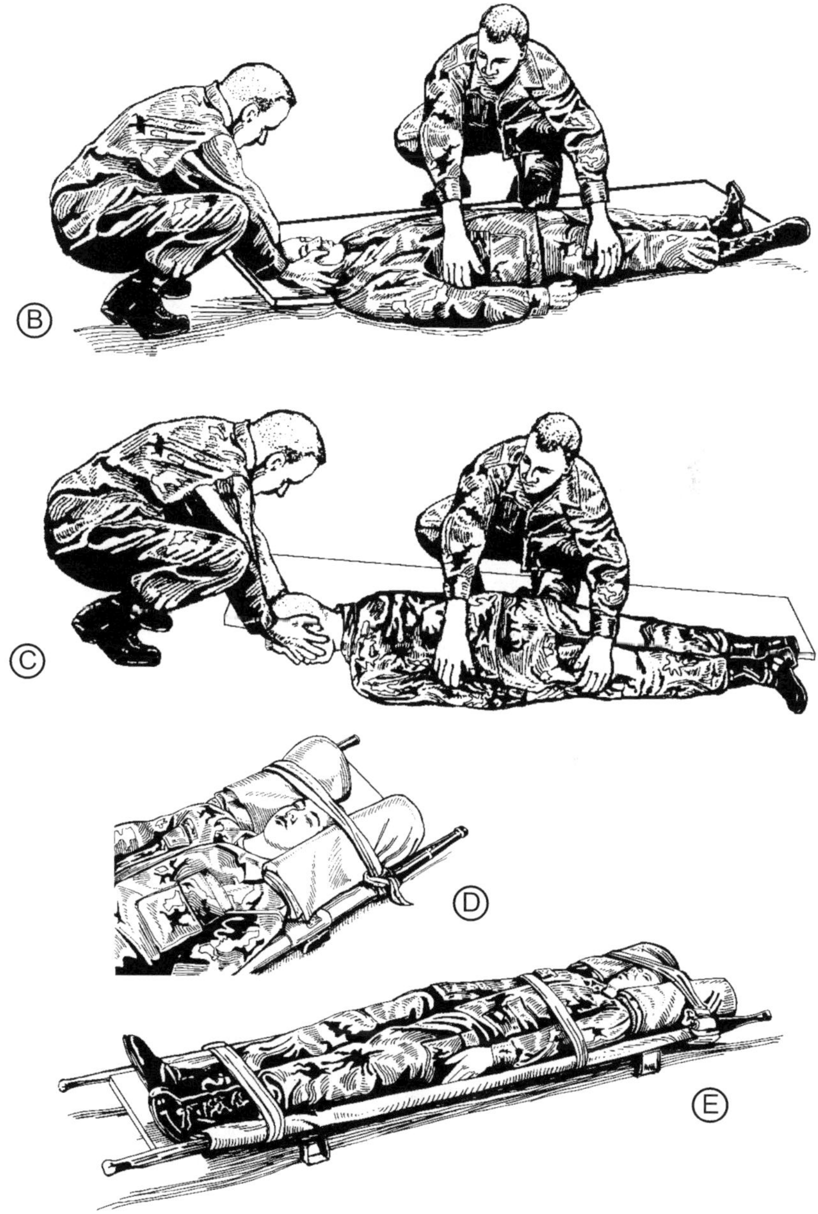

Figure 4-32. Preparing casualty with fractured neck for transportation (Illustrated A—E) (Continued).

CHAPTER 5

FIRST AID FOR CLIMATIC INJURIES

5-1. General

a. It is desirable, but not always possible, for an individual's body to become adjusted (acclimated) to an environment.

(1) The service members physical condition determines the amount of time their bodies need to adjust to the environment. Even those individuals in good physical condition need time before working or training in extremes of hot or cold weather. Climate-related injuries are usually preventable; prevention is both an individual and leadership responsibility.

(2) Several factors contribute to health and well-being in any environment—

- Diet.
- Sleep and rest.
- Exercise.
- Suitable clothing.

(3) Diet should be suited to an individual's needs in a particular climate. A special diet started for any purpose (such as weight reduction) should be done with appropriate medical supervision.

WARNING

Service members should use extreme caution when starting fad diets or taking over-the-counter herbal supplements. Medical records revealed that deaths and severe injuries occurred in individuals using dietary/ herbal supplements without medical monitoring.

NOTE

Weight loss and the use of weight loss supplements should be supervised by a trained health care provider.

(4) Specialized clothing and equipment (such as cold weather gear) for a specific environment should be obtained and used properly.

b. For information on the prevention of heat and cold injuries, refer to FM 21-10/Marine Corps Reference Publication (MCRP) 4-11.1D.

5-2. Heat Injuries

a. Heat injuries are environmental injuries. They may result when a service member—

- Is exposed to extreme heat, such as from the sun or from high temperatures.
- Does not wear proper clothing.
- Is in MOPP gear.
- Is inside closed spaces, such as inside an armored vehicle.
- Wears body armor.

b. Heat injury can be divided into three categories: heat cramps, heat exhaustion, and heatstroke.

c. Each service member must be able to recognize and give first aid for heat injuries.

WARNING

The heat casualty should be continually monitored for development of conditions which may require the performance of necessary basic lifesaving measures.

CAUTION

Do not use salt solutions in first aid procedures for heat injuries.

(1) Check the casualty for signs and symptoms of *cramping*.

- *Signs and symptoms.* Cramping is caused by an imbalance of chemicals (called electrolytes) in the body as a result of excessive sweating. This condition causes the casualty to exhibit:

- Cramping in the extremities (arms and legs).
- Abdominal (stomach) cramps.
- Excessive sweating.

NOTE

Thirst may or may not occur. Cramping can occur without the service member being thirsty.

- *First aid measures.*

- Move the casualty to a cool, shady area or improvise shade if none is available.

- Loosen his clothing (if not in a chemical environment).

NOTE

In a chemical environment, transport the heat casualty to a noncontaminated area as soon as the mission permits.

- Have him slowly drink at least one canteen full of water. (The body absorbs cool water faster than warm or cold water; therefore, cool water is preferred if it is available.)

- Seek medical assistance should cramps continue.

(2) Check the casualty for signs and symptoms of *heat exhaustion*.

- *Signs and symptoms*. Heat exhaustion is caused by loss of body fluids (dehydration) through sweating without adequate fluid replacement. It can occur in an otherwise fit individual who is involved in physical exertion in any hot environment especially if the service member is not acclimatized to that environment. These signs and symptoms are—

- Excessive sweating with pale, moist, cool skin.
- Headache.
- Weakness.

- Dizziness.
- Loss of appetite.
- Cramping.
- Nausea (with or without vomiting).
- Urge to defecate.
- Chills (gooseflesh).
- Rapid breathing.
- Tingling of hands and/or feet.
- Confusion.

- *First aid measures.*
  - Move the casualty to a cool, shady area or improvise shade if none is available.
  - Loosen or remove his clothing and boots (unless in a chemical environment); pour water on him and fan him.
  - Have him slowly drink at least one canteen of water.
  - Elevate his legs.
  - If possible, the casualty should not participate in strenuous activity for the remainder of the day.
  - Monitor the casualty until the symptoms are gone, or medical assistance arrives.

(3) Check the casualty for signs and symptoms of *heatstroke*.

WARNING

Heatstroke is a medical emergency which may result in death if care is delayed.

- *Signs and symptoms.* A service member suffering from heatstroke has been exposed to high temperatures (such as direct sunlight) or been dressed in protective overgarments, which causes the body temperature to rise. Heatstroke occurs more rapidly in service members who are engaged in work or other physical activity in a high heat environment. Heatstroke is caused by a failure of the body's cooling mechanism which includes a decrease in the body's ability to produce sweat. The casualty's skin is red (flushed), hot, and dry. He may experience weakness, dizziness, confusion, headaches, seizures, nausea, stomach pains or cramps, and his respiration and pulse may be rapid and weak. Unconsciousness and collapse may occur suddenly.

- *First aid measures.* Cool casualty immediately by—

  - Moving him to a cool, shady area or improvising shade if none is available.

  - Loosening or removing his clothing (except in a chemical environment).

  - Spraying or pouring water on him; fanning him to permit the coolant effect of evaporation.

  - Massaging his extremities and skin, which increases the blood flow to those body areas, thus aiding the cooling process.

  - Elevating his legs.

  - Having him slowly drink at least one canteen full of water if he is conscious.

NOTE

Start cooling casualty immediately. Continue cooling while awaiting transportation and during transport to an MTF.

- *Medical assistance.* Seek medical assistance because the casualty should be transported to an MTF as soon as possible. Do not interrupt the cooling process or lifesaving measures to seek help; if someone else is present send them for help. The casualty should be continually monitored for development of conditions that may require the performance of necessary basic lifesaving measures.

d. Table. See Table 5-1 for further information.

Table 5-1. Heat Injuries

INJURIES	SIGNS AND SYMPTOMS	FIRST AID[1]
HEAT CRAMPS	THE CASUALTY EXPERIENCES MUSCLE CRAMPS OF THE ARMS, LEGS, AND/OR STOMACH. THE CASUALTY MAY ALSO HAVE EXCESSIVE SWEATING.	1. MOVE THE CASUALTY TO A COOL SHADY AREA OR IMPROVISE SHADE AND LOOSEN CLOTHING.[2] 2. HAVE HIM SLOWLY DRINK AT LEAST ONE CANTEEN FULL OF COOL WATER SLOWLY. 3. MONITOR THE CASUALTY AND GIVE HIM MORE WATER AS TOLERATED.
HEAT EXHAUSTION	THE CASUALTY EXPERIENCES HEAVY SWEATING WITH PALE, MOIST, COOL SKIN; HEADACHE, WEAKNESS, DIZZINESS, AND/OR LOSS OF APPETITE, HEAT CRAMPS, NAUSEA (WITH OR WITHOUT VOMITING), URGE TO DEFECATE, CHILLS (GOOSE-FLESH), RAPID BREATHING, CONFUSION, AND TINGLING OF THE HANDS AND/OR FEET.	1. MOVE THE CASUALTY TO A COOL, SHADY AREA OR IMPROVISE SHADE AND LOOSEN OR REMOVE HIS CLOTHING.[2] 2. POUR WATER ON HIM AND FAN HIM TO PERMIT THE COOLANT EFFECT OF EVAPORATION. 3. HAVE HIM SLOWLY DRINK AT LEAST ONE CANTEEN FULL OF COOL WATER. 4. ELEVATE THE CASUALTY'S LEGS. 5. SEEK MEDICAL ASSISTANCE IF SYMPTOMS CONTINUE; MONITOR UNTIL SYMPTOMS ARE GONE OR MEDICAL ASSISTANCE ARRIVES.
HEATSTROKE[3] (SUNSTROKE)	THE CASUALTY STOPS SWEATING (RED [FLUSHED] HOT, DRY SKIN). HE FIRST MAY EXPERIENCE HEADACHE, DIZZINESS, NAUSEA, FAST PULSE AND RESPIRATION, SEIZURES, AND MENTAL CONFUSION. HE MAY COLLAPSE	1. MOVE THE CASUALTY TO A COOL, SHADY AREA OR IMPROVISE SHADE AND LOOSEN OR REMOVE HIS CLOTHING, REMOVE THE OUTER GARMENTS AND PROTECTIVE

Table 5-1. Heat Injuries (Continued)

INJURIES	SIGNS AND SYMPTOMS	FIRST AID[1]
	AND SUDDENLY BECOME UNCONSCIOUS. **THIS IS A MEDICAL EMERGENCY.**	CLOTHING IF THE SITUATION PERMITS.[2] 2. START COOLING THE CASUALTY IMMEDIATELY. SPRAY OR POUR WATER ON HIM. FAN HIM. MASSAGE HIS EXTREMITIES AND SKIN. 3. ELEVATE HIS LEGS. 4. IF CONSCIOUS, HAVE HIM SLOWLY DRINK AT LEAST ONE CANTEEN FULL OF COOL WATER. 5. SEEK MEDICAL AID. CONTINUE COOLING WHILE AWAITING TRANSPORT AND CONTINUE FIRST AID WHILE EN ROUTE.

LEGEND:

1 THE FIRST AID PROCEDURE FOR HEAT RELATED INJURIES CAUSED BY WEARING INDIVIDUAL PROTECTIVE EQUIPMENT (IPE) IS TO MOVE THE CASUALTY TO A CLEAN AREA AND GIVE HIM WATER TO DRINK.
2 WHEN IN A CHEMICAL ENVIRONMENT, DO NOT LOOSEN OR REMOVE THE CASUALTY'S CLOTHING.
3 CAN BE FATAL IF NOT PROVIDED FIRST AID AND MEDICAL TREATMENT PROMPTLY.

5-3. Cold Injuries

Cold injuries are most likely to occur when conditions are moderately cold, but accompanied by wet or windy conditions. Cold injuries can usually be prevented. Well-disciplined and well-trained service members can be protected even in the most adverse circumstances. They and their leaders must know the hazards of exposure to the cold. They must know the importance of personal hygiene, exercise, care of the feet and hands, and the use of protective clothing.

a. Contributing Factors.

(1) Temperature, humidity, precipitation, and wind greatly increase likelihood of cold injuries, and the service members with wet clothing are at great risk of cold injuries. Riverine operations (river, swamp, and stream crossings) increase likelihood of cold injuries. Low temperatures and low relative humidity (dry cold) promote frostbite. Higher temperatures, together with moisture, promote immersion syndrome. Windchill accelerates the loss of body heat and may aggravate cold injuries.

(2) Relatively stationary activities such as being in an observation post or on guard duty increase the service member's vulnerability to cold injury. Also, a service member is more likely to receive a cold injury if he is—

- In contact with the ground (such as marching, performing guard duty, or engaging in other outside activities).

- Immobile for long periods (such as while riding in an unheated or open vehicle).

- Standing in water, such as in a foxhole.

- Out in the cold for days without being warmed.

- Deprived of an adequate diet and rest.

- Not able to take care of his personal hygiene.

(3) Physical fatigue contributes to apathy, which leads to inactivity, personal neglect, carelessness, and reduced heat production. In turn, these increase the risk of cold injury. Service members with prior cold injuries have a higher-than-normal risk of subsequent cold injury; not necessarily involving the body part previously injured.

(4) Depressed or unresponsive service members are also vulnerable because they are less active. These service members tend to be careless about precautionary measures, especially warming activities, when cold injury is a threat.

(5) Excessive use of alcohol or drugs leading to faulty judgment or unconsciousness in a cold environment increases the risk of becoming a cold injury casualty.

b. *Signs and Symptoms.* Once a service member becomes familiar with the factors that contribute to cold injury, he must learn to recognize cold injury signs and symptoms.

(1) Many service members suffer cold injury without realizing what is happening to them. They may be cold and generally uncomfortable. These service members often do not notice the injured part because it is already numb from the cold.

(2) Superficial cold injury usually can be detected by numbness or tingling sensations. These signs and symptoms often can be relieved simply by loosening boots or other clothing and by exercising to improve circulation. In more advanced cases involving deep cold injury, the service member often is not aware that there is a problem until the affected part feels like a stump or block of wood.

(3) Outward signs of cold injury include discoloration of the skin at the site of injury. In light-skinned persons, the skin first reddens and then becomes pale or waxy white. In dark-skinned persons, grayness in the skin is usually evident. An injured foot or hand feels cold to the touch. Swelling may be an indication of deep injury. Also note that blisters may occur after rewarming the affected parts. Service members should work in pairs (buddy teams) to check each other for signs of discoloration and other symptoms.

c. *First Aid Measures.* First aid for cold injuries depends on whether they are superficial or deep. Rewarming the affected part using body heat can adequately treat cases of superficial cold injury. (For example, this can be done by covering cheeks with hands, putting fingertips in armpits, or placing the casualty's feet under the clothing of a buddy [next to his belly].) The injured part should **NOT** be massaged, exposed to a fire or stove, rubbed with snow, slapped, chafed, or soaked in cold water. Walking on injured feet should be avoided. Deep cold injury (frostbite) is very serious and requires prompt first aid to avoid or to minimize the loss of parts or all of the fingers, toes, hands, or feet. The sequence for treating cold injuries depends on whether the condition is life-threatening. The first priority in managing cold injuries is to remove the casualty from the cold environment (such as building an improvised shelter). Other injuries the casualty may have are provided first aid simultaneously while waiting for transportation or evacuation. If the casualty is to be transported in a nonmedical vehicle, first aid measures should be continued en route to the MTF.

d. *Conditions Caused by Cold.* Conditions caused by cold include chilblain, immersion syndrome (immersion foot and trench foot), frostbite, snow blindness, dehydration, and hypothermia.

(1) *Chilblain.*

- *Signs and symptoms.* Chilblain is caused by repeated prolonged exposure of bare skin at temperatures from 60° Fahrenheit (F) to

32°F, or 20°F for acclimated, dry, unwashed skin. The area may be acutely swollen, red, tender, and hot with itchy skin. There may be no loss of skin tissue in untreated cases but continued exposure may lead to infected, ulcerated, or bleeding lesions.

- *First aid measures.* Within minutes, the area usually responds to locally applied body heat. Rewarm the affected part by applying firm steady pressure with your hands, or placing the affected part under your arms or against the stomach of a buddy. **DO NOT** rub or massage affected areas.

NOTE

Medical personnel should evaluate the injury, because signs and symptoms of tissue damage may be slow to appear.

(2) *Immersion syndrome (immersion foot and trench foot).* Immersion foot and trench foot are injuries that result from fairly long exposure of the feet to wet conditions at temperatures from approximately 32°F to 50°F. Inactive feet in damp or wet socks and boots, or tightly laced boots which impair circulation, are even more susceptible to injury. This injury can be very serious; it can lead to loss of toes or parts of the feet. If exposure of the feet has been prolonged and severe, the feet may swell so much that pressure closes the blood vessels and cuts off circulation. Should an immersion injury occur, dry the feet thoroughly and transport the casualty to an MTF by the fastest means possible.

- *Signs and symptoms.* At first, the parts of the affected foot are cold and painless, the pulse is weak, and numbness may be present. Second, the parts may feel hot, and burning and shooting pains may begin. In later stages, the skin is pale with a bluish cast and the pulse decreases. Other signs and symptoms that may follow are blistering, swelling, redness, heat, hemorrhaging (bleeding), and gangrene.

- *First aid measures.* First aid measures are required for all stages of immersion syndrome injury. Rewarm the injured part gradually by exposing it to warm air. Protect it from trauma and secondary infections. Dry, loose clothing or several layers of warm coverings are preferable to extreme heat. Under no circumstances should the injured part be exposed to an open fire. Elevate the injured part to relieve the swelling. Transport the casualty to an MTF as soon as possible. When the part is rewarmed, the casualty often feels a burning sensation and pain. Symptoms may persist for days or weeks even after rewarming.

NOTE

When providing first aid for immersion foot and trench foot—**DO NOT** massage the injured part. **DO NOT** moisten the skin. **DO NOT** apply heat or ice.

(3) *Frostbite.* Frostbite is the injury of tissue caused from exposure to cold, usually below 32°F depending on the windchill factor, duration of exposure, and adequacy of protection. Individuals with a history of cold injury are likely to suffer an additional cold injury. The body parts most easily frostbitten are the cheeks, nose, ears, chin, forehead, wrists, hands, and feet. Frostbite may involve only the skin (superficial), or it may extend to a depth below the skin (deep). Deep frostbite is very serious and requires prompt first aid to avoid or to minimize the loss of parts or all of the fingers, toes, hands, or feet.

- *Signs and symptoms.*
  - Loss of sensation (numb feeling) in any part of the body.
  - Sudden blanching (whitening) of the skin of the affected part, followed by a momentary tingling sensation.
  - Redness of skin in light-skinned service members; grayish coloring in dark-skinned service members.
  - Blisters.
  - Swelling or tender areas.
  - Loss of previous sensation of pain in affected area.
  - Pale, yellowish, waxy-looking skin.
  - Frozen tissue that feels solid (or wooden) to the touch.

CAUTION

Deep frostbite is a very serious injury and requires immediate first aid and subsequent medical treatment to avoid or minimize loss of body parts.

- *First aid measures.*

  - *Face, ears, and nose.* Cover the casualty's affected area with his and/or your bare hands until sensation and color return.

  - *Hands.* Open the casualty's field jacket and shirt. (In a chemical environment, do not loosen or remove the clothing and protective overgarments.) Place the affected hands under the casualty's armpits. Close the field jacket and shirt to prevent additional exposure.

  - *Feet.* Remove the casualty's boots and socks if he does not need to walk any further to receive additional treatment. (Thawing the casualty's feet and forcing him to walk on them will cause additional pain and injury.) Place the affected feet under clothing and against the body of another service member.

WARNING

DO NOT attempt to thaw the casualty's feet or other frozen areas if he will be required to walk or travel to an MTF for additional medical treatment. The possibility of additional injury from walking is less when the feet are frozen than when they are thawed. (However, if possible avoid walking.) Thawing in the field increases the possibilities of infection, gangrene, or other injury.

NOTE

Thawing may occur spontaneously during transportation to the MTF; this cannot be avoided since the body in general must be kept warm.

In all of the above areas, ensure that the casualty is kept warm and that he is covered (to avoid further injury). Seek medical treatment as soon as possible. Reassure the casualty, protect the affected area from further injury by covering it lightly with a blanket or any dry clothing, and seek shelter out of the wind. Remove or loosen constricting clothing (except in a contaminated environment) and increase insulation. Ensure the casualty exercises as much as possible, avoiding trauma to the injured part, and is prepared for pain when thawing occurs. Protect the frostbitten part from additional injury. **DO NOT**—

- Rub the injured part with snow or apply cold water soaks.

- Warm the part by massage or exposure to open fire because the frozen part may be burned due to the lack of feeling.

- Use ointments or other salves.

- Manipulate the part in any way to increase circulation.

- Use alcohol or tobacco because this reduces the body's resistance to cold.

NOTE

Remember, when freezing extends to a depth below the skin, it is a much more serious injury. Extra care is required to reduce or avoid the chances of losing all or part of the toes or feet. This also applies to the fingers and hands.

(4) *Snow blindness.* Snow blindness is the effect that glare from an ice field or snowfield has on the eyes. It is more likely to occur in hazy, cloudy weather than when the sun is shining. Glare from the sun will cause an individual to instinctively protect his eyes. However, in cloudy weather, he may be overconfident and expose his eyes longer than when the threat is more obvious. He may also neglect precautions such as the use of protective eyewear. Waiting until discomfort (pain) is felt before using protective eyewear is dangerous because a deep burn of the eyes may already have occurred.

- *Signs and symptoms.* Symptoms of snow blindness are a sensation of grit in the eyes with pain in and over the eyes, made worse by moving the eyeball. Other signs and symptoms are watering, redness, headache, and increased pain on exposure to light.

- *First aid measures.* First aid measures consist of blindfolding or covering the eyes with a dark cloth which stops painful eye movement. Complete rest is desirable. If further exposure to light is not preventable, the eyes should be protected with dark bandages or the darkest glasses available. Once unprotected exposure to sunlight stops, the condition usually heals in a few days without permanent damage. The casualty should be evacuated to the nearest MTF.

(5) *Dehydration.* Dehydration occurs when the body loses too much fluid. A certain amount of body fluid is lost through normal body processes. A normal daily intake of liquids replaces these losses. When individuals are engaged in any strenuous exercises or activities, fluid is lost

through sweating and this loss creates an imbalance of fluids in the body, and if not matched by rehydration it can contribute to dehydration. The danger of dehydration is as prevalent in cold regions as it is in hot regions. In hot weather, the individual is aware of his body losing fluids through sweat. In cold weather, however, it is extremely difficult to realize that this condition exists since sweating is not as apparent as in a hot environment. The danger of dehydration in cold weather operations is a serious problem. In cold climates, sweat evaporates so rapidly or is absorbed so thoroughly by layers of heavy clothing that it is rarely visible on the skin. Dehydration also occurs during cold weather operations because drinking is inconvenient. Dehydration will weaken or incapacitate a casualty for a few hours, or sometimes, several days. Because rest is an important part of the recovery process, casualties must take care that limited movement during their recuperative period does not enhance the risk of becoming a cold injury casualty.

- *Signs and symptoms.* The symptoms of cold weather dehydration are similar to those encountered in heat exhaustion. The mouth, tongue, and throat become parched and dry, and swallowing becomes difficult. The casualty may have nausea (with or without vomiting) along with extreme dizziness and fainting. The casualty may also feel generally tired and weak and may experience muscle cramps. Focusing the eyes may also become difficult.

- *First aid measures.* The casualty should be kept warm and his clothes should be loosened (if not in a chemical environment) to allow proper circulation. Shelter from wind and cold must be provided. Fluid replacement should begin immediately and the service member transported to an MTF as soon as possible.

(6) *Hypothermia (general cooling).* When exposed to prolonged cold weather a service member may become both mentally and physically numb, thus neglecting essential tasks or requiring more time and effort to achieve them. Under some conditions (particularly cold water immersion), even a service member in excellent physical condition may die in a matter of minutes. The destructive influence of cold on the body is called *hypothermia.* This means bodies lose heat faster than they can produce it. Hypothermia can occur from exposure to temperatures either above or below freezing, especially from immersion in cold water, wet-cold conditions, or from the effect of wind. Physical exhaustion and insufficient food intake may also increase the risk of hypothermia. General cooling of the entire body to a temperature below 95°F is caused by continued exposure to low or rapidly dropping temperatures, cold moisture, snow, or ice. Fatigue, poor physical condition, dehydration, faulty blood circulation, alcohol or other drug use, trauma, and immersion can cause hypothermia. Remember, cold

may affect the body systems slowly and almost without notice. Service members exposed to low temperatures for extended periods may suffer ill effects even if they are well protected by clothing.

- *Signs and symptoms.* As the body cools, there are several stages of progressive discomfort and impairment. A sign that is noticed immediately is shivering. Shivering is an attempt by the body to generate heat. The pulse is faint or very difficult to detect. People with temperatures around 90°F may be drowsy and mentally slow. Their ability to move may be hampered, stiff, and uncoordinated, but they may be able to function minimally. Their speech may be slurred. As the body temperature drops further, shock becomes evident as the person's eyes assume a glassy state, breathing becomes slow and shallow, and the pulse becomes weaker or absent. The person becomes very stiff and uncoordinated. Unconsciousness may follow quickly. As the body temperature drops even lower, the extremities freeze, and a deep (or core) body temperature (below 85°F) increases the risk of irregular heart action. This irregular heart action or heart standstill can result in sudden death.

- *First aid measures.* Except in cases of the most severe hypothermia (marked by coma or unconsciousness and a weak pulse), first aid measures for hypothermia are directed towards protecting the casualty from further loss of body heat. For the casualty who is conscious, first aid measures are directed at rewarming the body evenly and without delay. Provide heat by using a hot water bottle or field expedient or another service member's body heat.

CAUTION

DO NOT expose the casualty to an open fire, as he may become burned.

NOTE

When using a hot water bottle or field expedient (canteen filled with warm water), the bottle or canteen must be wrapped in cloth prior to placing it next to the casualty. This will reduce the chance of burning the casualty's skin.

Always call or send for help as soon as possible and protect the casualty immediately with dry clothing or a sleeping bag. Then, move him to a warm place. Evaluate other injuries and provide first aid as required. First aid measures can be performed while the casualty is waiting transportation or while he is en route. In the case of an accidental breakthrough into ice water, or other hypothermic accident, strip the casualty of wet clothing immediately and bundle him into a sleeping bag. Rescue breathing should be started at once if the casualty's breathing has stopped or is irregular or shallow. Warm liquids (**NOT HOT**) may be given gradually if the casualty is conscious. **DO NOT** force liquids on an unconscious or semiconscious casualty because he may choke. The casualty should be transported on a litter because the exertion of walking may aggravate circulation problems. Medical personnel should immediately treat any hypothermia casualty. Hypothermia is life threatening until normal body temperature has been restored. The first aid measures for a casualty with severe hypothermia are based upon the following principles: attempt to avoid further heat loss, handle the casualty gently, and transport the casualty as soon as possible to the nearest MTF. If at all possible, the casualty should be evacuated by medical personnel.

WARNING

Rewarming a severely hypothermic casualty is extremely dangerous in the field due to the possibility of such complications as rewarming, shock and disturbances in the rhythm of the heartbeat. These conditions require treatment by medical personnel.

NOTE

Resuscitation of casualties with hypothermic complications is difficult if not impossible to do outside of an MTF setting.

CAUTION

The casualty is unable to generate his own body heat. Therefore, merely placing him in a blanket or sleeping bag is not sufficient.

e. *Table.* See Table 5-2 for further information.

Table 5-2. Injuries Caused by Cold and Wet Conditions

INJURIES	SIGNS/SYMPTOMS	FIRST AID
CHILBLAIN	RED SWOLLEN, HOT, TENDER, ITCHING SKIN. CONTINUED EXPOSURE MAY LEAD TO INFECTED (ULCERATED OR BLEEDING) SKIN LESIONS.	1. AREA USUALLY RESPONDS TO LOCALLY APPLIED REWARMING (BODY HEAT). 2. DO NOT RUB OR MASSAGE AREA. 3. SEEK MEDICAL AID.
IMMERSION SYNDROME (IMMERSION FOOT/TRENCH FOOT)	AFFECTED PARTS ARE COLD, NUMB, AND PAINLESS. PARTS MAY THEN BE HOT, WITH BURNING AND SHOOTING PAINS. ADVANCED STAGE: SKIN PALE WITH BLUISH CAST; PULSE DECREASES; BLISTERING, SWELLING, HEAT, HEMORRHAGING, AND GANGRENE MAY FOLLOW.	1. GRADUAL REWARMING BY EXPOSURE TO WARM AIR. 2. DO NOT MASSAGE OR MOISTEN SKIN. 3. PROTECT AFFECTED PARTS FROM TRAUMA. 4. DRY FEET THOROUGHLY, AVOID WALKING. 5. SEEK MEDICAL AID.
FROSTBITE	LOSS OF SENSATION (NUMB FEELING) IN ANY PART OF THE BODY. SUDDEN BLANCHING (WHITENING) OF THE SKIN OF THE AFFECTED PART, FOLLOWED BY A MOMENTARY TINGLING SENSATION. REDNESS OF SKIN IN LIGHT-SKINNED SERVICE MEMBERS; GRAYISH COLORING IN DARK-SKINNED SERVICE MEMBERS. BLISTERS. SWELLING OR TENDER AREAS. LOSS OF PREVIOUS SENSATION OF PAIN IN THE AFFECTED AREA. PALE YELLOWISH, WAXY-LOOKING SKIN. FROZEN TISSUE THAT FEELS SOLID (WOODEN) TO THE TOUCH.	1. WARM THE AREA AT THE FIRST SIGN OF FROSTBITE, USING FIRM, STEADY PRESSURE OF THE HAND, UNDERARM, OR ABDOMEN. 2. FACE, EARS, NOSE: COVER AREA WITH HANDS (CASUALTY'S OWN OR BUDDY'S). 3. HANDS: OPEN FIELD JACKET AND PLACE CASUALTY'S HANDS AGAINST HIS BODY, THEN CLOSE THE JACKET TO PREVENT HEAT LOSS. 4. FEET: REMOVE THE CASUALTY'S BOOTS AND SOCKS AND PLACE HIS FEET AGAINST THE BODY OF ANOTHER SERVICE MEMBER.

Table 5-2. Injuries Caused by Cold and Wet Conditions (Continued)

INJURIES	SIGNS/SYMPTOMS	FIRST AID
		5. WARNING: DO NOT ATTEMPT TO THAW THE CASUALTY'S FEET OR OTHER FROZEN AREAS IF HE WILL BE REQUIRED TO WALK OR TRAVEL TO AN MTF FOR ADDITIONAL TREATMENT. THE POSSIBILITY OF INJURY FROM WALKING IS LESS WHEN THE FEET ARE FROZEN THAN WHEN THEY HAVE BEEN THAWED. (HOWEVER, IF POSSIBLE AVOID WALKING.) THAWING IN THE FIELD INCREASES THE POSSIBILITY OF INFECTION, GANGRENE, OR INJURY. 6. LOOSEN OR REMOVE CONSTRICTING CLOTHING AND REMOVE ANY JEWELRY. 7. INCREASE INSULATION (COVER WITH BLANKET OR OTHER DRY MATERIAL). ENSURE CASUALTY EXERCISES AS MUCH AS POSSIBLE, AVOIDING TRAUMA TO INJURED PART.
SNOW BLINDNESS	EYES MAY FEEL SCRATCHY. WATERING, REDNESS, HEADACHE, AND INCREASED PAIN WITH EXPOSURE TO LIGHT CAN OCCUR.	1. COVER THE EYES WITH A DARK CLOTH. 2. SEEK MEDICAL AID.
DEHYDRATION	SIMILAR TO HEAT EXHAUSTION (REFER TO TABLE 5-1).	1. KEEP WARM. 2. CASUALTY NEEDS FLUID REPLACEMENT, REST, AND PROMPT MEDICAL AID.
HYPOTHERMIA	CASUALTY IS COLD. SHIVERING. CORE TEMPERATURE IS LOW. CONSCIOUSNESS MAY BE	**MILD HYPOTHERMIA** 1. REWARM BODY

Table 5-2. Injuries Caused by Cold and Wet Conditions (Continued)

INJURIES	SIGNS/SYMPTOMS	FIRST AID
	ALTERED. UNCOORDINATED MOVEMENTS MAY OCCUR. SHOCK AND COMA MAY RESULT AS BODY TEMPERATURE DROPS.	EVENLY AND WITHOUT DELAY. (NEED TO PROVIDE HEAT SOURCE; CASUALTY'S BODY UNABLE TO GENERATE HEAT.) 2. KEEP DRY, PROTECT FROM THE ELEMENTS. 3. WARM (NOT HOT) LIQUIDS MAY BE GIVEN GRADUALLY (TO CONSCIOUS CASUALTIES ONLY). 4. BE PREPARED TO START BASIC LIFE SUPPORT MEASURES FOR THE CASUALTY. 5. SEEK MEDICAL TREATMENT IMMEDIATELY. **SEVERE HYPOTHERMIA** 1. STABILIZE THE TEMPERATURE. 2. ATTEMPT TO AVOID FURTHER HEAT LOSS. 3. HANDLE THE CASUALTY GENTLY. 4. EVACUATE TO THE NEAREST MTF AS SOON AS POSSIBLE. **5. WARNING: HYPOTHERMIA IS A MEDICAL EMERGENCY. PROMPT MEDICAL ATTENTION IS NECESSARY.**

CHAPTER 6

FIRST AID FOR BITES AND STINGS

6-1. General

Snakebites, insect bites, or stings can cause intense pain and/or swelling. If not treated promptly and correctly, they can cause serious illness or death. The severity of a snakebite depends upon: whether the snake is poisonous or nonpoisonous, the type of snake, the location of the bite, and the amount of venom injected. Bites from humans and other animals, such as dogs, cats, bats, raccoons, and rats, can cause severe bruises and infection and tears or lacerations of tissue. Awareness of the potential sources of injuries can reduce or prevent them from occurring. Knowledge and prompt application of first-aid measures can lessen the severity of injuries from bites and stings and keep the service member from becoming a serious casualty.

6-2. Types of Snakes

a. Nonpoisonous Snakes. There are approximately 130 different varieties of nonpoisonous snakes in the United States. They have oval-shaped heads and round eyes. Unlike poisonous snakes, discussed below, nonpoisonous snakes do not have fangs with which to inject venom. Figure 6-1 depicts the characteristics of a nonpoisonous snake.

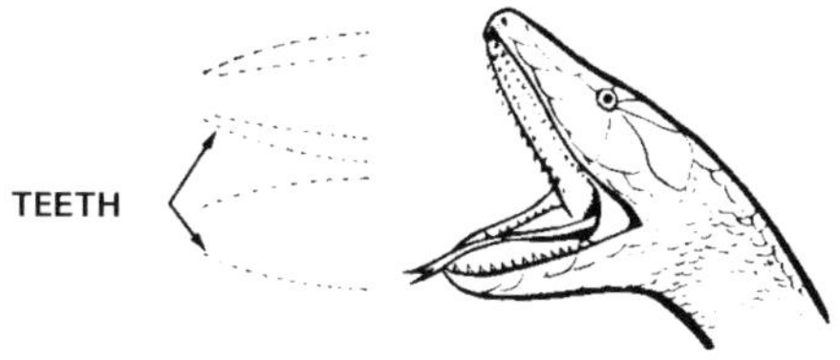

Figure 6-1. Characteristics of nonpoisonous snake.

b. Poisonous Snakes. Poisonous snakes are found throughout the world, primarily in tropical to moderate climates. Within the United States, there are four kinds: rattlesnakes, copperheads, water moccasins (cottonmouth), and coral snakes. Poisonous snakes in other parts of the world include sea snakes, the fer-de-lance, the bushmaster, and the tropical rattlesnake in tropical Central America; the Malayan pit viper in the tropical Far East; the cobra in Africa and Asia; the mamba (or black mamba) in central and southern Africa; and the krait in India and Southeast Asia. Refer to Figure 6-2 for characteristics of a poisonous pit viper.

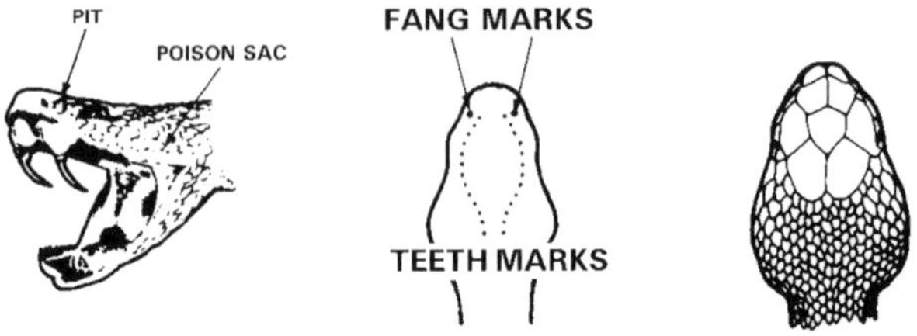

Figure 6-2. Characteristics of poisonous pit viper.

c. Pit Vipers (Poisonous). Figure 6-3 depicts a variety of poisonous snakes.

Figure 6-3. Poisonous snakes.

(1) Rattlesnakes, bushmasters, copperheads, fer-de-lance, Malayan pit vipers, and water moccasins (cottonmouth) are called pit vipers because of the small, deep pits between the nostrils and eyes on each side of the head (Figure 6-2). In addition to their long, hollow fangs, these snakes have other identifying features: thick bodies, slit-like pupils of the eyes, and flat, almost triangular-shaped heads. Color markings and other identifying characteristics, such as rattles or a noticeable white interior of the mouth (cottonmouth), also help distinguish these poisonous snakes. Further

identification is provided by examining the bite pattern of the wound for signs of fang entry. Occasionally there will be only one fang mark, as in the case of a bite on a finger or toe where there is no room for both fangs, or when the snake has broken off a fang.

(2) The casualty's condition provides the best information about the seriousness of the situation, or how much time has passed since the bite occurred. Pit viper bites are characterized by severe burning pain. Discoloration and swelling around the fang marks usually begins within 5 to 10 minutes after the bite. If only minimal swelling occurs within 30 minutes, the bite will almost certainly have been from a nonpoisonous snake or possibly from a poisonous snake which did not inject venom. The venom destroys blood cells, causing a general discoloration of the skin. Blisters and numbness in the affected area follow this reaction. Other signs, which can occur, are weakness, rapid pulse, nausea, shortness of breath, vomiting, and shock.

d. Corals, Cobras, Kraits, and Mambas. Corals (Figure 6-4), cobras (Figure 6-5), kraits, and mambas all belong to the same group even though they are found in different parts of the world. All four inject their venom through short, grooved fangs, leaving a characteristic bite pattern.

Figure 6-4. Coral snake.

(1) The small coral snake, found in the Southeastern US, is brightly colored with bands of red, yellow (or almost white), and black completely encircling the body. Other nonpoisonous snakes have the same coloring, but on the coral snake found in the US, the red ring always touches the yellow ring. To know the difference between a harmless snake and the coral snake found in the United States, remember the following:

"Red on yellow will kill a fellow,
Red on black, venom will lack."

Figure 6-5. Cobra snake.

(2) The venom of corals, cobras, kraits, and mambas produces symptoms different from those of pit vipers. Because there is only minimal pain and swelling, many people believe that the bite is not serious. Delayed reactions in the nervous system normally occur between 1 to 7 hours after the bite. Symptoms include blurred vision, drooping eyelids, slurred speech, drowsiness, and increased salivation and sweating. Nausea, vomiting, shock, respiratory difficulty, paralysis, convulsions, and coma will usually develop if the bite is not treated promptly.

e. Sea Snakes. Sea snakes (Figure 6-6) are found in the warm water areas of the Pacific and Indian oceans, along the coasts, and at the mouths of some larger rivers. Their venom is **VERY** poisonous, but their fangs are only 1/4 inch long. The first aid outlined for land snakes also applies to sea snakes.

Figure 6-6. Sea snake.

6-3. Snakebites

a. Poisonous snakes **DO NOT** always inject venom when they bite or strike a person. However, all snakes may carry tetanus (lockjaw); anyone bitten by a snake, whether poisonous or nonpoisonous, should immediately seek medical attention.

- Poison is injected from the venom sacs through grooved or hollow fangs. Depending on the species, these fangs are either long or short. Pit vipers have long hollow fangs. These fangs are folded against the roof of the mouth and extend when the snake strikes. This allows them to strike quickly and then withdraw. Cobras, coral snakes, kraits, mambas, and sea snakes have short, grooved fangs. These snakes are less effective in their attempts to bite, since they must chew after striking to inject enough venom (poison) to be effective. Figure 6-7 depicts the characteristics of a poisonous snakebite.

- In the event you are bitten, attempt to identify and/or kill the snake. Take it to medical personnel for inspection/identification. This provides valuable information to medical personnel who deal with snakebites. **TREAT ALL SNAKEBITES AS POISONOUS**.

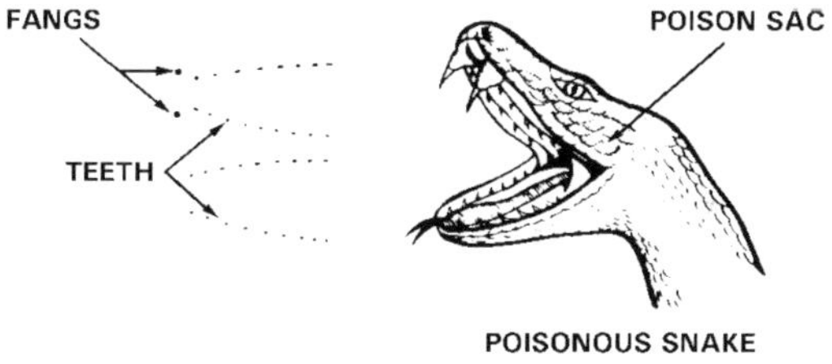

Figure 6-7. Characteristics of poisonous snakebite.

b. The venoms of different snakes cause different effects. Pit viper venom (hemotoxin [blood toxin]) destroys tissue and blood cells. Cobras, adders, and coral snakes inject powerful venom (neurotoxin [nerve toxin]) which affect the central nervous system, causing respiratory paralysis. Water moccasins and sea snakes have venom that is both hemotoxic and neurotoxic.

c. The identification of poisonous snakes is very important since medical treatment will be different for each type of venom. *Unless it can be positively identified, the snake should be killed and saved.* When this is not possible or when doing so is a serious threat to others, identification may sometimes be difficult since many venomous snakes resemble harmless varieties. When dealing with snakebite problems in foreign countries, seek advice, professional or otherwise, which may help identify species in the particular area of operations.

d. Get the casualty to an MTF as soon as possible and with minimum movement. Until evacuation or treatment is possible, have the casualty lie quietly and not move any more than necessary. If the casualty has been bitten on an extremity, **DO NOT** elevate the limb; keep the extremity level with the body. Keep the casualty comfortable and reassure him. If the casualty is alone when bitten, he should go to the medical facility himself rather than wait for someone to find him. Unless the snake has been positively identified, attempt to kill it and send it with the casualty. Be sure that retrieving the snake does not endanger anyone or delay transporting the casualty.

(1) If the bite is on an arm or leg, place a constricting band (narrow cravat [swathe], or narrow gauze bandage) one to two fingerbreadths above and below the bite (Figure 6-8). If the bite is on the hand or foot, place a single band above the wrist or ankle. The band should be tight enough to stop the flow of blood near the skin, but not tight enough to interfere with circulation. In other words, it should not have a tourniquet-like affect. If no swelling is seen, place the bands about 1 inch from either side of the bite. If swelling is present, put the bands on the unswollen part at the edge of the swelling. If the swelling extends beyond the band, move the band to the new edge of the swelling. (If possible, leave the old band on, place a new one at the new edge of the swelling, and then remove and save the old one in case the process has to be repeated.)

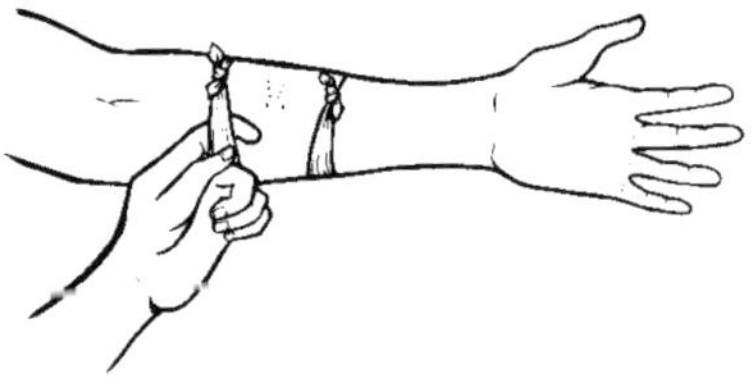

Figure 6-8. Constricting band.

CAUTION

DO NOT attempt to cut open the bite nor suck out the venom. If the venom should seep through any damaged or lacerated tissues in your mouth, you could immediately lose consciousness or even die.

(2) If the bite is located on an arm or leg, immobilize it at a level below the heart. **DO NOT** elevate an arm or leg even with or above the level of the heart.

> **CAUTION**
>
> When a splint is used to immobilize the arm or leg, take **EXTREME** care to ensure the splinting is done properly and does not bind. Watch it closely and adjust it if any changes in swelling occur.

(3) When possible, clean the area of the bite with soap and water. **DO NOT** use ointments of any kind.

(4) **NEVER** give the casualty food, alcohol, stimulants (coffee or tea), drugs, or tobacco.

(5) Remove rings, watches, or other jewelry from the affected limb.

6-4. Human or Animal Bites

Human or other land animal bites may cause lacerations or bruises. In addition to damaging tissue, bites always present the possibility of infection.

a. Human Bites. Human bites that break the skin may become seriously infected since the mouth is heavily contaminated with bacteria. Medical personnel **MUST** treat all human bites.

b. Animal Bites. Land animal bites can result in both infection and disease. Tetanus, rabies, and various types of fevers can follow an untreated animal bite. Because of these possible complications, the animal causing the bite should, if possible, be captured or killed (without damaging its head) so that it can be tested for disease.

c. First Aid.

(1) Cleanse the wound thoroughly with soap.

(2) Flush it well with water.

(3) Cover it with a sterile dressing.

(4) Immobilize the injured arm or leg, if appropriate.

(5) Transport the casualty immediately to an MTF.

NOTE

If unable to capture or kill the animal, provide medical personnel with any information that will help identify it.

6-5. Marine (Sea) Animals

With the exception of sharks and barracuda, most marine animals will not deliberately attack. The most frequent injuries from marine animals are wounds by biting, stinging, or puncturing. Wounds inflicted by marine animals can be very painful, but are rarely fatal.

a. Sharks, Barracuda, and Alligators. Wounds from these marine animals can involve major trauma as a result of bites and lacerations. Bites from large marine animals are potentially the most life threatening of all injuries from marine animals. Major wounds from these animals can be treated by controlling the bleeding, preventing shock, giving basic life support, splinting the injury, and by securing prompt medical aid.

b. Turtles, Moray Eels, and Corals. These animals normally inflict minor wounds. Treat by cleansing the wound(s) thoroughly and by splinting if necessary.

c. Jellyfish, Portuguese Man-of-War, Anemones, and Others. This group of marine animals inflict injury by means of stinging cells in their tentacles. Contact with the tentacles produces burning pain with a rash and small hemorrhages on the skin. Shock, muscular cramping, nausea, vomiting, and respiratory distress may also occur. Gently remove the clinging tentacles with a towel and wash or treat the area. Use diluted ammonia or alcohol, meat tenderizer, and talcum powder. If symptoms become severe or persist, seek medical assistance.

d. Spiny Fish, Urchins, Stingrays, and Cone Shells. These animals inject their venom by puncturing the skin with their spines. General signs and symptoms include swelling, nausea, vomiting, generalized cramps, diarrhea, muscular paralysis, and shock. Deaths are rare. Treatment consists of soaking the wounds in hot water (when available) for 30 to 60 minutes. This inactivates the heat sensitive toxin. In addition, further first aid measures (controlling bleeding, applying a dressing, and so forth) should be carried out as necessary.

> **CAUTION**
>
> Be careful not to scald the casualty with water that is too hot because the pain of the wound will mask the normal reaction to heat.

6-6. Insect (Arthropod) Bites and Stings

An insect bite or sting can cause great pain, allergic reaction, inflammation, and infection. If not treated correctly, some bites/stings may cause serious illness or even death. When an allergic reaction is not involved, first aid is a simple process. In any case, medical personnel should examine the casualty at the earliest possible time. It is important to properly identify the spider, bee, or creature that caused the bite/sting, especially in cases of allergic reaction.

a. Types of Insects. The insects found throughout the world that can produce a bite or sting are too numerous to mention in detail. Commonly encountered stinging or biting insects include brown recluse spiders (Figure 6-9), black widow spiders (Figure 6-10), tarantulas (Figure 6-11), scorpions (Figure 6-12), urticating caterpillars, bees, wasps, centipedes, conenose beetles (kissing bugs), ants, and wheel bugs. Upon being reassigned, especially to overseas areas, take the time to become acquainted with the types of insects to avoid.

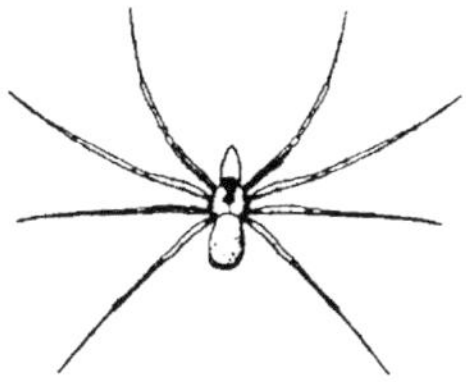

Figure 6-9. Brown recluse spider.

Figure 6-10. Black widow spider.

Figure 6-11. Tarantula.

Figure 6-12. Scorpion.

b. Signs and Symptoms. Discussed in paragraphs (1) and (2) below are the most common effects of insect bites/stings. They can occur alone or in combination with the others.

(1) *Less serious.* Commonly seen signs/symptoms are pain, irritation, swelling, heat, redness, and itching. Hives or wheals (raised areas of the skin that itch) may occur. These are the least severe of the allergic reactions that commonly occur from insect bites/stings. They are usually dangerous only if they affect the air passages (mouth, throat, nose, and so forth), which could interfere with breathing. The bites/stings of bees, wasps, ants, mosquitoes, fleas, and ticks are usually not serious and normally produce mild and localized symptoms. A tarantula's bite is usually no worse than that of a bee sting. Scorpions are rare and their stings (except for a specific species found only in the Southwest desert) are painful but usually not dangerous.

(2) *Serious.* Emergency allergic or hypersensitive reactions sometimes result from the stings of bees, wasps, and ants. Many people are allergic to the venom of these particular insects. Bites or stings from these insects may produce more serious reactions, to include generalized itching and hives, weakness, anxiety, headache, breathing difficulties, nausea, vomiting, and diarrhea. Very serious allergic reactions (called *anaphylactic shock*) can lead to complete collapse, shock, and even death. Spider bites (particularly from the black widow and brown recluse spiders) can also be

serious. Venom from the black widow spider affects the nervous system. This venom can cause muscle cramps, a rigid, nontender abdomen, breathing difficulties, sweating, nausea, and vomiting. The brown recluse spider generally produces local rather than system-wide problems; however, local tissue damage around the bite can be severe and can lead to an ulcer and even gangrene.

c. First Aid. There are certain principles that apply regardless of what caused the bite/sting. Some of these are—

- If there is a stinger present (for example, from a bee), remove the stinger by scraping the skin's surface with a fingernail or knife. **DO NOT** squeeze the sac attached to the stinger because it may inject more venom.

- Wash the area of the bite/sting with soap and water (alcohol or an antiseptic may also be used) to help reduce the chances of an infection and remove traces of venom.

- Remove jewelry from bitten extremities because swelling may occur.

- In most cases of insect bites the reaction will be mild and localized; use ice or cold compresses (if available) on the site of the bite/ sting. This will help reduce swelling, ease the pain, and slow the absorption of venom. Meat tenderizer (to neutralize the venom) or calamine lotion (to reduce itching) may be applied locally. If necessary, seek medical assistance.

- In more serious reactions (severe and rapid swelling, allergic symptoms, and so forth) treat the bite/sting like you would treat a snakebite; that is, apply constricting bands above and below the site.

- Be prepared to perform basic life-support measures, such as rescue breathing.

- Reassure the casualty and keep him calm.

- In serious reactions, attempt to capture the insect for positive identification; however, be careful not to become a casualty yourself.

- If the reaction to the bite/sting appears serious, seek medical assistance.

WARNING

Insect bites/stings may cause *anaphylactic shock* (a shock caused by a severe allergic reaction). This is a life-threatening event and a TRUE MEDICAL EMERGENCY. Be prepared to perform the basic life-support measures and to immediately transport the casualty to an MTF.

NOTE

Be aware that some allergic or hypersensitive individuals may carry identification or emergency insect bite treatment kits. If the casualty is having an allergic reaction and has such a kit, administer the medication in the kit according to the instructions which accompany the kit.

d. Supplemental Information. For additional information concerning biting insects, see FM 21-10.

6-7. First Aid for Bites and Stings

See the table below for information on bites and stings.

Table 6-1. First Aid Measures for Bites and Stings

TYPES	FIRST AID MEASURES
SNAKEBITE	1. MOVE CASUALTY AWAY FROM THE SNAKE. 2. REMOVE JEWELRY FROM THE AFFECTED AREA, IF APPLICABLE. 3. REASSURE CASUALTY AND KEEP HIM QUIET. 4. APPLY CONSTRICTING BAND, 1-2 FINGERBREADTHS FROM THE BITE. YOU SHOULD BE ABLE TO INSERT A FINGER BETWEEN THE BAND AND THE SKIN. a. *ARM OR LEG BITE.* PLACE ONE BAND ABOVE AND ONE BAND BELOW THE BITE SITE. b. *HAND OR FOOT BITE.* PLACE ONE BAND ABOVE THE WRIST OR ANKLE.

Table 6-1. First Aid Measures for Bites and Stings

TYPES	FIRST AID MEASURES
	5. IMMOBILIZE THE AFFECTED PART IN A POSITION BELOW THE LEVEL OF THE HEART. 6. KILL THE SNAKE (IF POSSIBLE, WITHOUT DAMAGING ITS HEAD OR ENDANGERING YOURSELF) AND SEND IT WITH THE CASUALTY. 7. SEEK MEDICAL ASSISTANCE IMMEDIATELY.
BROWN RECLUSE SPIDER OR BLACK WIDOW SPIDER BITE	1. KEEP CASUALTY QUIET. 2. REMOVE ALL JEWELRY FROM AFFECTED PART, IF APPLICABLE. 3. WASH THE AREA. 4. APPLY ICE OR FREEZE PACK, IF AVAILABLE. 5. SEEK MEDICAL ASSISTANCE.
TARANTULA BITE OR SCORPION STING OR ANT BITE	1. WASH THE AREA. 2. REMOVE ALL JEWELRY FROM AFFECTED PART, IF APPLICABLE. 3. APPLY ICE OR FREEZE PACK, IF AVAILABLE. 4. APPLY BAKING SODA, CALAMINE LOTION, OR MEAT TENDERIZER (IF AVAILABLE) TO BITE SITE TO RELIEVE PAIN AND ITCHING. 5. IF THE SITE OF THE BITE IS ON THE FACE, NECK (POSSIBLE AIRWAY PROBLEMS), OR GENITAL AREA, OR IF LOCAL REACTION SEEMS SEVERE, OR IF THE STING IS BY THE DANGEROUS TYPE OF SCORPION FOUND IN THE SOUTHWEST UNITED STATES DESERT, KEEP THE CASUALTY AS QUIET AS POSSIBLE. SEEK MEDICAL ASSISTANCE.
BEE STING	1. IF THE STINGER IS PRESENT, REMOVE BY SCRAPING WITH A KNIFE OR FINGERNAIL. DO NOT SQUEEZE VENOM SAC ON STINGER; MORE VENOM MAY BE INJECTED. 2. REMOVE ALL JEWELRY FROM AFFECTED PART, IF APPLICABLE.

Table 6-1. First Aid Measures for Bites and Stings

TYPES	FIRST AID MEASURES
	3. WASH THE AREA. 4. APPLY ICE OR FREEZE PACK, IF AVAILABLE. 5. IF ALLERGIC SIGNS OR SYMPTOMS APPEAR, BE PREPARED TO PERFORM BASIC LIFE SUPPORT MEASURES. SEEK IMMEDIATE MEDICAL ASSISTANCE.

CHAPTER 7

FIRST AID IN A NUCLEAR, BIOLOGICAL, AND CHEMICAL ENVIRONMENT

7-1. General

American forces have not been exposed to NBC weapons/agents on the battlefield since World War I. In future conflicts and wars we can expect the use of such agents. Nuclear, biological, and chemical weapons will rapidly degrade unit effectiveness by forcing troops to wear protective clothing and by creating confusion and fear. Through training in protective procedures and first aid, units can maintain their effectiveness on the integrated battlefield.

7-2. First Aid Materials

You may be issued the following materials to protect, decontaminate, and use as first aid for NBC exposure. You must know how to use the items; some items are described in *a* through *d* below. It is equally important that you know when to use them.

a. Nerve Agent Pyridostigmine Pretreatment (*NAPP*). You may be issued a blister pack of pretreatment tablets when your commander directs. The NAPP is a pretreatment; it is not an antidote. It improves the effectiveness of the nerve agent antidote. When ordered to take the pretreatment you must take one tablet every 8 hours, mission permitting. This must be taken prior to exposure to nerve agents, since it may take several hours to develop adequate blood levels.

NOTE

Commanders must follow investigational new drug protocols for use of the NAPP.

b. M291 Skin Decontaminating Kit. The M291 Skin Decontaminating Kit (Figure 7-1) contains six packets of XE-555 decontaminant resin.

WARNING

For external use only. May be slightly irritating to the eyes. Keep decontaminating powder out of eyes. Use water to wash toxic agent out of eyes.

c. *Nerve Agent Antidote Kit, MARK I.* Each service member is issued three MARK Is for use in first aid for nerve agent poisoning (Figure 7-2 and paragraph 7-6).

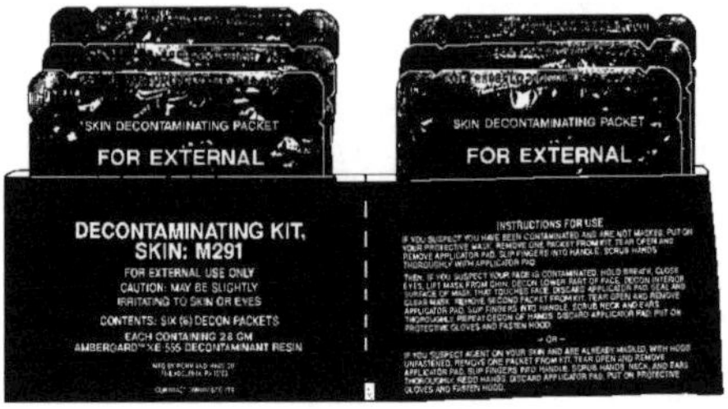

Figure 7-1. M291 Skin Decontamination Kit.

d. *Antidote Treatment, Nerve Agent, Autoinjector.* A new nerve agent antidote injection device, Antidote Treatment, Nerve Agent, Autoinjector (ATNAA) is currently under development that will replace the MARK I. The ATNAA is a multichambered device with the atropine and pralidoxime chloride in separate chambers. Both antidotes will be administered through a single needle.

7-3. Classification of Chemical and Biological Agents

a. Chemical agents are classified according to the primary physiological effects they produce, such as blistering, choking, vomiting, and incapacitating agents.

b. Biological warfare agents are classified according to the effect they have on man. The effects include their ability to incapacitate and cause death. Most biological warfare agents are delivered as aerosols that effect the respiratory tract; some can be delivered by releasing infected insects, by contaminating food and water, and by injection (injecting material in individuals by terrorist, not mass exposure). These agents are found in living organisms such as fungi, bacteria, and viruses.

WARNING

Swallowing water or food contaminated with nerve, blister, and other chemical agents and with some biological agents can be fatal. NEVER consume water or food that is suspected of being contaminated until it has been tested and found safe for consumption by medical personnel.

7-4. Conditions for Masking Without Order or Alarm

a. Once an attack with a chemical or biological agent is detected or suspected, or information is available that such an agent is about to be used, you must **STOP BREATHING** and mask immediately. **DO NOT WAIT** to receive an order or alarm under the following circumstances:

- Your position is hit by artillery missiles, rockets that produce vapors, smoke, and mists, and aerial sprays.
- Smoke or vapor cloud from an unknown source is present or approaching.
- A suspicious odor, liquid, or solid is present.
- A chemical or biological warfare agent attack is occurring.
- You are entering an area known or suspected of being contaminated.
- When casualties are being received from an area where chemical or biological agents have reportedly been used.
- You have one or more of the following symptoms:
  - An unexplained runny nose.
  - A sudden unexplained headache.
  - A feeling of choking or tightness in the chest or throat.
  - Dimness of vision.
  - Irritation of the eyes.
  - Difficulty in or increased rate of breathing without obvious reasons.
  - Sudden feeling of depression.
  - Dread, anxiety, or restlessness.
  - Dizziness or light-headedness.

- Slurred speech.
- Unexplained laughter or unusual behavior is noted in others.
- Numerous unexplained ill personnel.
- Service members suddenly collapsing without evident cause.
- Animals or birds exhibiting unusual behavior or suddenly dying.

b. For further information on protection and masking procedures, refer to FM 3-4, FM 4-02.7, FM 8-284, and FM 8-285.

7-5. First Aid for a Chemical Attack

Your field protective mask gives protection against biological and chemical warfare agents as well as radiological fallout. With practice you can mask in 9 seconds or less, or put on your mask with hood within 15 seconds.

a. Stop breathing. Don your mask, seal it properly, and clear and check it; then resume breathing. Give the alarm, and continue the mission. Keep your mask on until the "all clear" signal has been given.

NOTE

Keep your mask on until the area is no longer hazardous and you are told to unmask.

b. If symptoms of nerve agent poisoning (paragraph 7-7) appear, immediately give yourself one MARK I or ATNAA.

CAUTION

Do not inject a nerve agent antidote until you are sure you need it.

c. If your eyes and face become contaminated, you must immediately try to get under cover. You need shelter to prevent further contamination while performing decontamination procedures on your face. If no overhead cover is available, put your poncho over your head before beginning the decontamination process. Then you put on the remaining protective clothing. If vomiting occurs, the mask should be lifted momentarily and drained—with your eyes closed and while holding your breath—then replaced, cleared, and sealed.

d. If nerve agents are used, mission permitting, watch for persons needing nerve agent antidotes and immediately follow procedures outlined in paragraph 7-8*b* or *c*.

e. Decontaminate your skin immediately and clothing and equipment as soon as the mission permits.

7-6. Background Information on Nerve Agents

a. Nerve agents are among the deadliest of chemical agents. Nerve agents enter the body by inhalation, by ingestion, and through the skin. Depending on the route of entry and the amount, nerve agents can produce injury or death within minutes. Nerve agents can achieve their effects with small amounts. Nerve agents are absorbed rapidly, and the effects are felt immediately upon entry into the body. You will be issued three MARK Is or three ATNAAs and one Convulsant Antidote for Nerve Agent (CANA). Each MARK I consists of one atropine autoinjector and one pralidoxime chloride (2 PAM Cl) autoinjector (Figure 7-2A). Each ATNAA consist of a multichambered autoinjector with the atropine and pralidoxime chloride in separate chambers (Figure 7-2C). The CANA is a single autoinjector with flanges (Figure 7-2B). Procedures for use of both the MARK I and ATNAA are described below. You will use either the MARK I or the ATNAA in self-aid and buddy aid as issued.

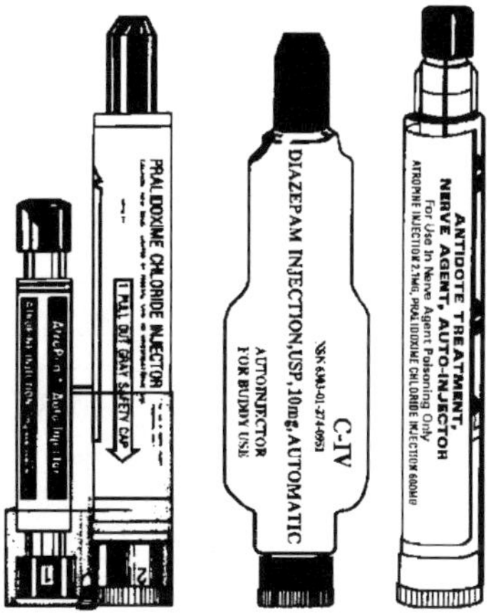

A B C

Figure 7-2. Nerve Agent Antidote Kit, MARK I, CANA, and ATNAA.

b. When you have the signs and symptoms of nerve agent poisoning, you should immediately put on the protective mask and then inject yourself with one set of the MARK I or ATNAA. Do not administer the CANA. You should inject yourself in the outer (lateral) thigh muscle (Figure 7-3) or if you are thin, in the upper outer (lateral) part of the buttocks (Figure 7-4).

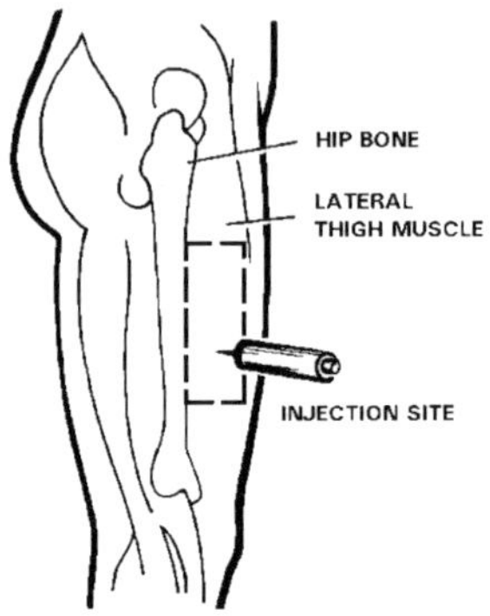

Figure 7-3. Thigh injection site.

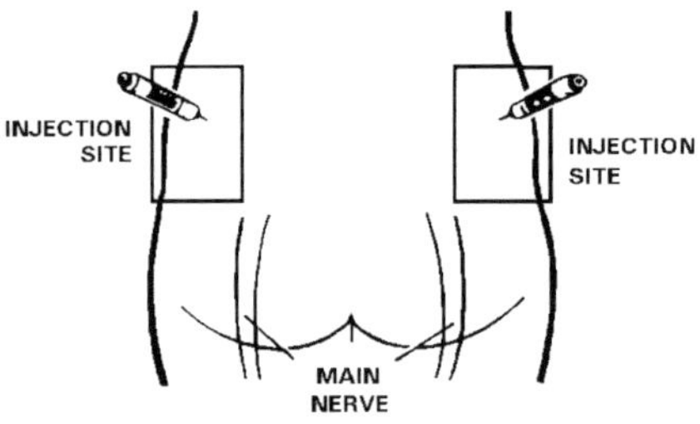

Figure 7-4. Buttocks injection site.

c. Also, you may come upon an unconscious chemical agent casualty who will be unable to care for himself and who will require first aid. You should be able to successfully—

(1) Mask him if he is unmasked.

(2) Inject him, if necessary, with all of **HIS** autoinjectors.

(3) Decontaminate his skin.

(4) Seek medical assistance.

7-7. Signs and Symptoms of Nerve Agent Poisoning

The symptoms of nerve agent poisoning are grouped as **MILD**—those that you recognize and for which you can perform self-aid, and **SEVERE**—those which require buddy aid.

a. MILD Signs and Symptoms.

- Unexplained runny nose.
- Unexplained sudden headache.
- Sudden drooling.
- Difficulty seeing (dimness of vision and miosis).
- Tightness in the chest or difficulty in breathing.
- Localized sweating and muscular twitching in the area of contaminated skin.
- Stomach cramps.
- Nausea.
- Tachycardia followed by bradycardia. (*Tachycardia* is an abnormally rapid heartbeat with a heart rate of over 100 beats per minute. *Bradycardia* is a slow heart rate of less than 60 beats per minute.)

b. SEVERE Signs and Symptoms.

- Strange or confused behavior.
- Wheezing, dyspnea (difficulty in breathing), and coughing.
- Severely pinpointed pupils.
- Red eyes with tearing.
- Vomiting.
- Severe muscular twitching and general weakness.
- Involuntary urination and defecation.

- Convulsions.
- Unconsciousness.
- Respiratory failure.
- Bradycardia.

7-8. First Aid for Nerve Agent Poisoning

First aid for nerve agent poisoning consists of administering the MARK I or ATNAA and CANA.

a. Injection Site. The injection site for administering the antidotes is normally in the outer thigh muscle. The thigh injection site is the area about a hand's width above the knee to a hand's width below the hip joint (Figure 7-3). It is important that the injection be given into a large muscle area. If the individual is thinly built, then the injections should be administered into the upper outer quarter (quadrant) of the buttock (Figure 7-4). Injecting in the buttocks of a thinly built individual avoids injury to the thighbone.

b. Self-Administer MARK I. If you experience any or all of the nerve agent **MILD** symptoms (paragraph 7-7*a*), you must **IMMEDIATELY** put on your protective mask and self-administer one MARK I (Figure 7-2A). Follow the procedure given in Table 7-1. The MARK I is carried in your protective mask carrier, pocket of the MOPP overgarment, or other location as specified in your unit tactical standing operating procedure (TSOP). (In cold weather, the MARK I should be stored in an inside pocket of your clothing to protect the antidote from freezing. A frozen MARK I cannot be immediately used to provide you with antidote, when needed. (However, the MARK I can still be used after complete thawing.)

Table 7-1. Self Aid for Nerve Agent Poisoning

MARK I*	ATNAA*
STEP 1. OBTAIN ONE MARK I.**	STEP 1. OBTAIN ONE ATNAA.**
STEP 2. CHECK INJECTION SITE.	STEP 2. CHECK INJECTION SITE.
STEP 3. HOLD MARK I AT EYE LEVEL WITH NONDOMINANT HAND WITH THE LARGE INJECTOR ON TOP (FIGURE 7-5A).	STEP 3. HOLD ATNAA WITH DOMINANT HAND (FIGURE 7-12A).

Table 7-1. Self Aid for Nerve Agent Poisoning (Continued)

MARK I*	ATNAA*
STEP 4. GRASP SMALL INJECTOR (ATROPINE) (FIGURE 7-5B) AND REMOVE FROM CLIP (FIGURE 7-5C).	STEP 4. GRASP SAFETY CAP WITH NONDOMINANT HAND AND REMOVE FROM INJECTOR (FIGURE 7-12B).
STEP 5. CLEAR HARD OBJECTS FROM INJECTION SITE.	STEP 5. CLEAR HARD OBJECTS FROM INJECTION SITE.
STEP 6. INJECT ATROPINE AT INJECTION SITE APPLYING EVEN PRESSURE TO THE INJECTOR (FIGURE 7-6 OR 7-7). HOLD IN PLACE FOR 10 SECONDS.	STEP 6. INJECT ATNAA AT INJECTION SITE APPLYING EVEN PRESSURE TO THE INJECTOR (FIGURE 7-14 OR 7-15). HOLD IN PLACE FOR 10 SECONDS.
STEP 7. HOLD USED INJECTOR WITH NONDOMINANT HAND.	STEP 7. BEND NEEDLE OF USED INJECTOR BY PRESSING ON A HARD SURFACE TO FORM A HOOK.
STEP 8. GRASP THE LARGE (2 PAM CI) INJECTOR (FIGURE 7-8B) AND PULL IT FROM CLIP (FIGURE 7-8C). DROP CLIP TO GROUND.	STEP 8. ATTACH USED INJECTOR TO BLOUSE POCKET FLAP OF BDO/JSLIST (FIGURE 7-16).
STEP 9. INJECT 2 PAM CI AT INJECTION SITE APPLYING EVEN PRESSURE TO THE INJECTOR (FIGURE 7-6 OR 7-7). HOLD IN PLACE FOR 10 SECONDS.	STEP 9. MASSAGE INJECTION SITE, MISSION PERMITTING.
STEP 10. BEND THE NEEDLES OF ALL USED INJECTORS BY PRESSING ON A HARD SURFACE TO FORM A HOOK.	
STEP 11. ATTACH ALL USED INJECTORS TO BLOUSE POCKET FLAP OF BDO/JSLIST (FIGURE 7-9).	
STEP 12. MASSAGE INJECTION SITE, MISSION PERMITTING.	

* USE STEPS LISTED FOR TYPE OF ANTIDOTE DEVICE ISSUED.
** ONLY ADMINISTER ONE MARK I OR ATNAA AS SELF-AID. DO NOT SELF-ADMINISTER CANA.

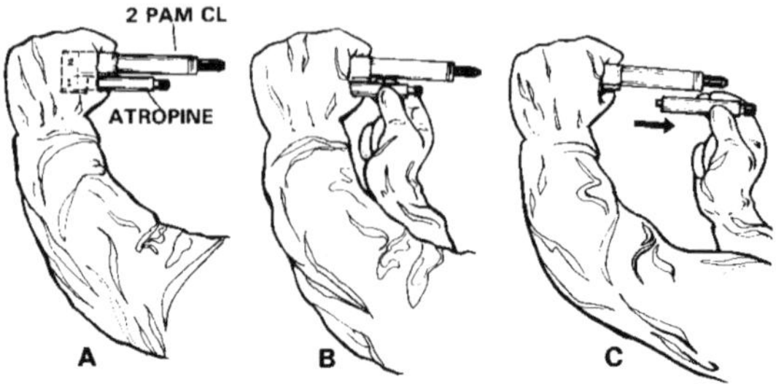

Figure 7-5. Removing the atropine autoinjector from the MARK I clip.

CAUTION

DO NOT cover or hold the needle end with your hand, thumb, or fingers—you might accidentally inject yourself. An accidental injection into the hand **WILL NOT** deliver an effective dose of the antidote, especially if the needle goes through the hand.

Figure 7-6. Thigh injection site for self-aid.

NOTE

If you are thinly built, inject yourself into the upper outer quadrant of the buttock (Figure 7-7). There is a nerve that crosses the buttocks; hitting this nerve can cause paralysis. Therefore, you must only inject into the *upper outer quadrant* of the buttock.

Figure 7-7. Buttocks injection site for self-aid.

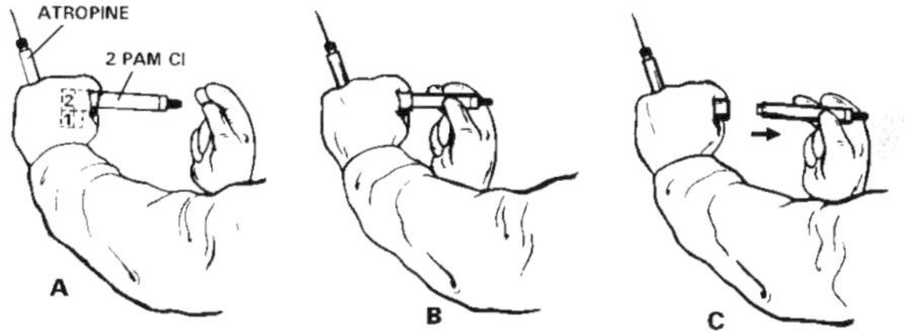

Figure 7-8. Removing the 2 PAM Cl autoinjector from the MARK I clip.

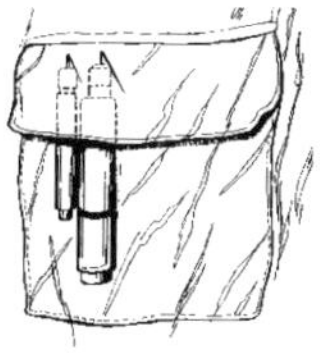

Figure 7-9. One set of used MARK I autoinjectors attached to pocket flap.

NOTES

1. **DO NOT** give yourself another set of injections. If you are able to walk without assistance, know who you are, and where you are, you **WILL NOT** need the second set of injections. (If not needed, giving yourself a second set of MARK I injections or ATNAA may create a nerve agent antidote overdose, which could cause incapacitation [inability to perform mission or defend yourself].)

2. If you continue to have symptoms of nerve agent poisoning, seek someone else (a buddy) to check your symptoms and administer the additional sets of injections, if required.

c. *Buddy Evaluation and Buddy Aid.* Service members may seek assistance after self-aid (self-administering one MARK I or ATNAA) or may become incapacitated after self-aid. A buddy must evaluate the individual to determine if additional antidotes are required to counter the effects of the nerve agent. Also, service members may experience **SEVERE** symptoms of nerve agent poisoning (paragraph 7-7*b*); they will not be able to treat themselves. In either case, other service members must perform buddy aid as quickly as possible. Before initiating buddy aid, determine if one set of MARK I autoinjectors has already been used so that no more than three sets of the antidote are administered. Buddy aid also includes administering the CANA with the third MARK I or ATNAA to prevent convulsions. Follow the procedures indicated in Table 7-2.

WARNING

Squat, DO NOT kneel, when masking the casualty or administering the nerve agent antidote to the casualty. Kneeling may force the chemical agent into or through your protective clothing.

CAUTION

DO NOT use your own MARK I, ATNAA, or CANA on a casualty. If you use your own, you may not have any antidote if needed for self-aid.

WARNING

DO NOT inject into areas close to the hip, knee, or thigh-bone.

Table 7-2. Buddy Aid/Combat Lifesaver Aid for Nerve Agent Casualty.

MARK I*	ATNAA*	CANA**
STEP 1. MASK THE CASUALTY AND POSITION HIM ON HIS SIDE (SWIMMER'S POSITION).	STEP 1. MASK THE CASUALTY AND POSITION HIM ON HIS SIDE (SWIMMER'S POSITION).	STEP 1. OBTAIN BUDDY'S CANA.
STEP 2. POSITION YOURSELF NEAR THE CASUALTY'S THIGH.	STEP 2. POSITION YOURSELF NEAR THE CASUALTY'S THIGH.	STEP 2. CHECK INJECTION SITE.

Table 7-2. Buddy Aid/Combat Lifesaver Aid for Nerve Agent Casualty (Continued).

MARK I*	ATNAA*	CANA**
STEP 3. OBTAIN BUDDY'S THREE OR REMAINING MARK Is.	STEP 3. OBTAIN BUDDY'S THREE OR REMAINING ATNAAs.	STEP 3. HOLD CANA IN A CLOSED FIST WITH DOMINANT HAND (FIGURE 7-12A).
STEP 4. CHECK INJECTION SITE.	STEP 4. CHECK INJECTION SITE.	STEP 4. GRASP SAFETY CAP WITH NONDOMINANT HAND AND REMOVE FROM INJECTOR (FIGURE 7-12B).
STEP 5. HOLD MARK I WITH NONDOMINANT HAND (FIGURE 7-5A).	STEP 5. HOLD ATNAA IN A CLOSED FIST WITH DOMINANT HAND (FIGURE 7-12A).	STEP 5. CLEAR HARD OBJECTS FROM INJECTION SITE.
STEP 6. GRASP SMALL INJECTOR (ATROPINE) AND REMOVE FROM CLIP (FIGURE 7-5B).	STEP 6. GRASP SAFETY CAP WITH NONDOMINANT HAND AND REMOVE FROM INJECTOR (FIGURE 7-12B).	STEP 6. INJECT CANA AT INJECTION SITE BY APPLYING EVEN PRESSURE TO THE INJECTOR, NOT A JABBING MOTION (FIGURE 7-14 OR 7-15). HOLD IN PLACE FOR 10 SECONDS.
STEP 7. CLEAR HARD OBJECTS FROM INJECTION SITE.	STEP 7. CLEAR HARD OBJECTS FROM INJECTION SITE.	STEP 7. BEND NEEDLE OF INJECTOR BY PRESSING ON A HARD SURFACE TO FORM A HOOK.
STEP 8. INJECT ATROPINE AT INJECTION SITE BY APPLYING EVEN PRESSURE TO THE INJECTOR, NOT A JABBING MOTION (FIGURE 7-10 OR 7-11). HOLD IN PLACE FOR 10 SECONDS.	STEP 8. INJECT ATNAA AT INJECTION SITE BY APPLYING EVEN PRESSURE TO THE INJECTOR, NOT A JABBING MOTION (FIGURE 7-14 OR 7-15). HOLD IN PLACE FOR 10 SECONDS.	STEP 8. ATTACH USED INJECTOR TO BLOUSE POCKET FLAP OF BDO/JSLIST (FIGURE 7-16).
STEP 9. HOLD USED INJECTOR BETWEEN LITTLE FINGER AND RING FINGER OF NONDOMINANT HAND (FIGURE 7-5A).	STEP 9. BEND NEEDLE OF INJECTOR BY PRESSING ON A HARD SURFACE TO FORM A HOOK.	STEP 9. MASSAGE INJECTION SITE, MISSION PERMITTING.
STEP 10. PULL LARGE INJECTOR (2 PAM CI) FROM CLIP (FIGURE 7-5C). DROP CLIP TO GROUND.	STEP 10. ATTACH ALL USED INJECTORS TO BLOUSE POCKET FLAP OF BDO/JSLIST (FIGURE 7-16).	

Table 7-2. Buddy Aid/Combat Lifesaver Aid for Nerve Agent Casualty (Continued).

MARK I*	ATNAA*	CANA**
STEP 11. INJECT 2 PAM CI AT INJECTION SITE BY APPLYING EVEN PRESSURE TO THE INJECTOR, NOT A JAB-BING MOTION (FIGURE 7-10 OR 7-11). HOLD IN PLACE FOR 10 SECONDS.	STEP 11. MASSAGE INJECTION SITE, MISSION PERMITTING.	
STEP 12. REPEAT STEPS ABOVE FOR REMAINING MARK Is.		
STEP 13. BEND THE NEEDLES OF ALL USED INJECTORS BY PRESSING ON A HARD SURFACE TO FORM A HOOK.		
STEP 14. ATTACH ALL USED INJECTORS TO BLOUSE POCKET FLAP OF BDO/JSLIST (FIGURE 7-13).		
STEP 15. MASSAGE INJECTION SITE, MISSION PERMITTING.		

* USE STEPS LISTED FOR TYPE OF ANTIDOTE DEVICE ISSUED.
** CANA IS USED IN BUDDY AID/CLS AID ONLY. DO NOT USE IN SELF-AID.

NOTE

If the casualty is thinly built, inject the antidote into the buttock. Only inject the antidote into the upper outer portion of the casualty's buttock (Figure 7-11). This avoids hitting the nerve that crosses the buttocks (Figure 7-4). Hitting this nerve can cause paralysis.

Figure 7-10. Injecting the casualty's thigh (Mark I or CANA).

Figure 7-11. Injecting the casualty's buttocks (Mark I or CANA).

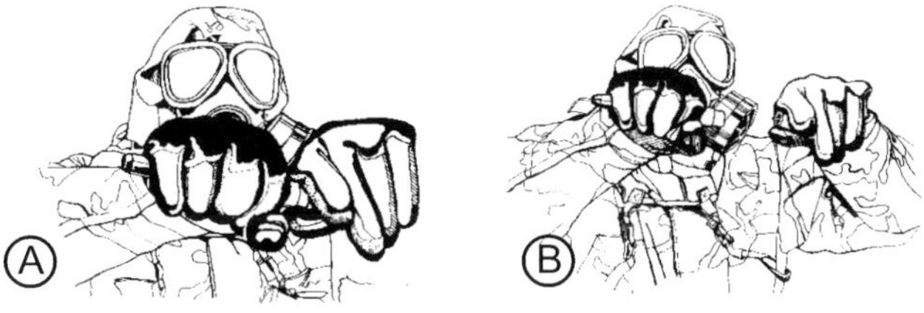

Figure 7-12. Preparing CANA or ATNAA for injection.

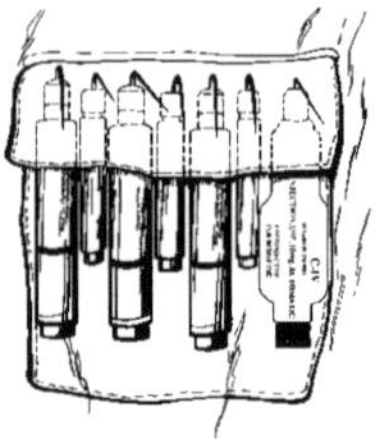

Figure 7-13. Three sets of used MARK I autoinjectors and one CANA autoinjector attached to pocket flap.

d. Self-Administer Antidote Treatment Nerve Agent Autoinjector. If you experience any or all of the nerve agent **MILD** symptoms (paragraph 7-7*b*), you must **IMMEDIATELY** self-administer one ATNAA following the procedure given Table 7-1.

NOTE

If you are thinly-built, inject yourself into the upper outer quarter (quadrant) of the buttock (Figure 7-15). There is a nerve that crosses the buttocks; hitting this nerve can cause paralysis. Therefore, you must only inject into the upper outer quarter (quadrant) of the buttocks.

Figure 7-14. Self-administration of ATNAA (thigh).

Figure 7-15. Self-administration of ATNAA (buttock).

NOTE

If you continue to have symptoms of nerve agent poisoning, seek someone else (a buddy) to check your symptoms and administer your remaining sets of injections, if required.

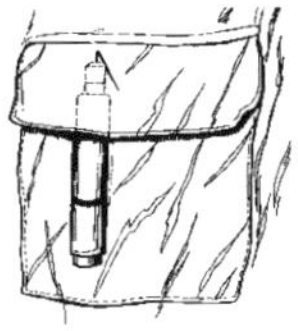

Figure 7-16. Used ATNAA attached to clothing.

e. Buddy Assistance. Service members may seek assistance after self-aid (self-administering one ATNAA) or may become incapacitated after self-aid. A buddy must evaluate the individual to determine if additional antidotes are required to counter the effects of the nerve agent. Also, service members may experience **SEVERE** symptoms of nerve agent poisoning (paragraph 7-7*b*); they will not be able to treat themselves. In either case, other service members must perform buddy aid as quickly as possible. Before initiating buddy aid, determine if one ATNAA has already been used so that no more than three ATNAA are administered. Buddy aid also includes administering the CANA with the third ATNAA to prevent convulsions. Follow the procedures indicated in Table 7-2.

WARNING

Squat, DO NOT kneel, when masking the casualty or administering the nerve agent antidotes to the casualty. Kneeling may force any chemical agent on your overgarment into or through your protective clothing.

Figure 7-17. Buddy injecting casualty's outer thigh (ATNAA or CANA).

NOTE

If the casualty is thinly built, inject the antidote into the buttocks (Figure 7-18). Only inject the antidote into the upper outer portion of the casualty's buttocks. This avoids hitting the nerve that crosses the buttocks (Figure 7-4). Hitting this nerve can cause paralysis.

WARNING

DO NOT inject into areas close to the hip, knee, or thighbone.

Figure 7-18. Buddy injecting casualty's buttocks (ATNAA or CANA).

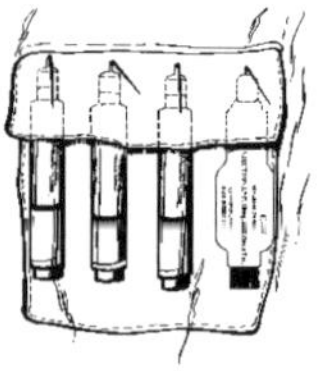

Figure 7-19. Three used ATNAAs and one CANA autoinjector attached to clothing.

f. Combat Lifesaver.

(1) The combat lifesaver must check to verify if the individual has received three sets of MARK I or ATNAAs. If not, the combat lifesaver performs first aid as described for buddy aid above. If the

individual has received the initial three sets of MARK I, then the combat lifesaver may administer additional atropine injections at approximately 15 minute intervals until atropinization is achieved (that is a heart rate above 90 beats per minute, reduced bronchial secretions, and reduced salivations). Administer additional atropine at intervals of 30 minutes to 4 hours to maintain atropinization or until the casualty is placed under the care of medical personnel. Check the heart rate by lifting the casualty's mask hood and feeling for a pulse at the carotid artery. Request medical assistance as soon as the tactical situation permits.

(2) The CLS should administer additional CANA to casualties suffering convulsions. Administer a second, and if needed, a third CANA at 5 to 10 minute intervals for a maximum of three injections (30 milligrams diazepam). Follow the steps and procedures described in buddy aid for administering the CANA. **DO NOT** give more than two additional injections for a total of three (one self-aid plus two by the CLS).

7-9. Blister Agents

Blister agents (vesicants) include mustard (H and HD), nitrogen mustards (HN), lewisite (L), and other arsenicals, mixtures of mustards and arsenicals, and phosgene oxime (CX). Blister agents may act on the eyes, mucous membranes, lungs, and skin. They burn and blister the skin or any other body parts they contact. Even relatively low doses may cause serious injury. Blister agents damage the respiratory tract (nose, sinuses, and windpipe) when inhaled and cause vomiting and diarrhea when absorbed. Lewisite and CX cause immediate pain on contact. However, mustard agents are deceptive as there is little or no pain at the time of exposure. Thus, in some cases, signs of injury may not appear for several hours after exposure.

a. Protective Measures. Your protective mask with hood and protective overgarment provide protection against blister agents. If it is known or suspected that blister agents are being used, **STOP BREATHING**, put on your mask and your protective overgarment.

CAUTION

Large drops of liquid vesicants on the protective overgarment ensemble may penetrate it if allowed to stand for an extended period. Remove large drops as soon as possible.

b. *Signs and Symptoms of Blister Agent Poisoning.*

(1) Immediate and intense pain upon contact with L, LH (lewisite and mustard) mixture, and CX. No initial pain upon contact with mustard.

(2) Inflammation and blisters (burns) resulting in tissue destruction. The severity of a chemical burn is directly related to the concentration of the agent and the duration of contact with the skin. The longer the agent is in contact with the tissue, the more serious the injury will be.

(3) Vomiting and diarrhea. Exposure to high concentrations of vesicants may cause vomiting or diarrhea.

(4) Death. The blister agent vapors absorbed during ordinary field exposure will probably not cause enough internal body (systemic) damage to result in death. However, death may occur from prolonged exposure to high concentrations of vapor or from extensive liquid contamination over wide areas of the skin, particularly *when decontamination is neglected or delayed.*

c. *First Aid Measures.*

(1) Use your M291 Skin Decontaminating Kit to decontaminate your skin and use water to flush contaminated eyes. Decontamination of vesicants must be done immediately (within 1 minute is best).

(2) If blisters form, cover them loosely with a field dressing and secure the dressing.

CAUTION

Blisters are actually burns. **DO NOT** attempt to decontaminate the skin where blisters have formed, as the agent has already been absorbed.

(3) If you receive blisters over a wide area of the body, you are considered seriously burned. Seek medical assistance immediately.

(4) If vomiting occurs, the mask should be lifted momentarily and drained—while the eyes are closed and the breath is held—and replaced, cleared, and sealed.

(5) Remember, if vomiting or diarrhea occurs after having been exposed to blister agents, seek medical assistance immediately.

7-10. Choking Agents (Lung-Damaging Agents)

Chemical agents that attack lung tissue, primarily causing fluid buildup (pulmonary edema), are classified as choking agents (lung-damaging agents). This group includes phosgene (CG), diphosgene (DP), chlorine (Cl), and chloropicrin (PS). Of these four agents, CG is the most dangerous and is more likely to be employed by the enemy in future conflict.

a. Protective Measures. Your protective mask gives adequate protection against choking agents.

b. Signs and Symptoms. During and immediately after exposure to choking agents (depending on agent concentration and length of exposure), you may experience some or all of the following signs and symptoms:

- Tears (lacrimation).
- Coughing.
- Choking.
- Tightness of chest.
- Nausea and vomiting.
- Headaches.

c. Self-Aid.

(1) The protective mask should be put on immediately when any of the conditions described in *b* above exist. Another indication of a CG attack is an odor like newly mown hay; however, **DO NOT** rely upon odor as indication of a chemical attack.

(2) If some CG is inhaled, normal combat duties should be continued unless there is difficulty in breathing, nausea, vomiting, or more than the usual shortness of breath during exertion. If any of the above symptoms occur and the mission permits, remain at quiet rest until medical evacuation is accomplished.

d. Death. With ordinary field exposure to choking agents, death will probably not occur. *However, prolonged exposure to high concentrations of the vapor and neglect or delay in masking can be fatal.*

7-11. Cyanogen (Blood) Agents

Cyanogen agents interfere with proper oxygen utilization in the body. Hydrogen cyanide (AC) and cyanogen chloride (CK) are the primary agents in this group.

a. Protective Measures. Your protective mask with a fresh filter gives adequate protection against field concentrations of cyanogen agent vapor. The protective overgarments, as well as the mask, are needed when exposed to liquid AC.

b. Signs and Symptoms. During and immediately after exposure to cyanogen agents (depending on agent concentration and length of exposure), you may experience some or all of the following signs and symptoms:

- Tearing (lacrimation).
- Eye, nose, and throat irritation.
- Sudden stimulation of breathing (unable to hold breath).
- Nausea.
- Coughing.
- Tightness of chest.
- Headache.
- Light-headedness (dizziness).
- Unconsciousness.

c. First Aid.

(1) *Hydrogen cyanide.* During any chemical attack, if you get a sudden stimulation of breath or detect an odor like bitter almonds, **PUT ON YOUR MASK IMMEDIATELY**. Speed is absolutely essential since this agent acts so rapidly that within a few seconds its effects will make it impossible for service members to put on their mask by themselves. Stop breathing until the mask is on, if at all possible. This may be very difficult since the agent strongly stimulates respiration.

(2) *Cyanogen chloride.* Put your mask on immediately if you experience any irritation of the eyes, nose, or throat. Service members

who are unable to mask should be masked by the nearest service member (buddy).

d. *Medical Assistance.* If you suspect that you have been exposed to blood agents, seek medical assistance immediately.

7-12. Incapacitating Agents

An incapacitating agent is a chemical agent which produces temporary, disabling conditions which persist for hours to days after exposure. Unlike riot control agents, which usually are momentary or fleeting in action, incapacitating agents have a persistent effect. It is likely that smoke-producing munitions or aerosols will disseminate such agents, thus making breathing their means of entry into the body. The protective mask is, therefore, essential.

a. There are no specific first aid measures to relieve the symptoms of incapacitating agents. Supportive first aid and physical restraint may be indicated. If the casualty is stuporous or comatose, be sure that respiration is unobstructed; then turn him on his side in case vomiting should occur. Complete cleansing of the skin with soap and water should be done as soon as possible; or, the M291 Skin Decontaminating Kit can be used if washing is impossible. Remove weapons and other potentially harmful items from service members who are suspected of having these symptoms. Harmful items include cigarettes, matches, medications, and small items that might be swallowed accidentally. Delirious (confused) persons have been known to attempt to eat items bearing only a superficial resemblance to food.

b. Incapacitating agents (anticholinergic drugs BZ type) may produce alarming dryness and coating of the lips and tongue; however, there is usually no danger of immediate dehydration. Fluids should be given sparingly, if at all, because of the danger of vomiting and because of the likelihood of temporary urinary retention due to paralysis of bladder muscles.

c. If the body temperature is elevated and mucous membranes are dry, immediate and vigorous cooling (as for heatstroke) is indicated. Methods that can be used to cool the skin are spraying with cool water and air circulation (fanning); applying alcohol soaked cloths and air circulation; and providing maximum exposure to air in a shaded area, along with maximum air circulation. Such cases are usually a result of anticholinergic poisoning. Rapid evacuation should be accomplished since medical treatment with the appropriate medication may be lifesaving.

CAUTION

DO NOT use **ice** for cooling the skin.

d. Reassurance and a firm, but friendly, attitude by individuals providing first aid will be beneficial if the casualty appears to comprehend what is being said. Conversation is a waste of time if the service member is incoherent or cannot understand what is being said. In such cases, the less said, the better it is—these casualties will benefit more from prompt and vigorous restraint and evacuation to an MTF.

7-13. Incendiaries

Incendiaries can be grouped as WP, thickened gasoline, metal, and oil and metal. You must learn to protect yourself against these incendiaries.

a. White phosphorus is used primarily as a smoke producer but can be used for its incendiary effect to ignite field expedients and combustible materials. The burns from WP are usually multiple, deep, and variable in size. When particles of WP get on the skin or clothing, they continue to burn until deprived of air. They also have a tendency to stick to a surface and must be brushed off or picked out.

(1) If burning particles of WP strike and stick to your clothing, quickly take off the contaminated clothing before the WP burns through to the skin.

(2) If burning WP strikes your skin, smother the flame with water, a wet cloth, or mud.

NOTE

Since WP is soluble in oil, **DO NOT** use grease, oily ointments, or eye ointments to smother the flame.

(3) Keep the WP particles covered with a wet material to exclude air until you can remove them or have them removed from your skin.

(4) Remove the WP particles from the skin by brushing them with a wet cloth and by picking them out with a knife, bayonet, stick, or other available object.

(5) Seek medical assistance when the mission permits.

b. Thickened fuel mixtures (napalm) have a tendency to cling to clothing and body surfaces, thereby producing prolonged exposure and severe burns. The first aid for these burns is the same as for other heat burns. The heat and irritating gases given off by these combustible mixtures may cause lung damage, which must be treated by medical personnel.

c. Metal incendiaries pose special problems. Thermite particles on the skin should be immediately cooled with water and then removed. The first aid for these burns is the same as for other heat burns. Particles of magnesium on the skin burn quickly and deeply. Like other metal incendiaries, they must be removed. Ordinarily, medical personnel should do the complete removal of these particles as soon as possible. Immediate medical treatment is required.

d. Oil and metal incendiaries have much the same effect on contact with the skin and clothing as those discussed (*b* and *c* above). First aid measures for burns are discussed in Chapter 3.

7-14. Biological Agents and First Aid

a. Biological attacks can result in combat ineffectiveness by introducing disease-causing organisms into a troop population.

b. Once a disease is identified, first aid or medical treatment is initiated, depending on the seriousness of the disease. First aid measures are concerned with observable symptoms of the disease such as diarrhea or vomiting.

7-15. Toxins

Toxins are alleged to have been used in past conflicts. Witnesses and victims have described the agent as toxic rain (or yellow rain) because it was reported to have been released from aircraft as a yellow powder or liquid that covered ground, structures, vegetation, and people.

a. *Signs and Symptoms.* The occurrence of the symptoms from toxins may appear in a period of a few minutes to several hours depending on the particular toxin, the service member's susceptibility, and the amount of toxin inhaled, ingested, or deposited on the skin. Symptoms from toxins usually involve the central nervous system but are often preceded by less prominent symptoms, such as nausea, vomiting, diarrhea, cramps, or stomach irritation and burning sensation. Typical neurological symptoms often develop rapidly in severe cases; for example, visual disturbances, inability to swallow, speech difficulty, lack of muscle coordination, and sensory abnormalities (numbness of mouth, throat, or extremities). Yellow rain (mycotoxins) also may have hemorrhagic symptoms, which could include any or all of the following:

- Dizziness.
- Severe itching or tingling of the skin.
- Formation of multiple, small, hard blisters.
- Coughing up blood.
- Shock (which could result in death).

b. Self-Aid. Upon recognition of an attack employing toxins, you must immediately take the following actions:

(1) Stop breathing, put on your protective mask with hood, and then resume breathing. Next, put on your protective clothing.

(2) Should severe itching of the face become unbearable, quickly—

- Loosen the cap on your canteen.
- Take and hold a deep breath and lift your mask.
- While holding your breath, close your eyes and flush your face with generous amounts of water.

CAUTION

DO NOT rub or scratch your eyes. Try not to let the water run onto your clothing or protective overgarment.

- Put your protective mask back on, seat it properly, clear it, and check it for a seal; then resume breathing.
- Decontaminate your skin by bathing with soap and water as soon as the mission permits.

- Change clothing and decontaminate your protective mask using soap and water. Replace the filters if directed.

(3) If vomiting occurs, the mask should be lifted momentarily and drained—while the eyes are closed and the breath is held—and replaced, cleared, and sealed.

c. *Medical Assistance.* If you suspect that you have been exposed to toxins, you should seek medical assistance immediately.

7-16. Nuclear Detonation

a. Three types of injuries may result from a nuclear detonation. These are thermal, blast, and radiation injuries. Many times the casualty will have a combination of these types of injuries. First aid for thermal and blast injuries is provided based on observable injuries, such as burns, hemorrhage, or fractures.

b. The signs and symptoms of radiation illness in the initial phase include the rapid onset of nausea, vomiting, and malaise (tiredness). The only first aid procedure for radiological casualties is decontamination.

CHAPTER 8

FIRST AID FOR PSYCHOLOGICAL REACTIONS

8-1. General

Psychological first aid is as natural and reasonable as physical first aid and is just as familiar. When you were hurt as a child, the understanding attitude of your parents did as much as the psychological effect of a bandage. Later, your disappointment or grief was eased by supportive words from a friend. Certainly, taking a walk and talking things out with a friend are familiar ways of dealing with an emotional crisis. The same natural feelings that make us want to help a person who is injured make us want to give a helping hand to a buddy who is upset. *Psychological first aid* really means nothing more complicated than assisting people with emotional distress whether it results from physical injury, disease, or excessive stress. Emotional distress is not always as visible as a wound or a broken bone. However, overexcitement, severe fear, excessive worry, deep depression, misdirected irritability, and anger are signs that stress has reached the point of interfering with effective coping. The more noticeable the symptoms become, the more urgent the need for you to be of help and the more important it is for you to know *how* to help.

8-2. Importance of Psychological First Aid

You must know how to give psychological first aid to be able to help yourself, your buddies, and your unit in order to keep performing the mission. Psychological first aid measures are simple and easy to understand. Your decision of what to do depends upon your ability to observe the service member and understand his needs. Making the best use of resources requires ingenuity on your part. A stress reaction resulting in poor judgment can cause injury or even death to yourself or others on the battlefield. It can be even more dangerous if other persons are affected by the judgment of an emotionally upset service member. If it is detected early enough, the affected service member stands a good chance of remaining in his unit as an effective member. If it is not detected early and if the service member becomes more emotionally upset, he may become a threat to himself and to others.

8-3. Situations Requiring Psychological First Aid

- Psychological first aid (buddy aid) is most needed at the first sign that a service member cannot perform the mission because of emotional distress. Stress is inevitable in combat, in hostage and terrorist situations, and in civilian disasters such as floods, hurricanes, or industrial accidents. Most emotional reactions to such situations are temporary, and the service member can still carry on with encouragement. Painful or disruptive symptoms may last for minutes, hours, or days. However, if the stress symptoms are seriously disabling, they may be psychologically contagious and endanger not only the emotionally upset service member but also the entire unit.

- Sometimes people continue to function well during a disastrous event, but suffer from emotional scars which impair their job performance or quality of life at a later time. Painful memories and dreams may recur for months and years and still be considered a normal reaction. However, if the memories are so painful that the person must avoid all situations which arouse them, becomes socially withdrawn, or shows symptoms of anxiety, depression, or substance abuse, he needs treatment. Experience with police, firemen, emergency medical technicians, and others who deal with disasters has proved that the routine application of psychological first aid to all the participants, including those who have functioned well, greatly reduces the likelihood of future serious post-traumatic stress disorders (PTSDs).

8-4. Interrelationship of Psychological and Physical First Aid

Psychological first aid should go hand in hand with physical first aid. The discovery of a physical injury or cause for an inability to function does not rule out the possibility of a psychological injury (or vice versa). The person suffering from pain, shock, fear of serious injury, or fear of death does not respond well to joking, indifference, or fearful-tearful attention. Fear and anxiety may take as high a toll of the service member's strength as does the loss of blood.

8-5. Goals of Psychological First Aid

The goals of psychological first aid are to—

- Be supportive; assist the service member in dealing with his stress reaction.

- Prevent, and if necessary control, behavior harmful to himself and to others.

- Return the service member to duty as soon as possible after dealing with the stress reaction.

8-6. Respect for Others' Feelings

a. Accept the service member you are trying to help without censorship or ridicule. Respect his right to his own feelings. Even though your feelings, beliefs, and behavior are different, DO NOT blame or make light of him for the way he feels or acts. Your purpose is to help him in this tough situation, not to be his critic. A person DOES NOT WANT to be upset and worried. When he seeks help, he needs and expects consideration of his fears, not abrupt dismissal or ridicule.

b. Realize that people are the products of a wide variety of factors. All people DO NOT react the same way to the same situations. Each individual has complex needs and motivations, both conscious and unconscious, that are uniquely his own. Often the one thing that finally causes the person to become overloaded by a stressful situation is not the stressor itself, but some other problem.

8-7. Emotional and Physical Disability

a. Accept emotional disability as being just as real as physical disability. If a service member's ankle is seriously sprained in a fall, no one expects him to run right away. A service member's emotions may be temporarily strained by the overwhelming stress of battle or other traumatic incident. DO NOT demand that he pull himself together immediately and carry on without a break. Some individuals can pull themselves together immediately, but others cannot. The service member whose emotional stability has been disrupted has a disability just as real as the service member who has sprained his ankle. There is an unfortunate tendency in many people to regard as real only what they can see, such as a wound or bleeding. Some people tend to assume that damage involving a person's mind and emotions is just imagined, that he is not really sick or injured, and that he could overcome his trouble by using his will power.

b. The terms *it's all in your head, snap out of it,* and *get control of yourself* are often used by people who believe they are being helpful. Actually, these terms are expressions of hostility because they show lack of understanding. They only emphasize weakness and inadequacy. Such terms are of no use in psychological first aid.

c. Every physically injured person has some emotional reaction to the fact that he is injured.

(1) It is normal for an injured person to feel upset. The more severe the injury, the more insecure and fearful he becomes, especially

if the injury is to a body part which is highly valued. For example, an injury to the eyes or the genitals, even though relatively minor, is likely to be extremely upsetting. An injury to some other part of the body may be especially disturbing to an individual for his own particular reason. For example, an injury of the hand may be a terrifying blow to a surgeon or an injury to the eye of a pilot.

(2) An injured service member always feels less secure, more anxious, and more afraid not only because of what has happened to him but because of what he imagines may happen as a result of his injury. This fear and insecurity may cause him to be irritable, uncooperative, or unreasonable. As you help him, always keep in mind that such behavior has little or nothing to do with you personally. He needs your patience, reassurance, encouragement, and support.

8-8. Combat and Other Operational Stress Reactions

Stress reaction is a temporary emotional disorder or inability to function, experienced by a previously normal service member as a reaction to the overwhelming or cumulative stress of combat. Stress reaction gets better with reassurance, rest, physical replenishment, and activities that restore confidence. All service members are likely to feel stress reaction under conditions of intense and/or prolonged stress. They may even become stress reaction casualties, unable to perform their mission for hours or days. Other combat and operational stress reactions (COSRs) may result in negative behavior, but are not termed *stress reaction,* as they need more intensive treatment. These negative COSRs may result in misconduct stress behaviors such as drug and alcohol abuse, criminal acts, looting, desertion, and self-inflicted wounds. These harmful COSRs can often be prevented by good psychological first aid. Service members who commit misconduct stress behaviors may require disciplinary action rather than medical treatment.

8-9. Reactions to Stress

Most service members react to stressful incidents after the situation has passed. All service members feel some fear. This fear may be greater than they have experienced at any other time, or they may be more aware of their fear. In such a situation, they should not be surprised if they feel shaky or become sweaty, nauseated, or confused. These reactions are normal and are not a cause for concern. However, some reactions, either short- or long-term, will cause problems if left unchecked. See paragraph 8-13 for more information.

a. Emotional Reactions.

(1) The most obvious combat stress reaction (CSR) is inefficient performance. This can be demonstrated by—

- Slow thinking (or reaction time).
- Difficulty recognizing priorities and seeing what needs to be done.
- Difficulty getting started.
- Indecisiveness and having trouble focusing attention.
- Tendency to do familiar tasks and be preoccupied with familiar details. (This can reach the point where the person is very passive, such as just sitting or wandering about not knowing what to do.)

(2) A less common reaction may be uncontrolled emotional outbursts; this can be demonstrated by crying, screaming, or laughing. Some service members will react in the opposite way. They will be very withdrawn and silent and try to isolate themselves from everyone. These service members should be encouraged to remain with their assigned unit. Uncontrolled reactions may appear by themselves or in any combination (the person may be crying uncontrollably one minute and then laughing the next). In this state, the person is restless and cannot keep still. He may run about, apparently without purpose. Inside, he feels a great rage or fear and his physical acts may show this. In his anger he may indiscriminately strike out at others.

b. *Loss of Adaptability.*

(1) In a desperate attempt to get away from the danger, which has overwhelmed him, a service member may panic and become confused. His mental ability may be so impaired he cannot think clearly or even follow simple commands. His judgment may be faulty and he may not be aware of his actions, such as standing up in his fighting position during an attack.

(2) In other cases, overwhelming stress may produce symptoms that are often associated with head injuries. For example, the service member may appear dazed or be found wandering around aimlessly. He may appear confused and disoriented and may seem to have a complete or partial loss of memory. In such cases, especially when no eyewitnesses can provide evidence that the service member has NOT suffered a head injury, it is necessary for him to be rapidly medically evacuated. **DO NOT** allow the

service member to expose himself to further personal danger until the cause of the problem has been determined.

c. *Sleep Disturbance and Repetition of Dreams.* A person who has been overwhelmed by stress often has difficulty sleeping. The service member may experience nightmares related to the stressors. Remember that nightmares, in themselves, are not considered abnormal when they occur soon after a period of intensive stress. As time passes, the nightmares usually become less frequent and less intense. In extreme cases, a service member, even when awake, may think repeatedly of the incident, feel as though it is happening again, and act out parts of his stress over and over again. For some persons, this repetitious reexperiencing of the stressful event may be necessary for eventual recovery; therefore, it should not be discouraged or viewed as abnormal. For the person reexperiencing the event, such reaction may be disruptive. The service member needs to be encouraged to *ventilate* about the incident. Ventilation is a technique where the service member is given the opportunity to talk extensively, often repetitiously about the experience.

8-10. Severe Stress or Stress Reaction

You do not need specialized training to recognize severe stress or stress reaction that will cause problems for the service member, the unit, or the mission. Reactions that are less severe, however, are more difficult to detect. To determine whether a person needs help, you must observe him to see whether he is doing something meaningful, performing his duties, taking care of himself, behaving in an unusual fashion, or acting out of character.

8-11. Application of Psychological First Aid

The emotionally disturbed service member has built a barrier against fear. He does this for his own protection, although he is probably not aware that he is doing it. If he finds that he does not have to be afraid and that there are normal, understandable things about him, he will feel safer in dropping this barrier. Persistent efforts to make him realize that you want to understand him will be reassuring, especially if you remain calm. Nothing can cause an emotionally disturbed person to become even more fearful than feeling that others are afraid of him. Try to remain calm. Familiar things, such as a cup of coffee, the use of his name, attention to a minor wound, being given a simple job to do, or the sight of familiar people and activities, will add to his ability to overcome his fear. He may not respond well if you get excited, angry, or abrupt.

a. *Ventilation.* After the service member becomes calmer, he is likely to have dreams about the stressful event. He also may think about it when he is awake or even repeat his personal reaction to the event. One benefit of this natural pattern is that it helps him master the stress by going over it just as one masters the initial fear of parachuting from an aircraft by doing it over and over again. Eventually, it is difficult to remember how frightening the event was initially. In giving first aid to the emotionally disturbed service member, you should let him follow this natural pattern. Encourage him to talk. Be a good listener. Let him tell, in his own words, what actually happened. If home front problems or worries have contributed to the stress, it will help him to talk about them. Your patient listening will prove to him that you are interested in him, and by describing his personal problem, he can work at mastering his fear. If he becomes overwhelmed in the telling, suggest a cup of coffee or a break. Whatever you do, assure him that you will listen again as soon as he is ready. Do try to help put the service member's perception of what happened back into realistic perspective; but DO NOT argue about it.

b. *Activity.*

(1) A person who is emotionally disturbed as the result of a combat action is a casualty of anxiety and fear. He is disabled because he has become temporarily overwhelmed by his anxiety. A good way to control fear is through activity. Almost all service members, for example, experience a considerable sense of anxiety and fear while they are poised, awaiting the opening of a big offensive; but this is normally relieved, and they actually feel better once they begin to move into action. They take pride in effective performance and pleasure in knowing that they are good service members, perhaps being completely unaware that overcoming their initial fear was their first major accomplishment.

(2) Useful activity is very beneficial to the emotionally disturbed service member who is not physically incapacitated. After you help a service member get over his initial fear, help him to regain some self-confidence. Make him realize his job is continuing by finding him something useful to do. Encourage him to be active. Get him to help load trucks, clean up debris, or dig fighting positions. If possible, get him back to his usual duty. Seek out his strong points and help him apply them. Avoid having him just sit around. You may have to provide direction by telling him what to do and where to do it. The instructions should be clear and simple and should be repeated. A person who has panicked is likely to argue. Respect his feelings, but point out more immediate, obtainable, and demanding needs. Channel his excessive energy and, above all, DO NOT argue. If you cannot get him interested in doing more profitable work, it may be necessary to enlist aid in controlling his overactivity before it spreads to the group and

results in more panic. Prevent the spread of such infectious feelings by restraining and segregating if necessary.

(3) Involvement in activity helps a service member in three ways; he—

- Forgets himself.
- Has an outlet for his excessive tensions.
- Proves to himself he can do something useful.

c. *Rest.* There are times, particularly in combat, when physical exhaustion is a principal cause for emotional reactions. A unit sleep plan should be established and implemented. When possible, service members should be given a safe and relatively comfortable area in which to sleep. Examples would be an area away from heavy traffic, noise, and congestion or a place that is clean and dry and protected from environmental conditions. The more uninterrupted sleep a service member gets the better he will be able to function in the tactical environment.

d. *Hygiene.* Field hygiene is an important ingredient in a service member's morale. A service member who is dirty and unkempt will not function as well as a service member who has had the opportunity to bathe and put on clean, dry clothing. During combat, unit leaders should stress the importance of personal hygiene. Good personal hygiene not only improves morale, it also is a preventive measure against disease and nonbattle injury (DNBI).

e. *Group Activity.* You have probably already noticed that a person works, faces danger, and handles serious problems better if he is a member of a closely-knit group. Each service member in the team supports the other team members. Esprit de corps is built because the service members have the same interests, goals, and mission, and as a result they are more productive; furthermore, they are less worried because everyone is involved. It is this spirit that takes a strategic hill in battle. It is so powerful that it is one of the most effective tools you have in your *psychological first aid bag*. Getting the service member back into the team or squad activities will reestablish his sense of belonging and security and will go far toward making him a useful member of the unit.

8-12. Reactions and Limitations

Up to this point the discussion has been primarily about the feelings of the emotionally distressed service member. What about your feelings toward him? Whatever the situation, you will have emotional reactions (conscious or unconscious) toward this service member. Your reactions can either help or hinder your ability to help him. When you are tired or worried, you may very easily become impatient with him if he is unusually slow or exaggerates. You may even feel resentful toward him. At times when many physically wounded lie about you, it will be especially natural for you to resent disabilities that you cannot see. Physical wounds can be seen and easily accepted. Emotional reactions are more difficult to accept as injuries. On the other hand, will you tend to be overly sympathetic? Excessive sympathy for an incapacitated person can be as harmful as negative feelings in your relationship with him. He needs strong help, but not your sorrow. To overwhelm him with pity will make him feel even more inadequate. You must expect your buddy to recover, to be able to return to duty, and to become a useful service member again. This expectation should be displayed in your behavior and attitude as well as in what you say. If he can see your calmness, confidence, and competence, he will be reassured and will feel a sense of greater security.

8-13. Stress Reactions

See Tables 8-1, 8-2, and 8-3 for more information.

Table 8-1. Mild Stress Reaction

PHYSICAL SIGNS*	EMOTIONAL SIGNS*
1. TREMBLING, TEARFUL 2. JUMPINESS, NERVOUSNESS 3. COLD SWEAT, DRY MOUTH 4. POUNDING HEART, DIZZINESS 5. INSOMNIA, NIGHTMARES 6. NAUSEA, VOMITING, DIARRHEA 7. FATIGUE 8. THOUSAND-YARD STARE 9. DIFFICULTY THINKING, SPEAKING, AND COMMUNICATING	1. ANXIETY, INDECISIVENESS 2. IRRITABLE, COMPLAINING 3. FORGETFUL, UNABLE TO CONCENTRATE 4. EASILY STARTLED BY NOISE, MOVEMENT 5. GRIEF, TEARFUL 6. ANGER, BEGINNING TO LOSE CONFIDENCE IN SELF AND UNIT

SELF- AND BUDDY AID
1. CONTINUE MISSION PERFORMANCE, FOCUS ON IMMEDIATE MISSION. 2. EXPECT SERVICE MEMBER TO PERFORM ASSIGNED DUTIES. 3. REMAIN CALM AT ALL TIMES; BE DIRECTIVE AND IN CONTROL. 4. LET SERVICE MEMBER KNOW HIS REACTION IS NORMAL, AND THAT THERE IS NOTHING SERIOUSLY WRONG WITH HIM. 5. KEEP SERVICE MEMBER INFORMED OF THE SITUATION, OBJECTIVES, EXPECTATIONS, AND SUPPORT. CONTROL RUMORS. 6. BUILD SERVICE MEMBER'S CONFIDENCE, TALK ABOUT SUCCEEDING. 7. KEEP SERVICE MEMBER PRODUCTIVE (WHEN NOT RESTING) THROUGH RECREATIONAL ACTIVITIES, EQUIPMENT MAINTENANCE.

8.	ENSURE SERVICE MEMBER MAINTAINS GOOD PERSONAL HYGIENE.
9.	ENSURE SERVICE MEMBER EATS, DRINKS, AND SLEEPS AS SOON AS POSSIBLE.
10.	LET SERVICE MEMBER TALK ABOUT HIS FEELINGS. DO NOT "PUT DOWN" HIS FEELINGS OF GRIEF OR WORRY. GIVE PRACTICAL ADVICE AND PUT EMOTIONS INTO PERSPECTIVE.
*	MOST OR ALL OF THESE SIGNS ARE PRESENT IN MILD STRESS REACTION. THEY CAN BE PRESENT IN ANY NORMAL SERVICE MEMBER IN COMBAT YET HE CAN STILL DO HIS JOB.

Table 8-2. More Serious Stress Reaction

PHYSICAL SIGNS*	EMOTIONAL SIGNS*
1. CONSTANTLY MOVES AROUND 2. FLINCHING OR DUCKING AT SUDDEN SOUNDS 3. SHAKING, TREMBLING (WHOLE BODY OR ARMS) 4. CANNOT USE PART OF BODY, NO PHYSICAL REASON (HAND, ARM, LEGS) 5. CANNOT SEE, HEAR, OR FEEL (PARTIAL OR COMPLETE LOSS) 6. PHYSICAL EXHAUSTION, CRYING 7. FREEZING UNDER FIRE, OR TOTAL IMMOBILITY 8. VACANT STARES, STAGGERS, SWAYS WHEN STANDS 9. PANIC RUNNING UNDER FIRE	1. RAPID AND/OR INAPPROPRIATE TALKING 2. ARGUMENTATIVE, RECKLESS MOVEMENTS/ACTIONS 3. INATTENTIVE TO PERSONAL HYGIENE 4. INDIFFERENT TO DANGER 5. MEMORY LOSS 6. SEVERE STUTTERING, MUMBLING, OR CANNOT SPEAK AT ALL 7. INSOMNIA, NIGHTMARES 8. SEEING OR HEARING THINGS THAT DO NOT EXIST 9. RAPID EMOTIONAL SHIFTS 10. SOCIAL WITHDRAWAL 11. APATHETIC 12. HYSTERICAL OUTBURSTS 13. FRANTIC OR STRANGE BEHAVIOR

TREATMENT PROCEDURES**

1. IF A SERVICE MEMBER'S BEHAVIOR ENDANGERS THE MISSION, SELF, OR OTHERS, DO WHATEVER IS NECESSARY TO CONTROL HIM.
2. IF THE SERVICE MEMBER IS UPSET, CALMLY TALK HIM INTO COOPERATING.
3. IF CONCERNED ABOUT THE SERVICE MEMBER'S RELIABILITY:
   - UNLOAD HIS WEAPON.
   - TAKE WEAPON IF SERIOUSLY CONCERNED.
   - PHYSICALLY RESTRAIN HIM ONLY WHEN NECESSARY FOR SAFETY OR TRANSPORTATION.
4. REASSURE EVERYONE THAT THE SIGNS ARE PROBABLY JUST STRESS REACTION AND WILL QUICKLY IMPROVE.
5. IF STRESS REACTION SIGNS CONTINUE:
   - GET THE SERVICE MEMBER TO A SAFER PLACE.
   - DO NOT LEAVE THE SERVICE MEMBER ALONE, KEEP SOMEONE HE KNOWS WITH HIM.
   - NOTIFY SENIOR NONCOMMISSIONED OFFICER (NCO) OR OFFICER.
   - HAVE THE SERVICE MEMBER EXAMINED BY MEDICAL PERSONNEL.

Table 8-2. More Serious Stress Reaction (Continued)

TREATMENT PROCEDURES**
6. GIVE THE SERVICE MEMBER EASY TASKS TO DO WHEN NOT SLEEPING, EATING, OR RESTING. 7. ASSURE THE SERVICE MEMBER HE WILL RETURN TO FULL DUTY IN 24 HOURS; AND, RETURN HIM TO NORMAL DUTIES AS SOON AS HE IS READY.
* THESE SIGNS ARE PRESENT IN ADDITION TO THE SIGNS OF MILD STRESS REACTION. ** DO THESE PROCEDURES IN ADDITION TO THE SELF- AND BUDDY AID CARE.

Table 8-3. Preventive Measures to Combat Stress Reaction

1. WELCOME NEW MEMBERS INTO YOUR TEAM, GET TO KNOW THEM QUICKLY. IF YOU ARE NEW, BE ACTIVE IN MAKING FRIENDS.
2. BE PHYSICALLY FIT (STRENGTH, ENDURANCE, AND AGILITY).
3. KNOW AND PRACTICE LIFESAVING SELF- AND BUDDY AID.
4. PRACTICE RAPID RELAXATION TECHNIQUES (FM 22-51).
5. HELP EACH OTHER OUT WHEN THINGS ARE TOUGH AT HOME OR IN THE UNIT.
6. KEEP INFORMED; ASK YOUR LEADER QUESTIONS, IGNORE RUMORS.
7. WORK TOGETHER TO GIVE EVERYONE FOOD, WATER, SHELTER, HYGIENE, AND SANITATION.
8. SLEEP WHEN MISSION AND SAFETY PERMIT; LET EVERYONE GET TIME TO SLEEP.
   - SLEEP ONLY IN SAFE PLACES AND BY STANDING OPERATING PROCEDURE (SOP).
   - IF POSSIBLE, SLEEP 6 TO 9 HOURS PER DAY.
   - TRY TO GET AT LEAST 4 HOURS SLEEP PER DAY.
   - GET GOOD SLEEP BEFORE GOING ON SUSTAINED OPERATIONS.
   - CATNAP WHEN YOU CAN, BUT ALLOW TIME TO WAKE UP FULLY.
   - CATCH UP ON SLEEP AFTER GOING WITHOUT.

APPENDIX A

FIRST AID CASE AND KITS, DRESSINGS, AND BANDAGES

A-1. First Aid Case with Field Dressings and Bandages

Every service member is issued a first aid case (Figure A-1A) with a field first aid dressing encased in a plastic wrapper (Figure A-1B). He carries it at all times for his use. The field first aid dressing is a standard sterile (germ-free) compress or pad with bandages attached (Figure A-1C). This dressing is used to cover the wound, to protect against further contamination, and to stop bleeding (pressure dressing). When a service member administers first aid to another person, he must remember to use the wounded person's dressing; he may need his own later. The service member must check his first aid case regularly and replace any used or missing dressing. The field first aid dressing may normally be obtained from his unit supply.

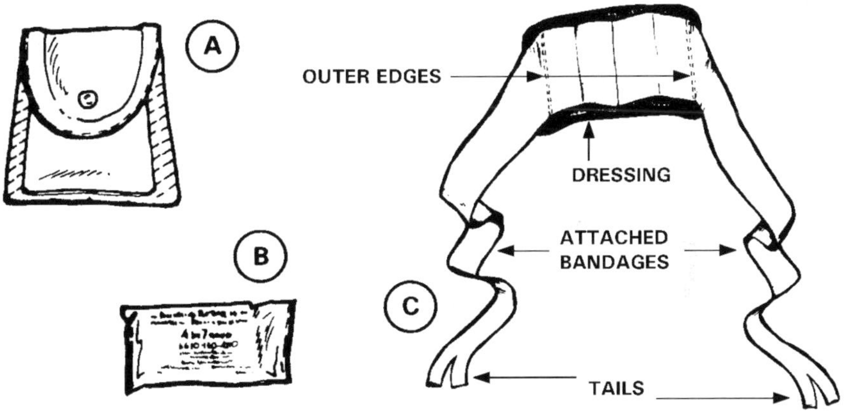

Figure A-1. Field first aid case and dressing (Illustrated A—C).

A-2. General Purpose First Aid Kits

General-purpose first aid kits are listed in the common table of allowances (CTA) 8-100. The operators, crew, and passengers carry these kits on Army vehicles, aircraft, and boats for use. Individuals designated by unit TSOP to be responsible for these kits are required to check them regularly and replace all items used. The general-purpose kit and its contents can be obtained through the unit supply system.

NOTE

Periodically check the dressings (for holes or tears in the packaging) and the medicines (for expiration date) that are in the first aid kits. If necessary, replace defective or outdated items.

A-3. Dressings

Dressings are sterile pads or compresses used to cover wounds. They usually are made of gauze or cotton wrapped in gauze (Figure A-1C). In addition to the standard field first aid dressing, other dressings such as sterile gauze compresses and small sterile compresses on adhesive strips may be available under CTA 8-100.

A-4. Standard Bandages

a. Standard bandages are made of gauze or muslin and are used over a sterile dressing to secure the dressing in place, to close off its edge from dirt and germs, and to create pressure on the wound and control bleeding. A bandage can also support an injured part or secure a splint.

b. Tailed bandages may be attached to the dressing as indicated on the field first aid dressing (Figure A-1C).

A-5. Triangular and Cravat (Swathe) Bandages

a. Triangular and cravat (or swathe) bandages (Figure A-2) are fashioned from a triangular piece of muslin (37 by 37 by 52 inches) provided in the general-purpose first aid kit. If it is folded into a strip, it is called a cravat. Two safety pins are packaged with each bandage. These bandages are valuable in an emergency since they are easily applied.

b. To improvise a triangular bandage, cut a square of available material, slightly larger than 3 feet by 3 feet, and *fold it diagonally*. If two bandages are needed, cut the material along the diagonal fold.

c. A cravat can be improvised from such common items as T-shirts, other shirts, bed linens, trouser legs, scarfs, or any other item made of pliable and durable material that can be folded, torn, or cut to the desired size.

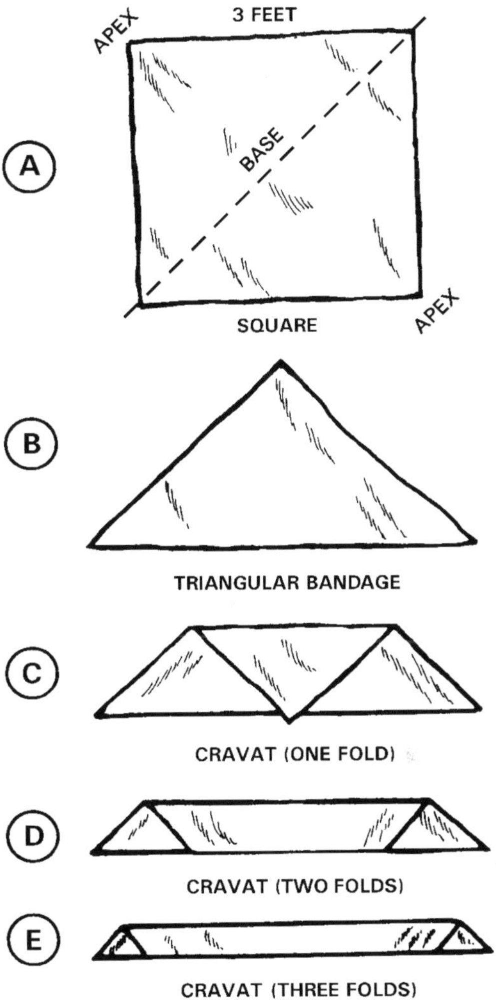

Figure A-2. Triangular and cravat bandages (Illustrated A—E).

APPENDIX B

RESCUE AND TRANSPORTATION PROCEDURES

B-1. General

A basic principle of first aid is to evaluate the casualty's injuries and administer first aid before moving him. However, adverse situations or conditions may jeopardize the lives of both the rescuer and the casualty if this is done. It may be necessary *first to rescue* the casualty before first aid can be effectively or safely given. The life and/or the well-being of the casualty will depend as much upon the manner in which he is *rescued and transported,* as it will upon the first aid and medical treatment he receives. Rescue actions must be done quickly and safely. Careless or rough handling of the casualty during rescue operations can aggravate his injuries.

B-2. Principles of Rescue Operations

a. When faced with the necessity of rescuing a casualty who is threatened by hostile action, fire, water, or any other immediate hazard, DO NOT take action without first determining the extent of the hazard and your ability to handle the situation. DO NOT become a casualty.

b. The rescuer must evaluate the situation and analyze the factors involved. This evaluation involves three major steps:

- Identify the task.
- Evaluate circumstances of the rescue.
- Plan the action.

B-3. Considerations

a. First determine if a rescue attempt is actually needed. It is a waste of time, equipment, and personnel to rescue someone not in need of rescuing. It is also a waste to look for someone who is not lost or needlessly risk the lives of the rescuer(s). In planning a rescue, attempt to obtain the following information:

- Who, what, where, when, why, and how the situation happened?

- How many casualties are involved and the nature of their injuries?

- What is the tactical situation?

- What are the terrain features and the location of the casualties?

- Will there be adequate assistance available to aid in the rescue/evacuation?

- Can first aid and/or medical treatment be provided at the scene; will the casualties require movement to a safer location?

- What specialized equipment will be required for the rescue operation?

- Is the rescue area contaminated? Will decontamination equipment and materiel be required for casualties, rescue personnel, and rescue equipment?

- How much time is available?

b. The time element can play a significant role in how the rescue is attempted. If the casualties are in imminent danger of losing their lives (such as near a burning vehicle or in a burning building) the time available will be relatively short and will sometimes cause a rescuer to compromise planning stages and/or the first aid which can be given. However, if the casualty is in a relatively secure area and his physical condition is strong, more deliberate planning can take place. A realistic estimate of time available must be made as quickly as possible to determine action time remaining. The key elements are the casualty's physical and mental condition, the tactical situation, and the environment.

B-4. Plan of Action

a. The casualty's ability to endure is of primary importance in estimating the time available. Age, physical condition, and extent of wounds and/or injuries will differ from casualty to casualty. Therefore, to determine the time available, you will have to consider—

- Endurance time of the casualty.

- Extent of injuries.

- Type of situation.
- Personnel and/or equipment availability.
- Weather.
- Terrain (natural and man-made).
- Environment (contaminated or uncontaminated).

b. In respect to terrain, you must consider altitude and visibility. In some cases, the casualty may be of assistance because he knows more about the particular terrain or situation than you do. Maximum use of secure/reliable trails or roads is essential.

c. When taking weather into account, ensure that blankets and/or rain gear are available. Even a mild rain can complicate a normally simple rescue. In high altitudes and/or extreme cold and gusting winds, the time available is critically shortened. Be prepared to provide shelter and warmth for the casualty as well as the rescuers.

B-5. Proper Handling of Casualties

a. You may have saved the casualty's life through the application of appropriate first aid measures. However, his life can be lost through rough handling or careless transportation procedures. Before you attempt to move the casualty—

- Evaluate the type and extent of his injuries.
- Ensure that dressings over wounds are adequately reinforced.
- Ensure that fractured bones are properly immobilized and supported to prevent them from cutting through muscle, blood vessels, and skin.

b. Based upon your evaluation of the type and extent of the casualty's injury and your knowledge of the various manual carries, you must select the best possible method of manual transportation. If the casualty is conscious, tell him how he is to be transported. This will help allay his fear of movement and gain his cooperation and confidence.

c. Buddy aid for chemical agent casualties includes those actions required to prevent an incapacitated casualty from receiving additional injury from the effects of chemical hazards. If a casualty is physically unable to decontaminate himself or administer the proper chemical agent antidote, the casualty's buddy assists him and assumes responsibility for his care. Buddy-aid includes—

- Administering the proper chemical agent antidote.
- Decontaminating the incapacitated casualty's exposed skin.
- Ensuring that his protective ensemble remains correctly emplaced.
- Maintaining respiration.
- Controlling bleeding.
- Providing other standard first aid measures
- Transporting the casualty out of the contaminated area.

B-6. Positioning the Casualty

The first step in any manual carry is to position the casualty to be lifted. If he is conscious, he should be told how he is to be positioned and transported. This helps lessen his fear of movement and to gain his cooperation. It may be necessary to roll the casualty onto his abdomen, or his back, depending upon the position in which he is lying and the particular carry to be used.

a. To roll a casualty onto his abdomen, kneel at the casualty's uninjured side.

(1) Place his arms above his head; cross his ankle which is farther from you over the one that is closer to you.

(2) Place your hands on the shoulder which is farther from you; place your other hand in the area of his hip or thigh (Figure B-1).

(3) Roll him gently toward you onto his abdomen (Figure B-2).

b. To roll a casualty onto his back, follow the same procedure described in *a* above, except gently roll the casualty onto his back, rather than onto his abdomen.

Figure B-1. Positioning the casualty.

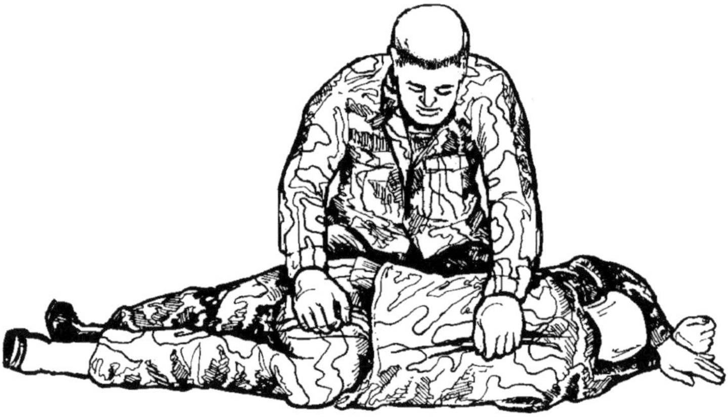

Figure B-2. Rolling casualty onto his abdomen.

B-7. Medical Evacuation and Transportation of Casualties

a. Medical evacuation of the sick and wounded (with en route medical care) is the responsibility of medical personnel who have been provided special training and equipment. Therefore, unless a good reason for you to transport a casualty arises, wait for some means of medical evacuation to be provided. When the situation is urgent and you are unable to obtain medical assistance or know that no medical evacuation assets are available, you will have to transport the casualty. For this reason, you must know how to transport him without increasing the seriousness of his condition.

b. Transporting a casualty by litter (FM 8-10-6) is safer and more comfortable for him than by manual means; it is also easier for you.

Manual transportation, however, may be the only feasible method because of the terrain or the combat situation; or it may be necessary to save a life. In these situations, the casualty should be transferred to a litter as soon as one can be made available or improvised.

B-8. Manual Carries

Casualties carried by manual means must be carefully and correctly handled, otherwise their injuries may become more serious or possibly fatal. Situation permitting, transport of a casualty should be organized and unhurried. Each movement should be performed as deliberately and gently as possible. Casualties should not be moved before the type and extent of injuries are evaluated and the required first aid is administered. The exception to this occurs when the situation dictates immediate movement for safety purposes (for example, it may be necessary to remove a casualty from a burning vehicle); that is, the situation dictates that the urgency of casualty movement outweighs the need to administer first aid. Manual carries are tiring for the bearers and involve the risk of increasing the severity of the casualty's injury. In some instances, however, they are essential to save the casualty's life. Although manual carries are accomplished by one or two bearers, the two-man carries are used whenever possible. They provide more comfort to the casualty, are less likely to aggravate his injuries, and are also less tiring for the bearers. The distance a casualty can be carried depends on many factors, such as—

- Nature of the casualty's injuries.
- Strength and endurance of the bearer(s).
- Weight of the casualty.
- Obstacles encountered during transport (natural or manmade).
- Type of terrain.

a. *One-man Carries*. These carries should be used when only one bearer is available to transport the casualty.

(1) The *fireman's carry* (Figure B-3) is one of the easiest ways for one individual to carry another. After an unconscious or disabled casualty has been properly positioned, he is raised from the ground, then supported and placed in the carrying position.

(*a*) After rolling the casualty onto his abdomen, straddle him. Extend your hands under his chest and lock them together.

(*b*) Lift the casualty to his knees as you move backward.

(*c*) Continue to move backward, thus straightening the casualty's legs and locking his knees.

(*d*) Walk forward, bringing the casualty to a standing position; tilt him slightly backward to prevent his knees from buckling.

(*e*) As you maintain constant support of the casualty with one arm, free your other arm, quickly grasp his wrist, and raise his arm high. Instantly pass your head under his raised arm, releasing it as you pass under it.

(*f*) Move swiftly to face the casualty and secure your arms around his waist. Immediately place your foot between his feet and spread them apart (approximately 6 to 8 inches).

(*g*) Grasp the casualty's wrist and raise his arm high over your head.

(*h*) Bend down and pull the casualty's arm over and down on your shoulder, bringing his body across your shoulders. At the same time, pass your arm between his legs.

(*i*) Grasp the casualty's wrist with one hand, and place your other hand on your knee for support.

(*j*) Rise with the casualty positioned correctly. Your other hand is free for use.

Ⓐ

Figure B-3. Fireman's carry (Illustrated A—J).

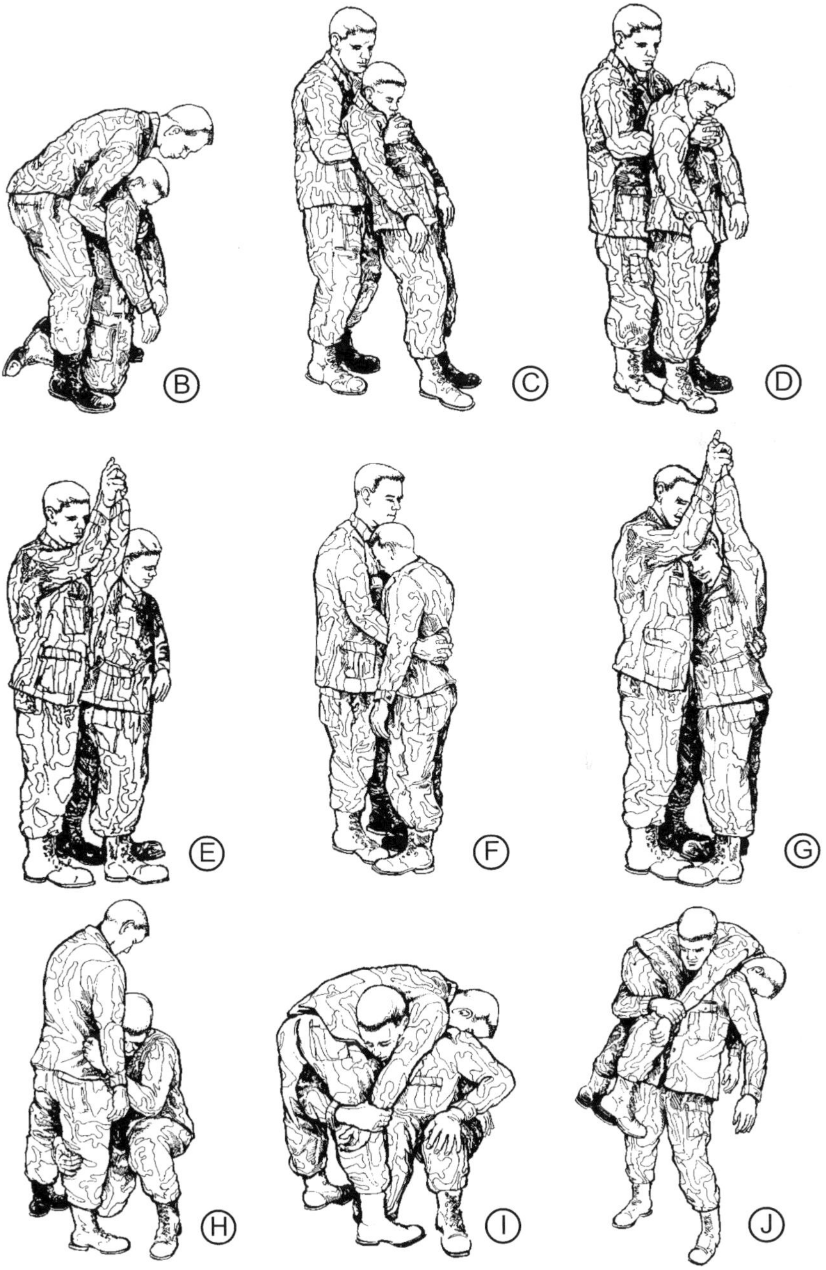

Figure B-3. Fireman's carry (Illustrated A—J) (Continued).

(2) The alternate method of the *fireman's carry* for raising a casualty from the ground is illustrated in Figure B-4; however, it should be used only when the bearer believes it to be safer for the casualty because of the location of his wounds. When the alternate method is used, care must be taken to prevent the casualty's head from snapping back and causing a neck injury. The steps for raising a casualty from the ground for the fireman's carry are also used in other one-man carries.

(*a*) Kneel on one knee at the casualty's head and face his feet. Extend your hands under his armpits, down his sides, and across his back.

(*b*) As you rise, lift the casualty to his knees. Then secure a lower hold and raise him to a standing position with his knees locked.

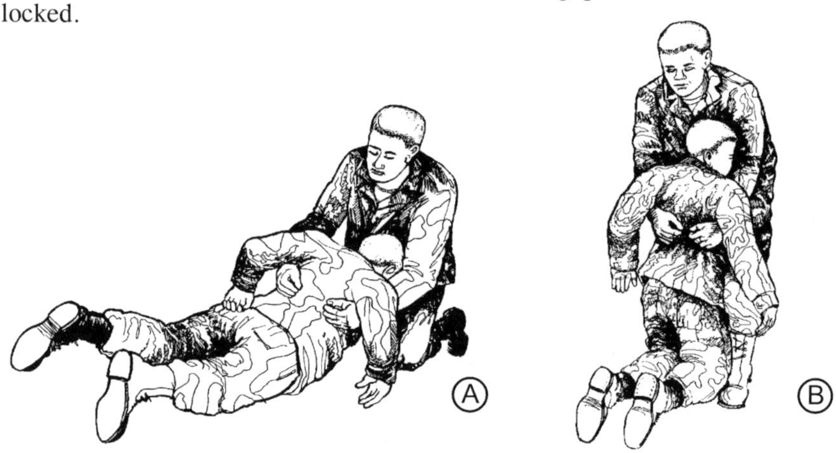

Figure B-4. Fireman's carry (alternate method) for lifting a casualty to a standing position (Illustrated A—B).

(3) In the *supporting carry* (Figure B-5), the casualty must be able to walk or at least hop on one leg, using the bearer as a crutch. This carry can be used to assist him as far as he is able to walk or hop.

(*a*) Raise the casualty from the ground to a standing position by using the fireman's carry.

(*b*) Grasp the casualty's wrist and draw his arm around your neck.

(*c*) Place your arm around his waist. The casualty is now able to walk or hop using you as a support.

Figure B-5. Supporting carry.

(4) The *arms carry* (Figure B-6) is useful in carrying a casualty for a short distance (up to 50 meters) and for placing him on a litter.

(*a*) Raise or lift the casualty from the ground to a standing position, as in the fireman's carry.

(*b*) Place one arm under the casualty's knees and your other arm around his back.

(*c*) Lift the casualty.

(*d*) Carry the casualty high to lessen fatigue.

Figure B-6. Arms carry.

(5) Only a conscious casualty can be transported by the *saddleback carry* (Figure B-7), because he must be able to hold onto the bearer's neck. To use this technique—

(*a*) Raise the casualty to an upright position, as in the fireman's carry.

(*b*) Support the casualty by placing an arm around his waist. Move to the casualty's side. Have the casualty put his arm around your neck and move in front of him with your back to support him.

(*c*) Have the casualty encircle his arms around your neck

(*d*) Stoop, raise him on your back and clasp your hands together beneath his thighs, if possible.

Figure B-7. Saddleback carry.

(6) In the *pack-strap carry* (Figure B-8), the casualty's weight rests high on the your back. This makes it easier for you to carry the casualty a moderate distance (50 to 300 meters). To eliminate the possibility of injury to the casualty's arms, you must hold his arms in a palms-down position.

(*a*) Lift the casualty from the ground to a standing position, as in the fireman's carry.

(*b*) Support the casualty with your arms around him and grasp his wrist closer to you.

(*c*) Place his arm over your head and across your shoulders.

(*d*) Move in front of him while still supporting his weight against your back.

(*e*) Grasp his other wrist and place this arm over your shoulder.

(*f*) Bend forward and raise or hoist the casualty as high on your back as possible so that his weight is resting on your back.

NOTE

Once the casualty is positioned on the bearer's back, the bearer remains as erect as possible to prevent straining or injuring his back.

Figure B-8. Pack-strap carry.

(7) The *pistol-belt carry* (Figure B-9) is the best one-man carry for a long distance (over 300 meters). The casualty is securely supported upon your shoulders by a belt. Both your hands and the casualty's (if conscious) are free for carrying a weapon or equipment, or climbing obstacles. With your hands free and the casualty secured in place, you are also able to creep through shrubs and under low-hanging branches.

(*a*) Link two pistol belts (or three, if necessary) together to form a sling. Place the sling under the casualty's thighs and lower back so that a loop extends from each side.

NOTE

If pistol belts are not available for use, other items such as a rifle sling, two cravat bandages, two litter straps, or any other suitable material, which will not cut or bind the casualty may be used.

(*b*) Lie face up between the casualty's outstretched legs. Thrust your arms through the loops and grasp his hands and trouser leg on his injured side.

(*c*) Roll toward the casualty's uninjured side onto your abdomen, bringing him onto your back. Adjust the sling, if necessary.

(*d*) Rise to a kneeling position. The belt will hold the casualty in place.

(*e*) Place one hand on your knee for support and rise to an upright position. (The casualty is supported on your shoulders.)

(*f*) Carry the casualty with your hands free for use in rifle firing, climbing, or surmounting obstacles.

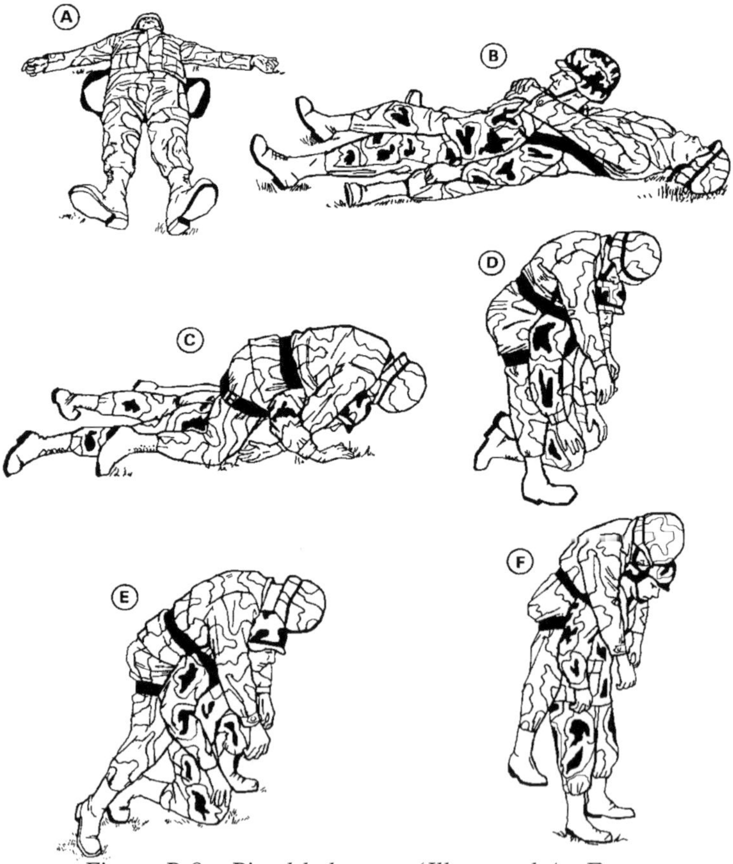

Figure B-9. Pistol-belt carry (Illustrated A—F).

(8) The *pistol-belt drag* (Figure B-10), as well as other drags, is generally used for short distances (up to 50 meters). This drag is useful in combat, since both the bearer and the casualty can remain closer to the ground than in any other drags.

(*a*) Extend two pistol belts or similar objects to their full length and join them together to make a continuous loop.

(*b*) Roll the casualty onto his back, as in the fireman's carry.

(*c*) Pass the loop over the casualty's head, and position it across his chest and under his armpits. Then cross the remaining portion of the loop, thus forming a figure eight. Keep tension on the belts so they do not come unhooked.

(*d*) Lie on your side facing the casualty.

(*e*) Slip the loop over your head and turn onto your abdomen. This enables you to drag the casualty as you crawl.

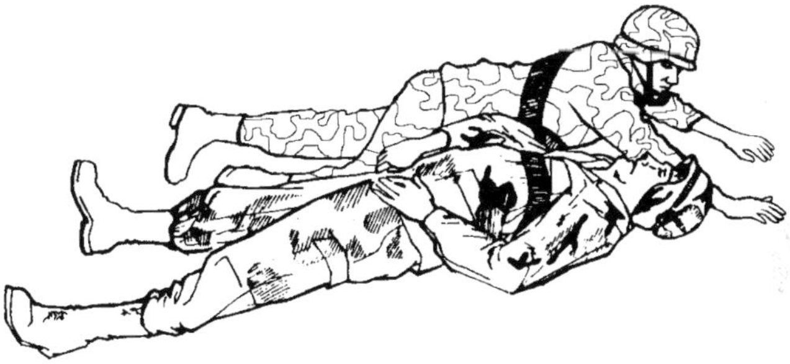

Figure B-10. Pistol-belt drag.

(9) The *neck drag* (Figure B-11) is useful in combat because the bearer can transport the casualty as he creeps behind a low wall or shrubbery, under a vehicle, or through a culvert. If the casualty is unconscious, his head must be protected from the ground. The neck drag cannot be used if the casualty has a broken arm.

NOTE

If the casualty is conscious, he may clasp his hands together around your neck.

(*a*) Tie the casualty's hands together at the wrists.

(*b*) Straddle the casualty in a kneeling face-to-face position.

(*c*) Loop the casualty's tied hands over and around your neck.

(*d*) Crawl forward dragging the casualty with you.

NOTE

If the casualty is unconscious, protect his head from the ground.

Figure B-11. Neck drag.

(10) The *cradle drop drag* (Figure B-12) is effective in moving a casualty up or down steps.

(*a*) Kneel at the casualty's head (with him lying on his back). Slide your hands, with palms up, under the casualty's shoulders and get a firm hold under his armpits.

(*b*) Rise (partially), supporting the casualty's head on one of your forearms. (You may bring your elbows together and let the casualty's head rest on both of your forearms.)

(*c*) Rise and drag the casualty backward. (The casualty is in a semisitting position.)

(*d*) Back down the steps, supporting the casualty's head and body and letting his hips and legs drop from step to step.

NOTE

If the casualty needs to be moved up the steps, you should back up the steps, using the same procedure.

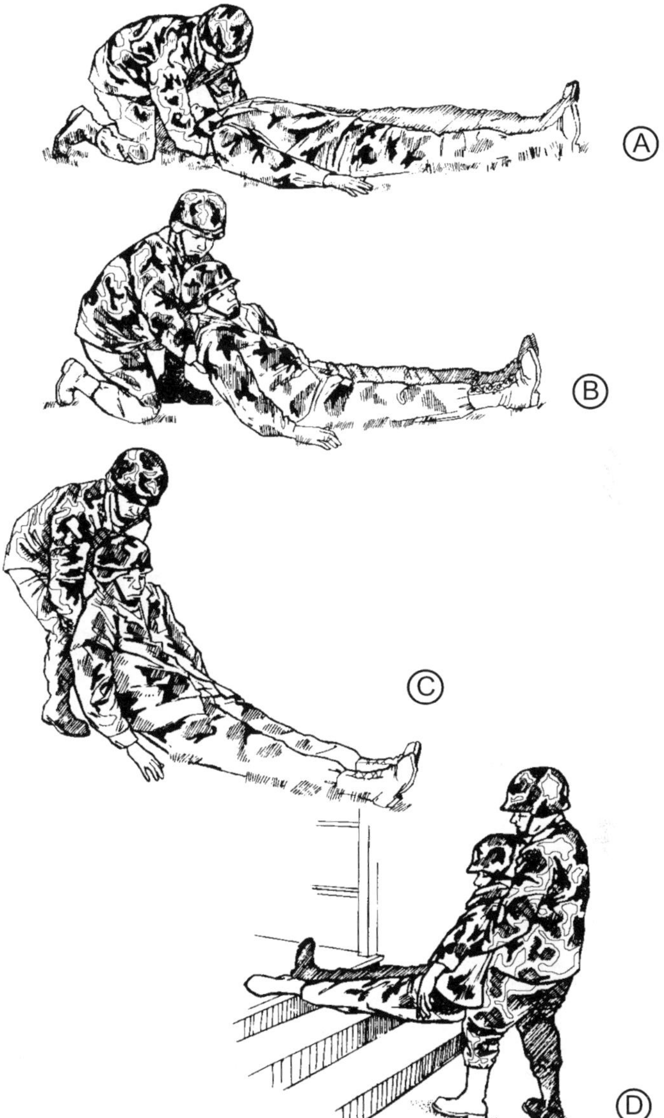

Figure B-12. Cradle-drop drag (Illustrated A—D).

(11) The *LBE carry using the bearer's LBE* can be used with a conscious casualty (Figure B-13).

(*a*) Loosen all suspenders on your LBE.

(*b*) Have the casualty place one leg into the loop formed by your suspenders and pistol belt.

(*c*) Squat in front of the standing casualty. Have him place his other leg into the loop, also.

(*d*) Have the casualty place his arms over your shoulders, lean forward onto your back, and lock his hands together.

(*e*) Stand up and lean forward into a comfortable position.

(*f*) Continue the mission.

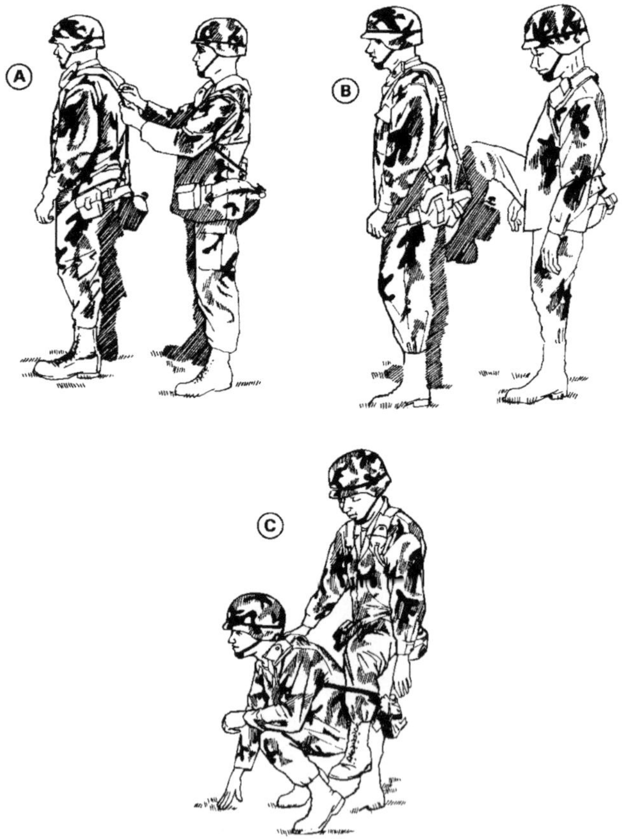

Figure B-13. Load bearing equipment carry using bearer's LBE (conscious casualty) (Illustrated A—F).

Figure B-13. Load bearing equipment carry using bearer's LBE (conscious casualty) (Illustrated A—F) (Continued).

(12) The *LBE carry using the bearer's LBE* can be used with an unconscious casualty or one who cannot stand (Figure B-14).

(*a*) Position the casualty on the flat of his back.

(*b*) Remove your LBE and loosen all suspender straps.

(*c*) Lift the casualty's leg and place it through the loop formed by your suspenders and pistol belt. Then place the other leg through the same loop. The LBE is moved up until the pistol belt is behind the casualty's thighs.

(*d*) Lay between the casualty's legs; work your arms through the LBE suspenders.

(*e*) Grasp the casualty's hand (on the injured side), and roll the casualty (on his uninjured side) onto your back.

(*f*) Rise to one knee and then push into a standing position.

(*g*) Bring the casualty's arms over your shoulders. Grasp his hands and secure them if the casualty is unconscious. If the casualty is conscious, have him lock his hands in front if he is able to do so.

(*h*) Lean forward into a comfortable position and continue the mission.

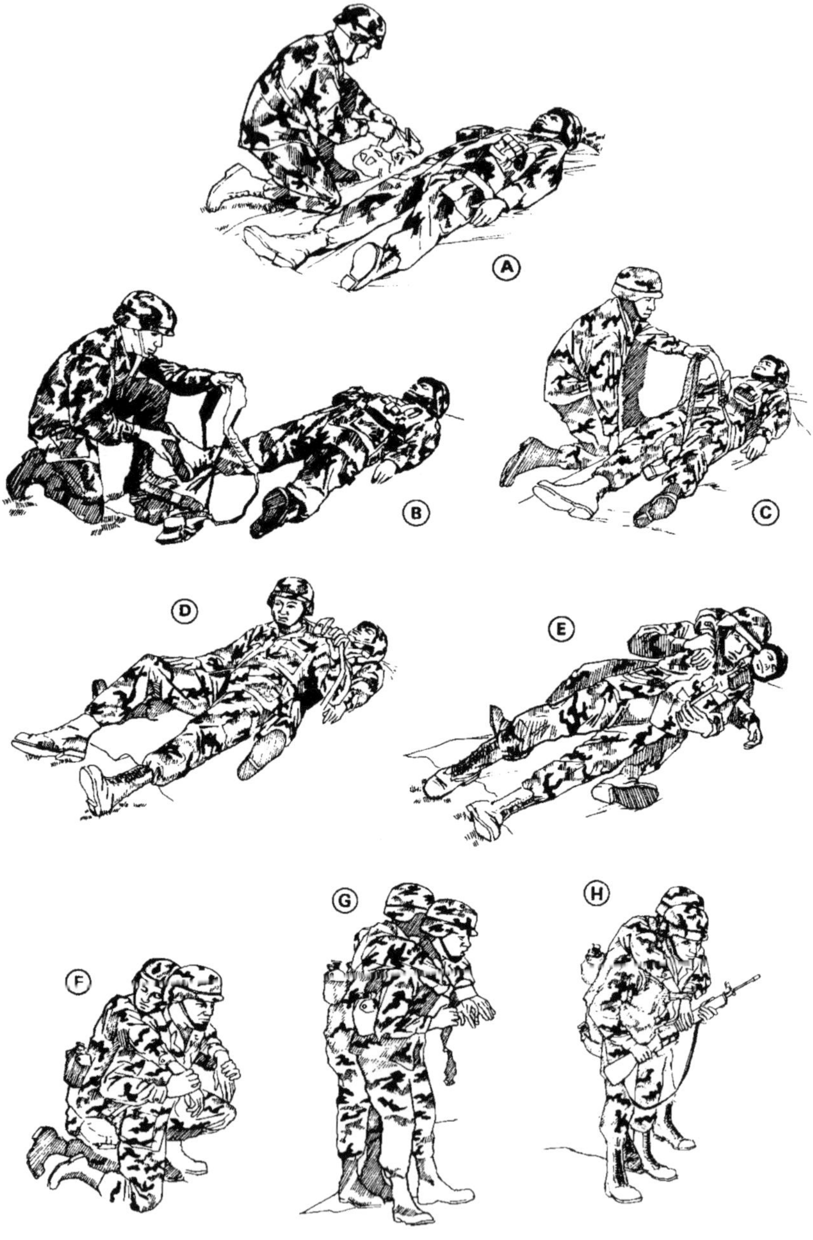

Figure B-14. Load bearing equipment carry using bearer's LBE (unconscious casualty or one that cannot stand) (Illustrated A—H).

(13) The *LBE carry using the casualty's LBE* (Figure B-15) can be used with a conscious or unconscious casualty.

(*a*) Position the casualty on his back with his LBE on.

(*b*) Loosen the casualty's two front suspenders.

(*c*) Position yourself between the casualty's legs, and slip your arms into the casualty's two front suspenders (up to your shoulders).

(*d*) Work his arms out of his LBE suspenders.

(*e*) Grasp the casualty's hand (on the injured side), and roll him (on his uninjured side) onto your back.

(*f*) Rise to one knee, then into a standing position.

(*g*) Grasp the casualty's hands and secure them, if the casualty is unconscious. Have the casualty lock his hands in front of you, if he is conscious.

(*h*) Lean forward into a comfortable position and continue the mission.

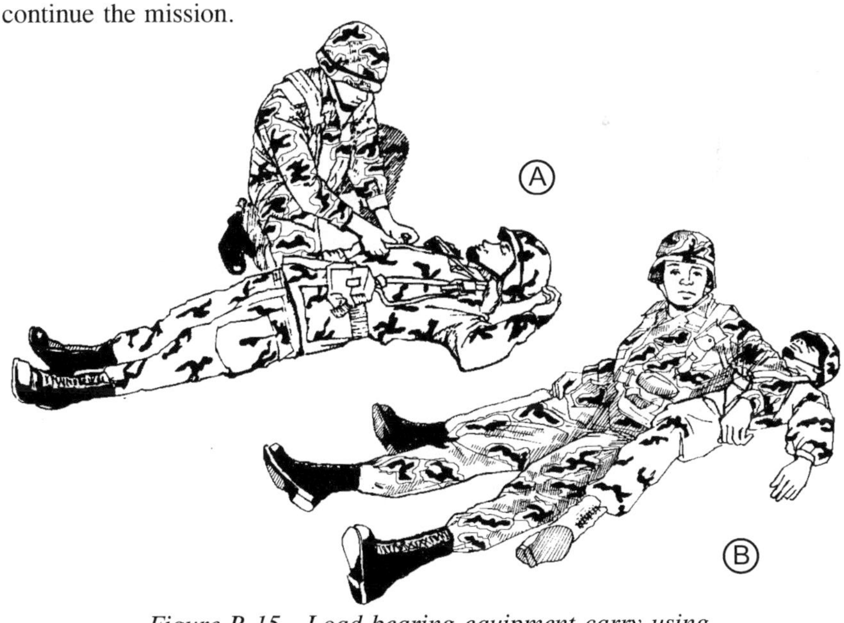

Figure B-15. Load bearing equipment carry using casualty's LBE (Illustrated A—G).

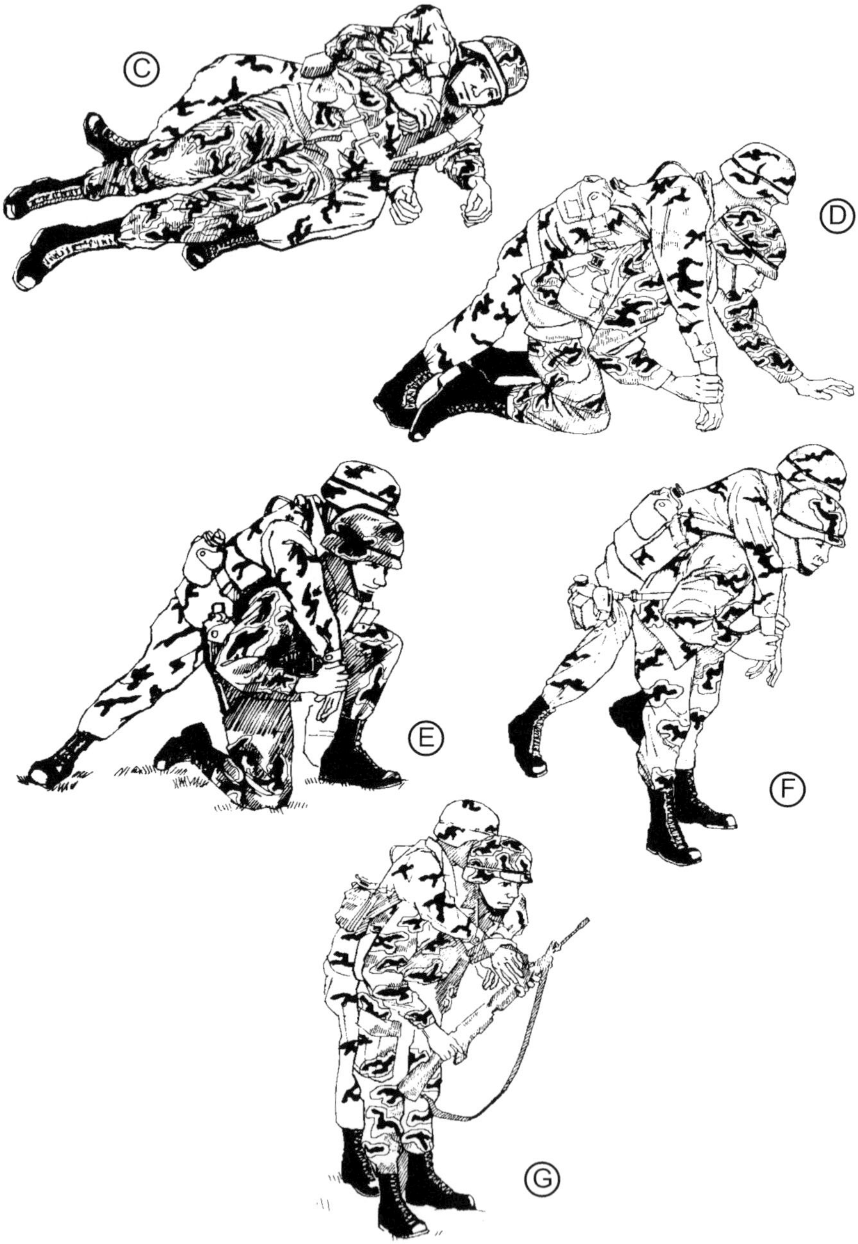

Figure B-15. Load bearing equipment carry using casualty's LBE (Illustrated A—G) (Continued).

b. *Two-man Carries*. These carries should be used whenever possible. They provide more casualty comfort, are less likely to aggravate injuries, and are less tiring for the bearers. Five different two-man carries can be used.

(1) The *two-man support carry* (Figure B-16) can be used in transporting either conscious or unconscious casualties. If the casualty is taller than the bearers, it may be necessary for the bearers to lift the casualty's legs and let them rest on their forearms. The bearers—

(*a*) Help the casualty to his feet and support him with their arms around his waist.

(*b*) Grasp the casualty's wrists and draw his arms around their necks.

Figure B-16. Two-man supporting carry.

(2) The *two-man arms carry* (Figure B-17) is useful in carrying a casualty for a moderate distance (50 to 300 meters) and placing him on a litter. To lessen fatigue, the bearers should carry the casualty high and as close to their chests as possible. In extreme emergencies when there is no time to obtain a spine board, this carry is the safest one for transporting a casualty with a back injury. If possible, two additional bearers should be used to keep the casualty's head and legs in alignment with his body. The bearers—

(*a*) Kneel at one side of the casualty; then they place their arms beneath the casualty's back, waist, hips, and knees.

(*b*) Lift the casualty while rising to their knees.

(*c*) Turn the casualty toward their chests, while rising to a standing position. Carry the casualty high to lessen fatigue.

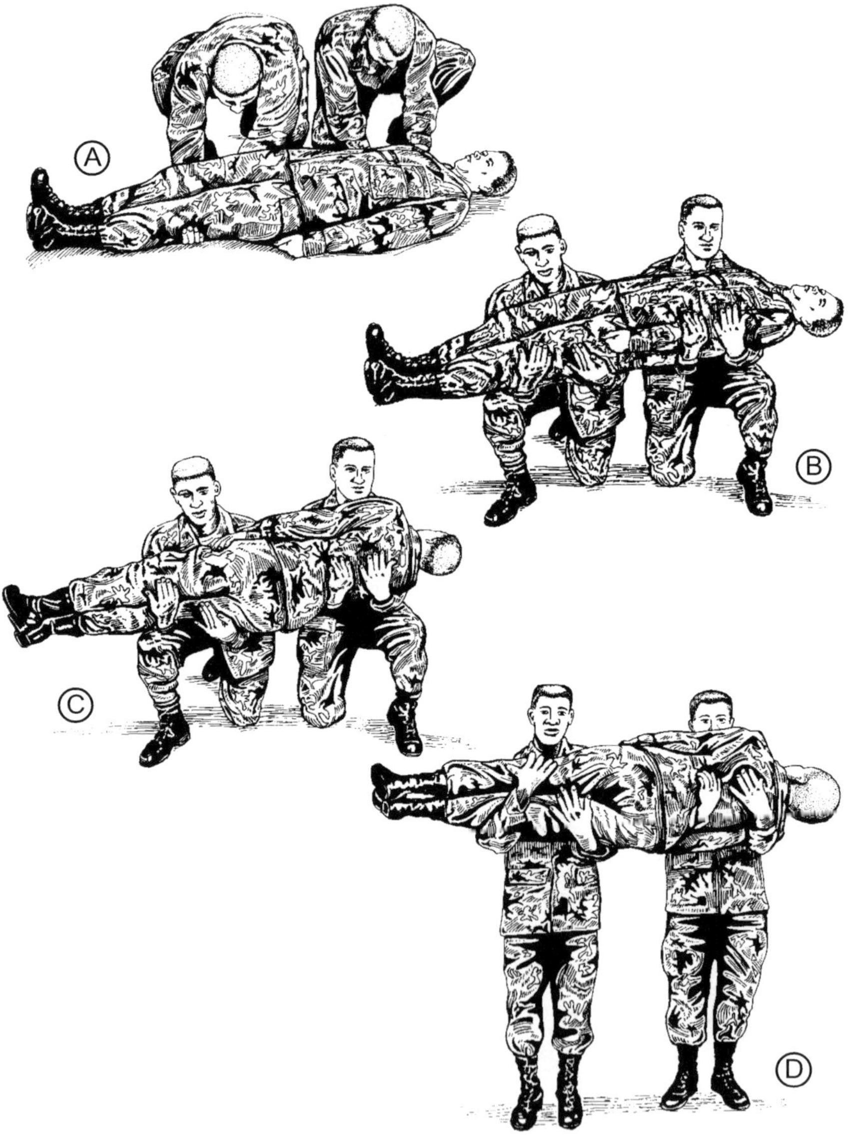

Figure B-17. Two-man arms carry (Illustrated A—D).

(3) The *two-man fore-and aft-carry* (Figure B-18) is a useful *two-man carry* for transporting a casualty for a long distance (over 300 meters). The taller of the two bearers should position himself at the casualty's head. By altering this carry so that both bearers face the casualty, it is useful for placing a casualty on a litter.

(*a*) The shorter bearer spreads the casualty's legs and kneels between them with his back to the casualty. He positions his hands behind the casualty's knees. The other bearer kneels at the casualty's head, slides his hands under the arms, across the chest, and locks his hands together.

(*b*) The two bearers rise together, lifting the casualty.

Figure B-18. Two-man fore-and-aft carry (Illustrated A—B).

(4) Only a conscious casualty can be transported with the *four-hand seat carry* (Figure B-19) because he must help support himself by placing his arms around the bearers' shoulders. This carry is especially useful in transporting a casualty with a head or foot injury for a moderate distance (50 to 300 meters). It is also useful for placing a casualty on a litter.

(*a*) Each bearer grasps one of his wrists and one of the other bearer's wrists, thus forming a packsaddle.

(*b*) The two bearers lower themselves sufficiently for the casualty to sit on the packsaddle; then, they have the casualty place his arms around their shoulders for support. The bearers then rise to an upright position.

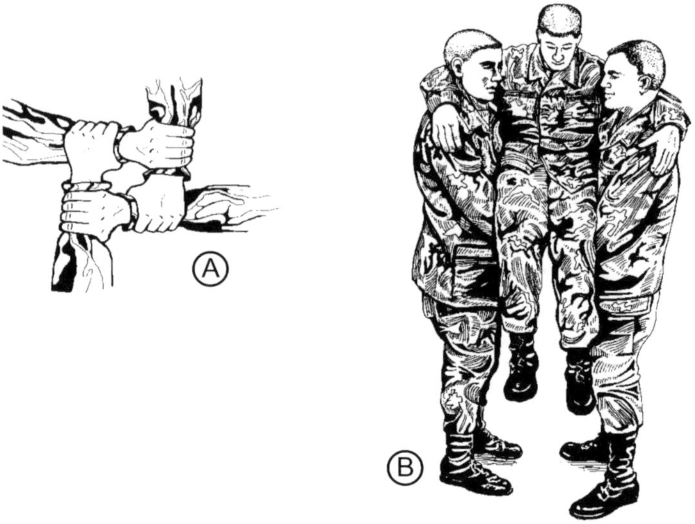

Figure B-19. Four-hand seat carry (Illustrated A—B).

(5) The *two-hand seat carry* (Figure B-20) is used when carrying a casualty for a short distance or for placing him on a litter. With the casualty lying on his back, a bearer kneels on each side of the casualty at his hips. Each bearer passes his arms under the casualty's thighs and back, and grasps the other bearer's wrists. The bearers rise lifting the casualty.

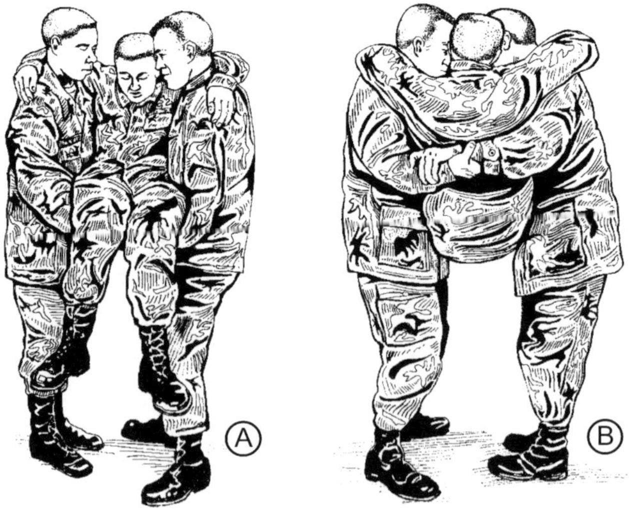

Figure B-20. Two-hand seat carry (Illustrated A—B).

B-9. Improvised Litters

Two men can support or carry a casualty without equipment for only short distances. By using available materials to improvise equipment, the casualty can be transported greater distances by two or more rescuers.

a. There are times when a casualty may have to be moved and a standard litter is not available. The distance may be too great for manual carries or the casualty may have an injury (such as a fractured neck, back, hip, or thigh) that would be aggravated by manual transportation. In these situations, litters can be improvised from materials at hand. Improvised litters must be as well constructed as possible to avoid risk of dropping or further injuring the casualty. Improvised litters are emergency measures and must be replaced by standard litters at the first opportunity.

b. Many different types of litters can be improvised, depending upon the materials available. A satisfactory litter can be made by securing poles inside such items as a blanket, poncho, shelter half, tarpaulin, mattress cover, jacket, shirt, or bed ticks, bags, and sacks (Figure B-18). Poles can be improvised from strong branches, tent supports, skis, lengths of pipe or other objects. If objects for improvising poles are not available, a blanket, poncho, or similar item can be rolled from both sides toward the center so the rolls can be gripped for carrying a patient. Most flat-surface objects of suitable size can be used as litters. Such objects include doors, boards, window shutters, benches, ladders, cots, and chairs. If possible, these objects should be padded for the casualty's comfort.

(1) To improvise a litter using a blanket and poles (Figure B-21), the following steps should be used.

Figure B-21. Litter made with blanket and poles.

(*a*) Open the blanket and lay one pole lengthwise across the center; then fold the blanket over the pole.

(*b*) Place the second pole across the center of the folded blanket.

(*c*) Fold the free edges of the blanket over the second pole and across the first pole.

(2) To improvise a litter using shirts or jackets (Figure B-22), button the shirt or jacket and turn it inside out, leaving the sleeves inside, (more than one shirt or jacket may be required), then pass the pole through the sleeves.

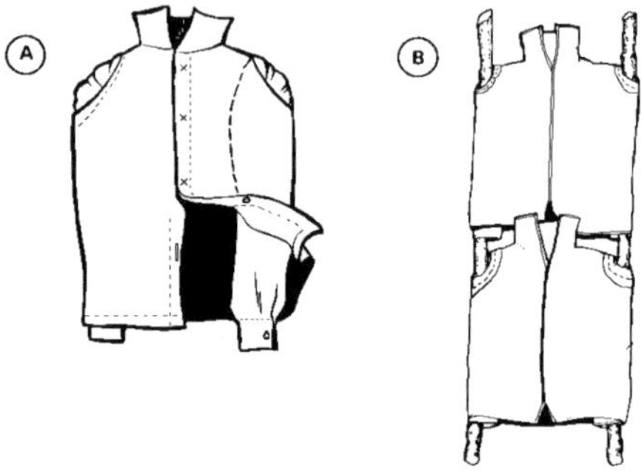

Figure B-22. Litter improvised from jackets and poles (Illustrated A—B).

(3) To improvise a litter from bed sacks and poles (Figure B-23), rip open the corners of bed ticks, bags, or sacks; then pass the poles through them.

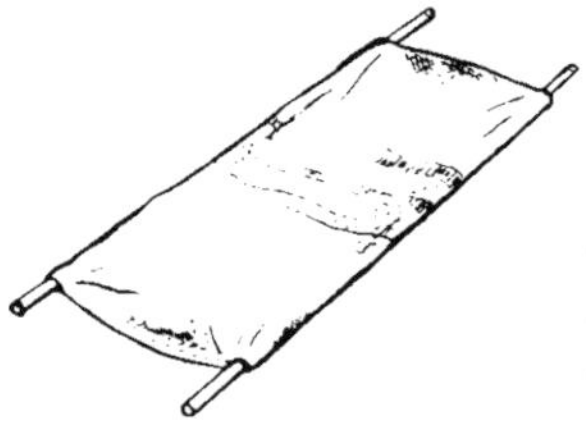

Figure B-23. Litter improvised from bed sacks and poles.

(4) If no poles are available, roll a blanket, shelter half, tarpaulin, or similar item from both sides toward the center (Figure B-24). Grip the rolls to carry the casualty.

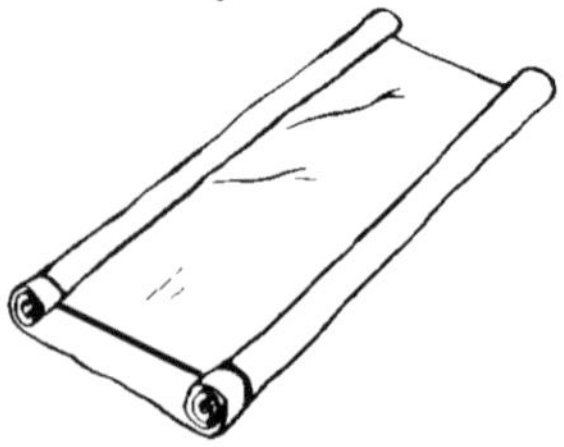

Figure B-24. Rolled blanket used as a litter.

c. Any of the appropriate carries may be used to place a casualty on a litter. These carries are:

- The one-man arms carry (Figure B-6).
- The two-man arms carry (Figure B-17).
- The two-man fore-and-aft carry (Figure B-18).
- The two-hand seat carry (Figure B-20).
- The four-hand seat carry (Figure B-19).

WARNING

Unless there is an immediate life-threatening situation (such as fire, explosion), DO NOT move a casualty with a suspected back or neck injury. Seek medical personnel for guidance on how to transport.

d. Either two or four service members (head/foot) may be used to lift a litter. To lift the litter, follow the procedure below.

(1) Raise the litter at the same time as the other carriers/ bearers.

(2) Keep the casualty as level as possible.

NOTE

Use caution when transporting on a sloping incline/hill.

GLOSSARY

ACRONYMS, ABBREVIATIONS, AND DEFINITIONS

AC hydrogen cyanide
AFMAN Air Force Manual
AOC area of concentration
AR Army regulation
ATM advanced trauma management
ATNAA Antidote Treatment, Nerve Agent, Autoinjector
attn attention

BDO battle dress overgarment
BDU battle dress uniform
BZ anticholinergic drugs

C Celsius
CANA Convulsant Antidote for Nerve Agent
CASEVAC casualty evacuation
cc cubic centimeter
CG phosgene
CHS combat health support
CK cyanogen chloride
Cl chlorine
CLS Combat Lifesaver
CNS central nervous system
CO_2 carbon dioxide
COSR combat and operational stress reactions
CSR combat stress reaction
CTA common table of allowance
CX phosgene oxime

DA Department of the Army
DD Department of Defense
DM diphenylaminochloroarsine (adamsite)
DNBI disease and nonbattle injury
DOD Department of Defense
DP diphosgene
DS direct support

EMT emergency medical treatment

F Fahrenheit
FM field manual

H mustard

HD mustard
HM Hospital Corpsman
HN nitrogen mustard
HSS health service support

IPE individual protective equipment
IV intravenous

JSLIST Joint Services Light Weight Integrated Suit Technology

L lewisite
lasers laser means Light Amplification by Stimulated Emission of Radiation and sources include range finders, weapons/guidance, communication systems, and weapons simulations such as MILES [Multiple Integrated Laser Engagement System].
LBE load bearing equipment
LX lewisite and mustard

MCRP Marine Corps Reference Publication
MILES Multiple Integrated Laser Engagement System
ml milliliter
MOPP mission-oriented protective posture
MOS military occupational specialty
MTF medical treatment facility

NAPP Nerve Agent Pyridostigmine Pretreatment
NATO North Atlantic Treaty Organization
NBC nuclear, biological, and chemical
NCO noncommissioned officer
NTRP Navy Tactical Reference Publication

occlusive dressing air tight transparent dressing used to seal and cover wounds
oz ounce

PAM pamphlet
PS chloropicrin
PTSD post-traumatic stress disorder

QSTAG Quadripartite Standardization Agreement

SOP standing operating procedure
STANAG standardization agreement
STP soldier training publication

2 PAM Cl pralidoxime chloride
TB MED technical bulletin medical
TM technical manual
TSOP tactical standing operating procedure

US United States

WP white phosphorus

REFERENCES

DOCUMENTS NEEDED

These documents must be available to the intended users of this publication.

NATO STANAGs

These agreements are available on request using DD Form 1425 from Standardization Document Order Desk, 700 Robin Avenue, Building 4, Section D, Philadelphia, Pennsylvania 19111-5094.

2122. *Medical Training in First Aid, Basic Hygiene and Emergency Care*. 10 December 1975.
2126. *First Aid Kits and Emergency Medical Care Kits*. 27 September 1983.
2358. *First Aid and Hygiene Training in NBC Operations*. 3 March 1989.
2871. *First Aid Material for Chemical Injuries*. 10 March 1986.

ABCA QSTAGs

These agreements are available on request using DD Form 1425 from Standardization Document Order Desk, 700 Robin Avenue, Building 4, Section D, Philadelphia, Pennsylvania 19111-5094.

535. *Medical Training in First Aid, Basic Hygiene and Emergency Care.* 12 November 1979.

Joint and Multiservice Publications

FM 21-10. *Field Hygiene and Sanitation.* MCRP 4-11.1D. 21 June 2000.

Army Publications

AR 350-41. *Training In Units.* 19 March 1993.
DA PAM 350-59. *Army Correspondence Course Program Catalog.* 26 October 2001.
FM 3-4. *NBC Protection.* FMFM 11-9. 29 May 1992 (Reprinted with basic including Change 1, 28 October 1992; Change 2, 26 February 1996.)
FM 3-5. *NBC Decontamination.* MCWP 3-37.3. 28 July 2000. (Change 1, 31 January 2002.)

FM 3-100. *Chemical Operations Principles and Fundamentals*. MCWP 3-3.7.1. 8 May 1996.

FM 4-02.33 (8-33). *Control of Communicable Diseases Manual* (17th Edition). 3 January 2000.

FM 8-10-6 (4-02.2). *Medical Evacuation in a Theater of Operations—Tactics, Techniques, and Procedures*. 14 April 2000.

FM 4-02.7 (8-10-7). *Health Service Support in a Nuclear, Biological, and Chemical Environment*. 1 October 2002.

FM 8-284 (4-02.284). *Treatment of Biological Warfare Agent Casualties*. NAVMED P-5042; AFMAN (I) 44-156; MCRP 4-11.1C. 17 July 2000. (Change 1, 8 July 2002.)

FM 8-285 (4-02.285). *Treatment of Chemical Agent Casualties and Conventional Military Chemical Injuries*. NAVMED P-5041; AFJMAN 44-149; FMFM 11-11. 22 December 1995.

FM 22-51 (4-02.22). *Leaders' Manual for Combat Stress Control*. 29 September 1994.

CTA 8-100. *Army Medical Department Expendable/Durable Items*. 31 August 1994.

CTA 50-900. *Clothing and Individual Equipment*. 1 September 1994.

STP 21-1-SMCT. *Soldier's Manual of Common Tasks Skill Level I*. 1 October 2001.

Department of Defense Forms

DD Form 1425. *Specifications and Standards Requisition*. March 1986.

READINGS RECOMMENDED

These readings contain relevant supplemental information.

Joint and Multiservice Publications

FM 8-9 (4-02.11). *NATO Handbook on the Medical Aspects of NBC Defensive Operations AMEDP-6 (B), Part I—Nuclear, Part II—Biological, Part III—Chemical*. NAVMED P-5059; AFJMAN 44-151V1V2V3. 1 February 1996.

TB MED 81. *Cold Injury*. NAVMED P-5052-29; AFP 161-11. 30 September 1976.

TB MED 507. *Occupational and Environmental Health Prevention, Treatment and Control of Heat Injury*. NAVMED P-5052-5; AFP 160-1. 25 July 1980.

Army Publications

AR 310-25. *Dictionary of United States Army Terms (Short Title: AD)*. 15 October 1983. (Reprinted with basic including Change 1, 21 May 1986.)

AR 310-50. *Authorized Abbreviations and Brevity Codes*. 15 November 1985.

TM 3-4230-216-10. *Operator's Manual for Decontaminating Kit, Skin: M258A1 (NSN 4230-01-101-3984) and Training Aid, Skin Decontaminating: M58A1 (6910-01-101-1768)*. 17 May 1985. (Change 1, 22 January 1997.)

INDEX

References are to paragraph numbers except where specified otherwise.

Military Operations on Urbanized Terrain

Table of Contents

Chapter 1

Modern Urban Warfare

> **"... the likelihood is high that in the future, the National Command Authorities will again commit Marines to missions in urban areas."**
>
> *A Concept for Future Military Operations on Urbanized Terrain*

1001. Importance of Urban Areas. Throughout history, military planners have viewed cities as centers of gravity. As such, in war, cities are something to be either protected or taken away, depending upon one's perspective (MCDP 1, *Warfighting*). Cities house the population centers, transportation hubs, seats of government, sources of wealth, centers for industry, information networks, and key nodes of communication within a nation. Recent forecasts based on population statistics and the worldwide migration trend from agrarian to industrialized societies predict that 85 percent of the world's population will reside in urbanized areas by the year 2025. As the world trend toward urbanization increases, the military significance of cities is likely to increase proportionally.

Urbanized areas, themselves, may be significant sources of future conflict. Cities historically are where radical ideas ferment, dissenters find allies, mixtures of people cause ethnic friction, and discontented groups receive media attention. Adversaries may focus on the capture of radio and television stations in an attempt to influence public opinion and attain their political goals. Our political leaders may take advantage to neutralize or stabilize some extremely volatile political situations, or to provide assistance to allies in need of support, by deploying U.S. forces into urban environments,

1002. The Marine Corps Role in Urban Warfare. As the Nation's force in readiness, forward deployed with expeditionary forces, Marines must be prepared to fight on urbanized terrain. In the past two decades, MAGTFs ranging in size from MEFs (Saudi Arabia, Desert Shield/Desert Storm; Somalia, Restore Hope) through Marine expeditionary units (MEUs) (Beirut, Lebanon; Grenada, Urgent Fury; Somalia, Eastern Exit and Restore Hope) have participated in MOUT. The task-organization and combined-arms aspect of the MAGTF makes it well suited for combat on urbanized terrain.

The results of geographical studies show that 60 percent of politically significant urban areas outside allied or former Warsaw Pact territory are located along or within 25 miles of a coastline; 75 percent are within 150 miles; 87 percent are within 300 miles; 95 percent are within 600 miles; and all are within 800 miles. U.S. embassies and diplomatic facilities are primarily located in cities where the host country's political and economic leadership is concentrated. The Marine Corps will continue to play a prominent role in future evacuations of U.S. citizens, as well as the conduct of peace, counterinsurgency, and contingency operations centered on urbanized areas.

Today's Marine air-ground task forces (MAGTFs) are deployed as part of naval expeditionary forces (NEFs) that maintain a *global forward presence for rapid crisis response*. These integrated combined-arms forces are part of the Nation's proven contingency and naval power projection force. Therefore, Marines may find themselves rapidly deployed and employed in actions across the spectrum of military operations. Many of these trouble spots will likely be located in or around large urban centers.

In the years since World War II, the United States has employed military force more than 200 times. Of these, four out of five involved naval forces, and the majority of the naval efforts included Marines embarked in amphibious ships. The reasons are straightforward: availability and adaptability. Availability derives from the loiter time of forward deployed forces embarked on amphibious shipping. Adaptability comes from the Marine Corps' MAGTF organization, doctrine, training, and equipment, which prepare us for expeditionary missions from the sea in support of a variety of missions, including forcible entry. Enhancing our adaptability are the maritime prepositioning forces (MPFs). MPFs provide a rapid buildup of combat and logistics equipment that is joined with Marines on a distant shore, creating a substantial combat force. Despite our availability and adaptability, the prospect of urban warfare combined with an amphibious assault is a complex task which requires special preparation.

At the outset of a developing situation, forward-deployed expeditionary forces can move quickly within range of a crisis that threatens the political stability of a country. Urban intervention operations must often be planned and executed in a matter of hours or days (rather than weeks or months) to take advantage of the internal turmoil surrounding a developing crisis. Navy and Marine forces should anticipate deployment to urbanized areas on a day-to-day basis.

1003. Distinguishing Features of Urbanized Terrain. Urbanized terrain is a complex and challenging environment. It possesses all of the characteristics of the natural landscape, coupled with manmade construction, resulting in an incredibly complicated and fluid environment that influences the conduct of military operations in unique ways.

Military operations on urbanized terrain (MOUT) is defined as *all military actions planned and conducted on a topographical complex and its adjacent terrain where manmade construction is the dominant feature. It includes combat in cities, which is that portion of MOUT involving house-to-house and street-by-street fighting in towns and cities* (Marine Corps Reference Publication (MCRP) 5-12A, *Operational Terms and Graphics*). MOUT effects the tactical options available to a commander. A <u>built-up area</u> is *a concentration of structures, facilities, and populations, such as villages, cities, and towns,* that form the economic and cultural focus for the surrounding area. (MCRP 5-12A)

a. Cities. Cities are centers of finance, politics, transportation, communication, industry, and culture. They generally have large population concentrations ranging from tens of thousands to millions of people. Because of their psychological, political, or logistical value, control of cities have often been the scenes of pitched battles.

(1) Operations in built-up areas are normally conducted to capitalize on the operational or tactical significance of a particular city. In developing nations, control of only a few cities is often the key to the control of national resources. The side that controls a major city usually has a psychological advantage, which can be enough to significantly affect the outcome of a countrywide conflict.

(2) The abundance of guerrilla and terrorist operations conducted in built-up areas (e.g., Santo Domingo, Caracas, Belfast, Managua, and Beirut) demonstrates the importance many insurgent groups place on urban warfare.

(3) In the past 40 years, many cities have expanded dramatically, losing their well-defined boundaries as they extended into the countryside. New transportation systems (highways, canals, and railroads) have been built to connect population centers. Industries have grown along those connectors, creating "strip areas." Rural areas, although retaining much of their farmlike character, are connected to the towns by a network of secondary roads.

b. Multiple Avenues of Approach. Urbanized terrain is a unique battlespace that provides both attacker and defender with numerous and varied avenues of approach and fields of fire. The urban battlespace is divided into four basic levels: *building*, *street*, *subterranean,* and *air*. Operations can be conducted from above ground, on ground level, inside buildings, or below the ground. Most operations will include fighting on all levels simultaneously.

(1) Building Level. Buildings provide cover and concealment; limit or increase fields of observation and fire; and canalize, restrict, or block movement of forces, especially mechanized forces. They provide optimum perches for snipers and antiair weapons. Buildings also provide antitank weapons optimum positioning to allow engagement from above, exploiting an inherent weakness found in most armored vehicles.

(2) Street Level. While streets provide the means for rapid advance or withdawal, forces moving along streets are often canalized by buildings and have little space for off-road maneuver. Because they are more difficult to bypass, obstacles on streets in urbanized areas are usually more effective than those on roads in open terrain.

(3) Subterranean Level. Subterranean systems are easily overlooked but can be important to the outcome of operations. These areas may be substantial and include subways, sewers, cellars, and utility systems (Figure 1-1 on page 1-4). The city of Los Angeles alone has more than 200 miles of storm sewers located under the city streets. Both attacker and defender can use subterranean avenues to maneuver to the rear or the flanks of an enemy. These avenues also facilitate the conduct of ambushes, counterattacks, and infiltrations. (See Appendix E)

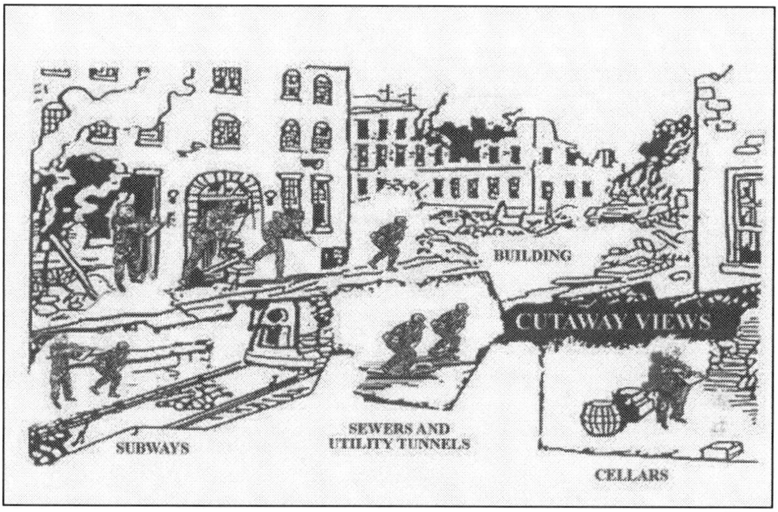

Figure 1-1. Subterranean Systems

(4) Air Level. The air provides another avenue of approach in urbanized areas. Aviation assets can be used for high speed insertion or extraction of troops, supplies, and equipment. While aviation assets are not affected by obstacles on the streets, they are affected by light towers, signs, power lines, and other aerial obstructions. They are also vulnerable to the man-portable surface-to-air missile threat, crew served weapons, and small arms fire.

c. Categories of Built-Up Areas. Built-up areas are generally classified as:

- Villages (populations of 3,000 or less)
- Strip areas (industrialized zones built along roads connecting towns or cities)
- Towns or small cities (populations of up to 100,000 and not part of a major urban complex)
- Large cities with associated urban sprawl (populations in the millions, covering hundreds of square kilometers).

d. Characteristics of Urbanized Areas. A typical urban area consists of combinations of the *city core*, *commercial ribbon*, *core periphery*, *residential sprawl*, *outlying industrial areas*, and *outlying high-rise areas*. Each of the urban area's regions has distinctive characteristics that may weigh heavily in planning for MOUT. Most urbanized areas resemble the generalized model shown in Figure 1-2.

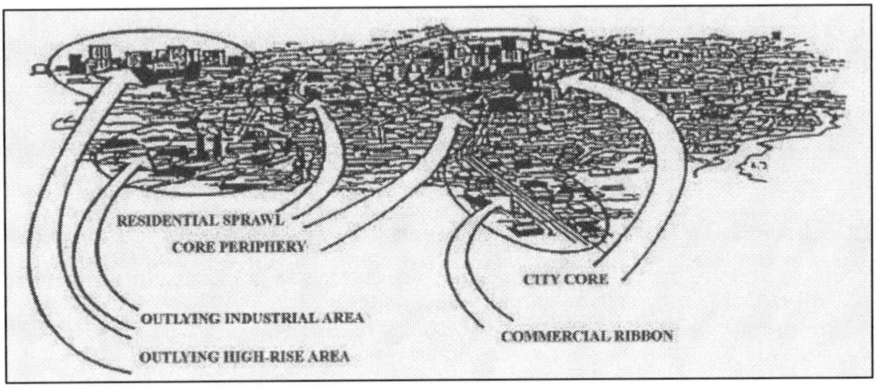

Figure 1-2. Typical Urban Area

(1) City Core. In most cities, the city core has undergone more recent development than the core periphery. As a result, the two regions are often quite different. Typical city cores are made up of high-rise buildings which vary in height. Modern urban planning for built-up areas allows for more open spaces between buildings than in old city cores or in core peripheries (Figure 1-3).

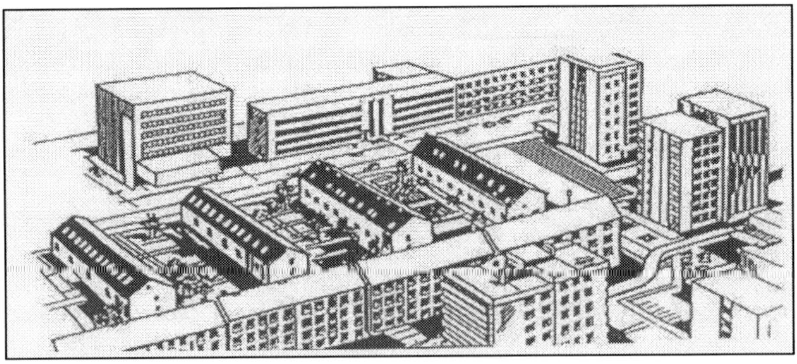

Figure 1-3. City Core

(2) Commercial Ribbon. Commercial ribbons are composed of rows of stores, shops, and restaurants that are built along both sides of major streets through built-up areas.

Typically, such streets arc 25 meters wide or more. The buildings in the outer areas are uniformly two to three stories tall—about one story taller than the dwellings on the streets behind them (Figure 1-4).

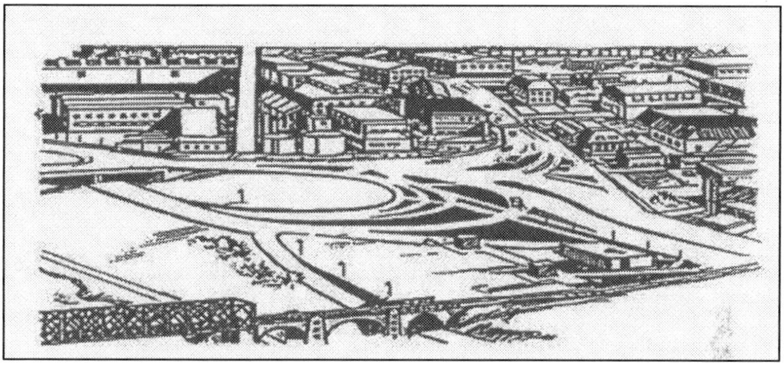

Figure 1-4. Commercial Ribbons

(3) Core Periphery. The core periphery generally consists of streets 12 - 20 meters wide with continuous fronts of brick or concrete buildings. The building heights are fairly uniform—2 or 3 stories in small towns, 5 to 10 or more stories in large cities (Figure 1-5).

Figure 1-5. Core Periphery

(4) Residential Sprawl. Residential sprawl areas consist mainly of low houses or apartments that are one to three stories tall. The area is primarily composed of detached dwellings that are usually arranged in irregular patterns along streets, with many smaller open areas between structures (Figure 1-6).

Figure 1-6. Residential Sprawl

(5) Outlying Industrial Areas. These areas generally consist of clusters of industrial buildings varying from one to five stories in height. Buildings generally vary dramatically in size and composition to match the needs of the particular businesses they house. Industrial parks are good examples of this category (Figure 1-7).

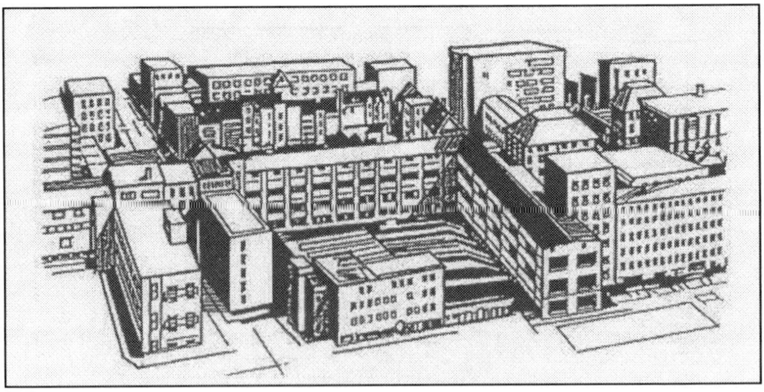

Figure 1-7. Outlying Industrial Areas

(6) Outlying High-Rise Areas. These areas are similar in composition to city core areas, but may be composed of clusters of more modern multistory high-rise buildings in outlying parts of the city. Building height and size may vary dramatically (Figure 1-8). Generally, there is more open space between buildings located in the outlying high-rise areas than is found within the city core area.

Figure 1-8. Outlying High-Rise Area

1004. Modern Battles for Urbanized Terrain. Urban warfare is as old as war itself. Since man began building villages, he has fought battles in and around them. Geography, politics, and economics dictate that cities will continue to be an objective of armies in warfare. From the armies that invaded and liberated Europe twice during the 20th century, to the forces that fought in Korea and Vietnam, to our most recent urban battles in the Middle East and Southwest Asia, *the basic principles of combat in built-up areas have essentially remained unchanged in this century*. While the principles remain the same, the introduction of helicopters, fixed-wing aircraft, armor, and precision-guided munitions (PGMs) has altered some of the techniques associated with urban combat.

Twenty two modern urban battles are discussed to illustrate the trends, dominant factors, and principles of combat in urbanized areas. (The majority of this material is extracted from the 1987 study, *Modern Experience in City Combat,* produced by Abbott Associates.)

• Aachen	• Manila
• Arnhem	• Ortona
• Ashrafiyeh	• Quang Tri City I
• Ban Me Thuot	• Quang Tri City II
• Beirut Port/Hotel (I)	• Seoul
• Beirut 1982 (II)	• Sidon
• Berlin	• Stalingrad
• Cherbourg	• Suez City
• Hue	• Tel Zaatar
• Jerusalem	• Tyre
• Khorramshahr	• Zahle

A brief description of each battle is provided to illustrate that battle's significance.

a. Stalingrad (1942 - 1943). The tenacious Soviet defense of Stalingrad cost the attacking Germans dearly in every way and set up conditions for a decisive counteroffensive. This classic urban battle involved large forces and resulted in innovative urban combat techniques and the creation of the highly successful storm groups (task-organized assault units). (Length of battle: greater than 30 days) (Casualties: 1,630,000+)

b. Ortona (1943). In this Italian town, determined resistance by a battalion of the elite German 3rd Parachute Regiment against Canadian Army attackers demonstrated the difficulty of overcoming a well-prepared defense. The Canadians were unfamiliar with urban combat and had to develop urban fighting techniques during the battle. After the town was largely destroyed and the defender had extracted a high cost in time and casualties from Canadian forces, the German parachute battalion withdrew. (Length of battle: 6 - 13 days) (Casualties: estimated in the hundreds)

c. Aachen (1944). The battle for Aachen, Germany, in the fall of 1944 developed during the U.S. First Army's offensive to breach the Westwall fortifications and the vaunted Siegfried Line. Aachen, the ancient capital of Charlemagne, had symbolic political and psychological significance to both the Germans and Americans. Furthermore, it was the first city on German soil to face an assault by the Allies. This first major battle on German soil foreshadowed bitter resistance against the American attackers in subsequent battles. The German defenders surrendered only after the city was destroyed. Although the U.S. Army had achieved a clear tactical victory, the German defense of Aachen cost the U.S. First Army valuable time and delayed the planned attack to the Rhine river. (Length of battle: 14 - 30 days) (Casualties: 8,000+)

d. Arnhem (1944). On September 17, 1944, Operation Market-Garden, the largest airborne operation in history, was launched in the Netherlands. The plan was to land three airborne divisions to seize key bridges along a 100-kilometer-long corridor through which allied mechanized forces would pass as the first step in the final offensive into Germany. The British 1st Airborne Division made a surprise landing near the Dutch city of Arnhem in order to seize a bridge over the Rhine river for advancing British forces. An unexpected German armor force counterattacked and eliminated all footholds, virtually destroying the British division before a linkup could occur. (Length of battle: 6 - 13 days) (Casualties: estimated in the thousands)

e. Cherbourg (1944). By June 17, 1944, U.S. forces advancing toward Cherbourg from the Normandy beachhead succeeded in cutting off defending German forces in the Cotentin Peninsula. Four German divisions withdrew to a perimeter surrounding Cherbourg. After much fighting, particularly in strongpoints outside the city, the German garrison surrendered to the Americans. Unfortunately for the Americans, the port facilities were destroyed which denied their early use by Allied forces. (Length of battle: 6 - 13 days) (Casualties: estimated in the thousands)

f. Berlin (1945). The long, bloody Soviet offensive to seize the German capital city effectively concluded the last battle of World War II in Europe. Bitter fighting occurred, but the defense was never well coordinated due in part to poor preparation by the Germans. (Length of battle: 14 - 30 days) (Casualties: estimated in the thousands)

g. Manila (1945). Japanese Army troops evacuated Manila under pressure from advancing American forces, but the local Japanese naval commander independently decided to hold the city at all costs. Despite defending Manila with poorly trained and equipped personnel, the determined resistance resulted in a high number of casualties to attacking U.S. forces as well as the destruction of the city and much of its population. (Length of battle: 14 - 30 days) (Casualties: 22,000+)

h. Seoul (1950). Following the Inchon landing, U.S. and Republic of Korea (ROK) forces recaptured the South Korean capital from the North Koreans. The fighting was unusual in that combat was largely centered on seizure of street barricades rather than buildings. (Length of battle: 6 - 13 days) (Casualties: Marines, 2,383; others, estimated in the thousands)

i. Jerusalem (1967). Israeli forces seized Jerusalem in a well prepared and well executed operation. Despite an uncoordinated Jordanian defense, Israeli casualties in this battle were the highest of those encountered during the Six Day War. Regular Jordanian forces withdrew during the latter stages of the battle, effectively ending organized resistance. (Length of battle: 48 hours to 5 days) (Casualties: Israeli forces, 400+; Jordanian forces, estimated in the hundreds)

j. Hue (1968). On January 31, 1968, the North Vietnamese Army (NVA) and Viet Cong (VC) forces launched their Tet Offensive at targets throughout South Vietnam. As part of this operation, two NVA/VC regiments and two sapper battalions conducted a surprise attack and seized part of the walled city (Citadel) of Hue. The NVA/VC held this part of the walled city for about three weeks against determined U.S. and South Vietnamese forces before finally succumbing. The battle for Hue is considered one of the most intense and savage battles of the Vietnam War. (Length of battle: 14 - 30 days) (Casualties: Marines, 433; others, 5000+)

k. Quang Tri City I and II (1972). An objective of the North Vietnamese 1972 winter-spring offensive was the capture of Quang Tri, the northernmost major city in South Vietnam. The NVA overwhelmed the Army, Republic of Vietnam (ARVN) defenders (I). Later, the city was recaptured (II) by a smaller ARVN force using extensive artillery and air support. The large conventional forces involved on both sides made Quang Tri I and II the major urban battles of the Vietnam War. (Length of battle: Quang Tri I, 6 - 13 days; Quang Tri II, 30 days or greater) (Casualties: battles combined, 30,000+)

l. Suez City (1973). Israeli Defense Forces (IDF) attempted to seize this Egyptian city before the anticipated United Nations (U.N.) cease-fire to end the Yom Kippur War. IDF armored shock tactics led to disaster against a well-prepared Egyptian defense. High casualties forced the IDF to withdraw. (Length of battle: less than 24 hours) (Casualties: Israeli forces, estimated 100 - 500; others, unknown)

m. Ban Me Thuot (1975). This South Vietnamese highlands town was the first strategic city to fall in the final North Vietnamese general offensive in 1975 South Vietnamese forces were surprised and overwhelmed. The fall of Ban Me Thuot resulted in a rout that the North Vietnamese exploited to achieve total victory in Vietnam. (Length of battle: 24 - 48 hours) (Casualties: estimated in the hundreds)

n. Beirut I (1976). When the Lebanese civil war broke out in the spring of 1975, combat in the capital city of Beirut assumed a central role. The battle for Beirut I was a series of small, local operations between largely irregular Christian and Muslim forces fighting over control of the hotel and port districts. Combat was not decisive, but led to changes in the boundary, called the "Green Line." This separated the antagonists and lead to the stagnation of the Lebanese conflict. (Length of battle: greater than 30 days) (Casualties: estimated in the hundreds)

o. Tel Zaatar (1976). Lebanese Christian attackers encircled and methodically besieged this Palestinian camp before overcoming its defenders with a final assault. (Length of battle: greater than 30 days) (Casualties: estimated in the hundreds)

p. Ashrafiyeh (1978). The Syrian forces occupying portions of Lebanon faced a complex political situation in which the power of the Christian militia was seen as a clear threat to stability. In an attempt to weaken the Christian militia by an attack on their center of power, the Syrians laid siege to the Christian militia stronghold of East Beirut (Ashrafiyeh). This urban battle was essentially an artillery bombardment without air attacks. Syria failed to break the will of the defenders and final positions remained unchanged. (Length of battle: greater than 30 days) (Casualties: estimated in the hundreds)

q. Khorramshahr (1980). Iranian regular forces initially evacuated this port city in the face of an Iraqi offensive. Irregular Iranian forces, however, continued to fight. They offered prolonged resistance and inflicted heavy casualties. Iraq eventually won this battle, but at a high cost in time and resources that ultimately served to halt the entire offensive against Iran. The intensity of fighting during the battle for Khorramshahr earned the city the nickname, "City of Blood." (Length of battle: 14 - 30 days) (Casualties: Iraqi, 3,000 - 9,000; Iranian, estimated in the thousands)

r. Zahle (1981). Syria laid siege to the Lebanese regular forces and militia in this key crossroads town. Fighting was inconclusive and ended in a negotiated settlement whereby the Lebanese defenders evacuated the town. (Length of battle: greater than 30 days) (Casualties: estimated in the hundreds)

s. Beirut II (1982). The siege of Beirut culminated the Israeli campaign to evict the Palestine Liberation Organization (PLO) from Lebanon. Fighting under domestic and world political pressures, the IDF besieged the PLO, selectively applying heavy ground and air firepower in conjunction with psychological warfare and limited-objective ground operations. The fighting resulted in a negotiated PLO evacuation from the city. (Length of battle: greater than 30 days) (Casualties: 2,300+)

t. Sidon (1982). Israeli forces easily seized this PLO southern headquarters during the invasion of Lebanon. The IDF was fully prepared for major urban combat using lessons learned from earlier battles, but resistance was unexpectedly light as PLO forces had largely withdrawn from the city. (Length of battle: 48 hours to 5 days) (Casualties: unknown)

u. Tyre (1982). The Israeli attack on the PLO in this Lebanese coastal city was well planned, with excellent intelligence on the target. All branches of the IDF participated in an operation that included naval fire support and amphibious landings. PLO resistance was uncoordinated and easily overcome. (Length of battle: 48 hours to 5 days) (Casualties: Israeli forces, 120+; others unknown)

1005. Modern Urban Battle Analysis and Observations. The 20th-century urban warfare experience has caused us to reevaluate old factors and consider new developments that affect the way we fight in this environment. The following discussion, based on the historical analysis of these select urban battles, illustrates the importance of maneuver warfare and combined arms philosophies in the urban environment. Factors that have had an impact on the manner in which urban warfare has been conducted are:

Intelligence is imperative to success in urban warfare. "Maneuver warfare requires a firm focus on the enemy" (MCDP 2, *Intelligence*). Few subsequent tactical changes can overcome the far-reaching impact of a major intelligence error.

a. Intelligence. The historical review of modern urban battles discloses that the attacker will usually win. Failures to win generally reflect classic military errors not characteristically unique to cities. *However, of the battles studied, battles lost were attributed to errors in* *initial intelligence.* The battles of **Arnhem** and **Suez City** probably would never have occurred had the attacker known the strength and locations of the defender's forces. At **Stalingrad**, the attacking Germans were aware of the defending forces facing them in the Sixth Army's zone, but the buildup of Soviet forces in other areas of the front was not anticipated and resulted in tactical surprise at those points, diluting the German offensive push to seize the city.

Surprise is a combat multiplier and can substantially reduce the cost of urban warfare.

b. Surprise. Surprise is a combat multiplier and can substantially reduce the cost of urban warfare. It can be achieved through deception, stealth, and ambiguity (MCDP 1-3, *Tactics*). Surprise was achieved by the attacker at **Aachen** and **Ban Me Thuot** and by the defender at **Suez City**. Surprise can be an important asset to increase leverage, but, as the failure at **Arnhem** shows, not necessarily a decisive one. When surprise is employed as a means to overcome other disadvantages, it is important to maintain accurate intelligence. In urban areas, tactical surprise by the attacker can be used to preempt effective defensive preparation of a city.

c. Combined Arms. The MAGTF must capitalize on one of the key means for gaining advantage in maneuver warfare — the use of combined arms (MCDP 1-3, *Tactics*). The use of combined arms places the enemy in a dilemma. Any action the enemy takes to avoid one combat arm makes him more vulnerable to another. An analysis of categories of weapons systems found in a MAGTF helps to illustrate it is a warfighting organization well suited for MOUT:

Combat in urban areas is primarily a small unit, infantry intensive operation

(1) Infantry. Combat in urban areas is primarily a small-unit, infantry intensive operation. Restrictions on maneuver, particularly for mechanized units, increases oportunities for infiltration. Urban combat requires small-unit leadership, initiative, and skill. Decentralized actions and difficulties in command , control, and communications are typically encountered. Built-up areas, like close terrain found in other operational environments, are generally considered to be most suited for operations conducted by infantry. Infantry units can be organized, trained and equipped to negotiate urbanized terrain that restricts observation, fields of fire, and mechanized movement.

(2) Armor. The role of armor in urban warfare can be significant. Of the 22 battles studied, armor participated in 21. In three-fourths of these battles, organic tank support was a central element when special assault teams were employed. *Overall, special assault units supported by tanks were more successful than any other task organization.*

The use of tanks to the attacker inside a city has been effective only when they were protected by infantry. Tanks in support of infantry act as an "assault gun" that delivers concentrated, sustained fires to reduce held strongpoints.

The U.S. experience in **Hue** demonstrated the key role that armor can play on a combined-arms team fighting inside the city. The Marines' most effective weapons during the battle were the M48A1 Patton tank and the M-50 Ontos. Both were protected by infantry. The M48A1, with its 90-mm main gun, was used extensively to reduce fortified positions. The Ontos, an armor-protected tracked vehicle mounting six 106-mm recoilless rifles, was highly effective against concrete and steel structures. The munitions of these armored systems provided breaches that the infantry could exploit.

Armor providing direct-fire support inside a built-up area requires protection by infantry.

In contrast, during the **Suez City** battle, Israeli armor forces attacked on "armor thrust avenues" into the city, outpacing their armored personnel carrier (APC) mounted paratroop/infantry support. The Egyptian defenders lacked organic artillery (except limited antiaircraft artillery (AAA) and mortars) and had no air support and virtually no armor support. The Egyptians prepared "kill zones" on the principal avenues down which the IDF armored forces attacked. As the lead IDF armor battalion entered the second of the three road intersection objectives, the Egyptians engaged with Sagger missiles, RPGs, ZSU-23 antiaircraft guns, antitank grenades thrown from balconies, and small arms. All of the tank commanders in the lead battalion were killed or wounded. Disabled vehicles blocked the road. Vehicles veering into the narrow side streets became trapped and were destroyed. *The lack of infantry to protect the armor proved disastrous to the attacking armor battalion.*

(3) Artillery. Artillery has played an important role in most major urban conflicts. At **Aachen,** U.S. forces combined infantry with antitank teams and artillery (in a direct-fire role) down to the squad and fire team level. Also, artillery firing shells with "delay" fuzes in an indirect-fire role were used to penetrate one or more floors before exploding, thus driving the enemy to the ground where infantry and armor could attack. Artillery was also positioned to fire *perpendicular* to the direction of movement of assaulting forces. Thus, fratricide from artillery range errors was alleviated.

Artillery has two distinct roles: outside the built-up area to isolate or prevent isolation with indirect-fire; and within the built-up area to provide direct-fire support.

Artillery proved most useful for interdicting enemy supplies, enemy evacuation, and the movement of reinforcements in the enemy's rear; for physically and psychologically harassing the enemy; and in direct-fire roles within a city.

Artillery employed in the indirect-fire role has been effective in disrupting defenders in half of the studied battles. However, some indirect-fire roles have proved more effective than others in urban combat. Artillery was most effective in the interdiction of supplies, enemy evacuation, movement of reinforcements in the enemy's rear (outside the city), and for indiscriminate physical and psychological harassment of the enemy. Artillery used as an indirect-fire siege weapon, as was done at **Ashrafiyeh** and **Zahle**, proved ineffective. Artillery can also cause problems for the attacker. The rubble resulting from indirect artillery fires can create considerable obstacles for the attacker while providing the defender with obstacles, materials, cover, and concealment.

Artillery employed in the direct-fire role proved useful in the reduction of strongpoints. Self-propelled artillery was used effectively inside **Aachen** and **Stalingrad** and more recently by the IDF at **Beirut II**.

(4) Mortars. The mortar is the most used indirect-fire weapon in urban combat. The mortar's high angle of fire allows the round to reach the street level accurately without being masked by surrounding structures. During the battle for **Hue** city, the most effective fire support provided to Marines was the indirect fire from 60-mm, 81-mm, and 4.2-inch mortars. Approximately 20,000 rounds of high explosive (HE) 60-mm and 81-mm mortar ammunition were expended during the battle. The 4.2-inch mortar was used primarily to deliver riot control agent munitions. It was discovered that 4.2-inch riot control agent munitions could be fired with great effectiveness through the tops of buildings to drive the enemy out. When the enemy emerged from their concealed positions, HE rounds were fired to complete the attack.

(5) Antiaircraft Artillery. Antiaircraft artillery (AAA) was extremely useful in a ground fire role in some urban battles. AAA was used only rarely in World War II, and then generally against assaulting personnel rather than against structures. AAA has been used more frequently in more recent events, but against buildings rather than people. *The high rates of fire of modern AAA make it excellent in terms of shock and destructive potential.* However, ammunition supply can be a problem because the volume of fire. During **Hue**, the Duster, an AAA vehicle mounting twin 40-mm guns, was provided to the Marines by the U.S. Army. These guns were extremely effective in the suppression of enemy positions due to the lethality of the 40-mm round and the quantities in which it could be delivered. During the **Suez City** battle, the Egyptians used the ZSU-23 antiaircraft gun in the direct-fire role against armored columns. This gun proved to be frightening and effective. Its high volume of explosive power created shock among IDF armor personnel akin to the shock that the IDF armor thrust tactics were designed to create. During **Sidon,** the IDF used 20-mm antiaircraft Vulcans in the direct-fire role against ground targets with great success.

(6) Aviation. Historically, aviation assets have played an important role in helping to isolate the objective and to interdict the flow of the defender's supplies and reinforcements. However, aviation has been relatively ineffective when not used in conjunction with ground forces. In past wars, bombing operations have been used in attempts to reduce the defender's will to resist and destroy their physical capabilities. In the majority of urban battles, aerial bombing, by itself, did not erode the defender's will to resist, nor did it significantly degrade the defender's military capabilities.

Aviation plays an important role in interdicting movement of the defender's supplies and reinforcement into the built-up area.

The Marine air-ground task force is employed as a combined arms team. As such, the aviation combat element will not usually be called upon to conduct independent operations in MOUT. It will function as part of a MAGTF and, regardless of operating on or over urbanized terrain, will still execute the six functions of Marine aviation (i.e., offensive air support, assault support, antiair warfare, electronic warfare, air reconnaissance, and control of aircraft and missiles) in support of the MAGTF.

In future urban warfare, aviation will be even more effective due to advances in fixed- and rotary-wing aircraft, unmanned aerial vehicles, precision guided munitions, improved munitions, communications, sensors, and targeting systems. Our battle study already indicates a trend toward more extensive aviation participation in MOUT. As an example, the IDF in **Beirut II** employed bombing by fixed-wing aircraft using cluster bomb units, "smart" bombs, phosphorous, and other munitions. Attack helicopters operated on the outskirts of the built up area with impunity, and medical evacuation (MEDEVAC) proceeded swiftly and efficiently using helicopter support.

d. Combat Forces. Whether attacking or defending, the size of the force relative to the enemy can be critical to success. When provided with adequate forces, the attacker can isolate and encircle the defender and prevent a breakout or linkup. The defender can use them to create a mobile defense or to create strong reserves for counterattacks.

Other factors that impact on the size of forces required are the degree of surprise achieved and the firepower utilized (aviation, armor, artillery, mortars, etc.). They should be weighed against the sophistication of the prepared defense. Ample consideration should be given to the local population, degree of external support, and utilization of existing services (communications, water, etc.).

Regardless of the size or quality of defensive forces, the defender usually extracts large costs from the attacker in time, resources, and casualties.

In the historical examples reviewed, the defender was usually outnumbered by the attacker, the quality of the defender's available forces was inferior, and defeat of the forces defending the city was usually certain. However, regardless of the size or quality of defensive forces, the defender can extract enormous costs to the attacker in time, resources, and casualties. As was seen at **Khorramshahr**, the Iranian defenders, outnumbered 4 to 1, still held the city for 26 days. Although the Iranian defenders eventually lost the city, its defense allowed the remaining Iranian forces time to organize and redeploy. Furthermore, the winter rains that followed the battle turned much of the region into a sea of mud and largely halted further Iraqi efforts. The Iraqi army's offensive thrusts into Iran lost momentum as a result of the defense of Khorramshahr.

e. Special Assault Teams. In these battle studies, "shock units" or "special assault teams" have been used by attackers (and often by defenders) with great success. These organizations were characterized by the integration of combined arms at the battalion level and below. Control was decentralized to lower echelons (down to the squad level in some cases). Assault teams typically contained infantry with various combinations of armor, artillery, and engineers. Hence, all combat arms should plan, train, and develop common tactics, techniques, and procedures for use on urbanized terrain. Before the battle of **Aachen**, U.S. ground forces conducted intensive urban training with antitank teams and artillery pieces down to the squad and fire team levels. In contrast, the defender, in most cases having to fight with whatever forces were locally available, had no opportunity to conduct combined-arms training.

The use of combined arms warfare on urbanized terrain is imperative.

f. Time. In most cases, successful conclusion of an urban battle took two to three times longer than the initial estimates. Consequently, the additional tome resulted in the expenditure of more logistics and the loss of more personnel then initially anticipated. This often had adverse affects on the overall campaign. Well-planned urban defense, even if the defender is isolated or lacking in aviation, armor, or artillery weapons, can be time consuming to the attacker. Time can allow the defender to reorganize, redeploy, or marshal resources in other areas.

Generally, urban warfare is time consuming.

Three battles where time played an unanticipated critical role in the attacker's strategic timetable were **Aachen**, **Khorramshahr**, and **Stalingrad**. In these battles, the defenders delayed the attacker longer than was estimated, resulting in the modification of operational or strategic plans.

g. Isolation. *The attacker won all urban battles where the defender was totally isolated.* Even the partial isolation of the defenders resulted in attackers enjoying a success rate of 80 percent. Conversely, attackers won only 50 percent of the battles in which defenders were not significantly isolated, and those victories came at great cost.

No single factor is more important to the attacker's success than isolation of the urban area.

In most urban battles, some form of isolation occurred as a result of the attacker's actions. While it is unrealistic to envision complete isolation of a city until enemy forces to the rear of the city are pushed far beyond its outermost boundaries, *total isolation does not appear necessary.* The key to the attacker's success is in stemming the *unimpeded* flow of manpower, supplies, and weapons to replace the defender's losses.

h. Cost. The cost of conducting urban warfare is relative to the percentage of total expended resources, the time elapsed, and the results achieved. The cost to the attacker was considered high in the majority of urban battles. A high cost does not necessarily imply that the results were not worth the price. *The attacker and defender must thoroughly evaluate the overall cost prior to committing to an urban battle.*

From the offensive point of view, several factors can be associated with cost. First, and most importantly, isolating the urban area is critical to the attacker's success. Second, overwhelming superiority is needed if all costs are to be minimized. Third, the operation should be carefully planned. Fourth, intelligence is invaluable. Knowing where and how the city has been prepared for defense is also important. Fifth, attacking forces should understand the unique nature of urban combat. Clear tactics, techniques, and procedures for urban combat is required. Every aspect, from taking a building to using destructive technology and coordinating combined arms, must be thoroughly understood. Careful consideration of these five factors can minimize the cost of urban warfare to the attacker.

Attacker cost was generally high in casualties, time, and resources in the majority of urban battles studied.

From the standpoint of the defender, the critical variable is the defensive preparation of the city. Defensive preparations should include measures to prevent isolation of the city by the attackers. The capture of a prepared city can be made to be extremely costly. Preparations can include creating kill zones, clearing fields of fire, constructing canalizing obstacles, establishing reinforcing and fall-back positions, decentralizing command and control, and organizing multiple movement routes above ground, between rooftops, and below ground in subways and sewers. Artillery and aviation support can also be valuable force multipliers in defense.

i. Rules of Engagement (ROE). The nature of the military operation may restrict our use of weapons. The majority of urban battles since 1967 (such as **Beirut II**, **Hue**, **Jerusalem**) have had one or more of the following restrictions imposed on the attacking force:

(1) Minimizing civilian casualties and/or collateral destruction in order to:

- Avoiding alienation of the local population
- Reducing the risk of adverse world or domestic opinion
- Preserving facilities for future use
- Preserving cultural facilities and grounds.

(2) Limiting the use of specific ground or air weapons.

j. Logistics. Timely combat service support, particularly in the areas of ammunition resupply and casualty treatment and evacuation, is a critical element in MOUT.

(1) Logistics Support. Historically, combat in urban environments has seen a dramatic increase in the amounts of Class IV (shoring, sandbags, concertina wire), Class V (ammunition), and Class VIII (medical material). Unique items, such as rope, grappling hooks, and ladders are required for operations on urbanized terrain. Intense close-quarter combat requires a continuous flow of ammunition, particularly small-arms, tank, antitank, mortar, and artillery ammunition, as well as mines, grenades, and demolition explosives. Medical supplies must be readily available to treat the anticipated increase in casualties. Once battle the has been initiated, combat forces will require continuous supply.

Urban warfare requires a flexible, balanced logistics system capable of sustaining the close-quarter fighting evolutions within the built-up area.

One method of providing continuous logistics support is to establish a "push system." Essentially, this system pushes supplies to fighting units without their having to request them. A "push system" should negate the inherent delays of a "pull system," which requires units to request supplies then await their arrival. Optimally however, a balanced approach should be taken to sustain engaged forces (MCDP 4.). Combat service support, in terms of its timeliness and anticipation of the true needs of the combat units, should sustain the tempo of operations. It should avoid the delivery of unwanted or unnecessary supplies. To achieve this balanced approach, logisticians must develop flexible support plans that readily adapt to the ebb and flow of urban combat.

Mobile aid stations and surgical teams should be positioned as far forward as the tactical situation allows and be provided with dedicated evacuation vehicles.

(2) Health Service Support (HSS). Responsive treatment and evacuation plans should be established to handle the expected increase in casualties. Both immediate treatment and evacuation are critical in maintaining the morale and confidence of forces engaged in urban warfare. To meet casualty treatment and evacuation needs, mobile aid stations with surgical teams should be placed as far forward as the tactical situation allows. Dedicated vehicles and aircraft for the evacuation of casualties should be provided.

In addition to casualties resulting from physical injury on the battlefield, there are often considerable psychological casualties in modern battles, particularly urban battles. This can be attributed to the intense, sustained, and close-quarter combat associated with urban warfare. In many battles, attacking troops were not properly trained for urban warfare and felt inadequate in their ability to fight on equal terms with the defender. Many felt moral or ethical dilemmas associated with collateral damage and civilian casualties. Finally, the constant threat of being killed from any quarter coupled with the extreme fatigue of fighting in an urban environment can create psychological casualties. Leaders at all levels should be attuned to the symptoms associated with psychological casualties in order to get effected individuals prompt treatment so that they can be returned to their units.

1006. Implications of Urban Warfare. The commander charged with making decisions needs to understand the operational and strategic implications of a tactical struggle in an urban area (MCDP 1-1, *Strategy*, and MCDP 1-2, *Campaigning*.). Three urban battles (**Stalingrad**, **Hue**, and **Beirut II**) illustrate the importance of seeing beyond the tactical nature of the battle.

Commanders must see beyond the immediate tactical implications of fighting an urban battle.

The battle for **Stalingrad** had major operational and strategic implications. The entanglement of German forces at Stalingrad bought time for the Soviets to mobilize and prevent a major combined effort of two German Army Groups in the south. While Stalingrad resulted operationally in the destruction of the German Sixth Army, the strategic results were even greater. Stalingrad caused a complete change of German strategy in the east. Hitler made major changes in his General Staff, and from this point on in the war, he was a man estranged from his military leadership. The loss of enough men and equipment to field one-fourth of the German Army shook the foundations of the Third Reich.

The Battle for **Hue**, although only one of over one hundred different attacks of the Tet Offensive of 1968, had a negative impact on the will of both the American people and their political leadership. Hue marked a revolution in the coverage of war by modern mass media. It was the first time Americans could and watch an ongoing battle from their living room on the evening news. Hue was a television bonanza for almost a month. When North Vietnamese leadership directed that Hue be held for at least seven days, it was clearly *not* their intent to win a tactical battle, but to strike at the *strategic center of gravity*—in this case, the will of the American people. Although the battle for Hue was a tactical victory for the U.S., the North Vietnamese had achieved their strategic goal of making the American public question the costs associated with the war.

During the battle of **Beirut II**, the IDF's objective was to drive the PLO from Lebanon. The ability of the PLO to leverage the media to gain an advantage was one of the most significant weapons in their arsenal. Despite an Israeli tactical victory, the costs in image, prestige, allies, and, most importantly, its own national will were enormous. The results of this battle eventually caused a change of political leadership at the highest levels of government.

1007. Key Insights. Analysis of modern urban battles provides insight into the major factors affecting the course and outcome of combat. Consideration of these key insights are required in order to effectively and efficiently plan and execute MOUT:

a. MOUT is infantry intensive.

b. A tactical battle may have far-reaching implications. No longer are the strategic, operational, and tactical levels overlap, creating the situation where tactical actions can have operational and possibly strategic repercussions (MCDP 1-1, *Strategy*, and MCDP 1-2, *Campaigning*).

c. Commanders at all levels must understand the impact that media representation will have on the accomplishment of operational and strategic objectives.

d. Maneuver warfare doctrine must be applied to the environment.

e. Intelligence is imperative to success in urban warfare.

f. Surprise is a combat multiplier.

g. Armor, artillery, and aviation are effective at the outer perimeter of built-up areas for causing isolation or preventing reinforcement.

h. Armor operating inside a built-up area must be protected by infantry.

i. Artillery providing direct fire inside a built-up area can be effective in the reduction of strongpoints.

j. As force ratio increases in favor of the attacker, combat duration decreases.

k. Urban warfare is time consuming.

l. Isolation of an urban defender ultimately ensures his defeat.

m. Attack of an urban area is costly to the attacker in terms of resources and casualties.

1008. Necessity for Preparation. Since MAGTFs are usually forward-deployed forces, it is inevitable that MAGTFs will be tasked to deploy to urban areas and conduct military operations. Fighting in this environment will be violent, close, and personal. Fighting in villages, towns, and cities are likely to become more frequent for military forces as the United States responds to global crises. Forces will be expected to exercise adaptability in responding to a wide variety of missions in the world's built-up areas. Marines may soon find themselves operating in the conceptual "three block battlefield" within a city grid: feeding refugees in one block, separating combatants in the adjoining block, and returning hostile fire in a third block.

MOUT is infantry intensive. The remaining chapters and appendices of this publication focus on how the GCE fights and integrates supporting elements in an urban environment.

Chapter 2

Offensive Operations

Urban combat isolates and separates units. Operations are reduced to a series of small-unit actions, placing a premium on small-unit leadership, initiative, and skill.

2001. Introduction. The Marine Corps' manuever warfare philosophy guides the conduct of offensive operations in MOUT. This warfighting philosophy serves to guide commanders through the development, planning, and execution of missions. A detailed discussion of maneuver warfare can be found in MCDP 1, *Warfighting*, MCDP 1-3, *Tactics*, MCWP 3-1, *Ground Combat Operations* (under development), and other Marine Corps doctrinal publications.

This chapter focuses only on offensive operations in an urban environment where collateral damage is of minimal concern. It describes tactics, techniques and procedures to be employed for seizing and clearing buildings and other urbanized areas. Chapter 6 addresses some considerations associated with fighting under constraints and restraints on urbanized terrain.

Section I

Planning

2101. Considerations. Identification of the adversary's centers of gravity and critical vulnerabilities will allow the commander to focus his efforts on those portions of the built-up area essential for mission accomplishment.

a. Reasons for Attacking a Built-Up Area. A commander considers the following before deciding to attack a built-up area:

(1) Tactical Advantage. Cities control key routes of commerce and provide a tactical advantage to the commander who controls them. Control of features such as bridges, railways, and road networks can have a significant impact on future operations. Urbanized areas may be used by the enemy as a base of operations from which they launch their own offensive operations. It may be advantageous to attack those bases and separate the enemy from their support infrastructure.

(2) Political Advantage. The political importance of a built-up area may justify the use of time and resources to liberate it. Capturing a city could destroy the seat of local and national government. At the very least, it could deal the enemy a decisive psychological blow.

(3) Economical Advantage. The destruction or capture of key industrial and commercial cities with the resulting denial of production and distribution of equipment and supplies strikes at the enemy's future ability to wage war. The requirement for a logistics base, especially a port or airfield, may play a pivotal role in the enemy's ability to continue the conflict. Capture of such cities may prove extremely beneficial to the attackers, who can use these resources to their advantage.

(4) Potential Threats to Operations

(a) Enemy Threat Too Great To Bypass. Though the terrain around a built-up area may facilitate its bypass, the enemy within that urbanized area may remain a threat capable of interdicting lines of communications. This may require the enemy force to be contained or destroyed.

(b) Terrain Does Not Allow Bypass. The urbanized area may sit between two natural slopes on the avenue of approach and thus require capture in order to secure the main supply route. Additionally, the urbanized area, itself, may sit on dominating terrain that threatens combat support and CSS elements.

b. Reasons for Not Attacking a Built-Up Area. The commander considers the following reasons for not attacking a built-up area:

(1) Built-Up Area Not Required To Support Future Operations. The attacker may have adequate supply support and resources established at another site from which combat forces can be supported. Therefore, during the estimate process, commanders may assess that the urbanized area is not necessary to support future operations.

(2) Enemy Not a Threat. The commander may decide to bypass upon determination that no substantial threat exists in the built-up area that could affect the unit's ability to accomplish its mission.

(3) Time or Risk Unacceptable. The commander's intent may dictate that speed of movement is essential to the mission. Because MOUT can be time-consuming, the commander may choose to bypass the urbanized area to maintain tempo. Furthermore, the potential for numerous casualties, the expenditure of critical resources, or the restrictions placed on attacking forces may result in unacceptable risks to the commander's primary mission.

(4) Declared an Open City. The area may have been declared an "open city" because it is undefended or of religious or historical significance. By international agreements, open cities are demilitarized and must be neither defended nor attacked. (See HR, 25, Annex to Hague Convention No. IV, *Embodying Regulations Respecting the Laws and Customs of War on Land*, 18 October 1907.) The attacking force must assume civil administrative control and treat the civilians as noncombatants in an occupied country. The defender must immediately evacuate and cannot arm the civilian population. A city can be declared open only before it is attacked. Other reasons for not defending could be the presence of large numbers of noncombatants, hospitals, or wounded personnel or the city's cultural, religious, or historical significance.

2102. Commander's Estimate. Once assigned the mission to attack a built-up area, a thorough mission analysis is conducted. Marines follow the same planning process in MOUT as for other operational environments by evaluating the mission, enemy, terrain and weather, troops and support available, and time available (METT-T). Based on METT-T and application of maneuver warfare concepts, the commander can decide on a plan that defeats the enemy by exploiting identified critical vulnerabilities. The commander may not have sufficient forces to fight everywhere and so must focus on the enemy and concentrate resources to decisive ends. This may involve making conscious decisions to accept risk at some place or time and to seize opportunities at others. The challenges of an urbanized environment dictate an even greater need to fully understand maneuver warfare.

This section addresses METT-T considerations as they pertain to the offense in urban warfare. METT-T is covered in detail in MCRP 3-11.5A, *Marine Troop Leader's Guide.*

a. Mission. Mission analysis is the first step of the estimate process. One of the important mission analysis considerations in urban conflict is the scope of clearance required to accomplish the mission. Commanders and planners should ask themselves the following questions:

- Do I need to clear every building?
- Should I clear only certain blocks?
- Should I only control certain areas?
- What level of protection is required for my lines of communication?

b. Enemy. Maneuver warfare focuses on the enemy. The intelligence preparation of the battlefield (IPB) process should be used to analyze the enemy and terrain in detail. (See Field Manual (FM) 34-130, *Intelligence Preparation of the Battlefield.*) The commander decides whether threat forces are conventional or unconventional.

(1) Conventional Forces. Most potential adversaries have adopted techniques of urban combat from either the United States or the former Soviet Union. Therefore, potential adversaries will build their urban defense to counter the attack of a combined arms force. Potential enemy forces will normally organize their defense in depth. Prepared strong points will usually form the perimeter of a larger defense while reserves locate in a separate position within the perimeter. You can expect ambushes to fill gaps in the perimeter while dummy strongpoints are established to deceive the attacker. Positions for securing the entrances to and exits from underground structures and routes are normally established. Security positions will normally be located forward of first-echelon defensive positions.

(2) Unconventional Forces. Urban areas have become a haven for unconventional forces. The large noncombatant population provides cover and concealment for unconventional force operations. Conventional forces operating in MOUT will normally be placed under restrictive ROE to minimize collateral damage. Unconventional forces will often use our restrictive ROE and the noncombatant population to their advantage when devising an urban defense. (See FM 34-130, FM 7-98, *Operations in a Low-Intensity Conflict*, and Chapter 6 of this manual.)

c. Terrain and Weather

(1) Terrain. Offensive operations should be tailored to the urban environment based on a detailed analysis of urbanized terrain (see Appendix I). Commanders and subordinate leaders use **KOCOA** (**k**ey terrain, **o**bservation and fields of fire, **c**over and concealment, **o**bstacles, and **a**venues of approach) to identify important terrain factors:

(a) K: Key Terrain. Key terrain surrounding an urban area can facilitate entry or deny escape. Within the city, airports or airfields, stadiums, parks, sports fields, school playgrounds, public buildings, road junctions, bridges, or industrial facilities may be key terrain.

Critical public buildings are identified during the terrain-analysis phase of an IPB. Hospitals, clinics, and surgical facilities are important because the laws of war prohibit their attack when not being used for military purposes other than medical support. The locations of civil defense, air raid shelters, and food supplies are critical in dealing with civilian affairs. Additionally, population size, locations, and density; density of the built-up area; firefighting capabilities; the location of hazardous materials; police and security capabilities; civil evacuation plans; and key public buildings should be identified.

(b) O: Observation and Fields of Fire. Urbanized terrain is characterized by restrictive observation and fields of fire. Weapon ranges can be greatly reduced because of buildings and other manmade structures. On the other hand, high ground or tall buildings can provide perches which enhance line of sight (LOS) for observation and communications as well as for individual and crew-served weapons. This includes man-portable surface-to-air missiles.

(c) C: Cover and Concealment. Buildings, sewers, and subways can provide excellent cover and concealment for enemy and friendly forces. They also provide covered and/or concealed maneuver routes within the built-up area. The civilian population can also offer cover and concealment to enemy forces.

(d) O: Obstacles. Natural or manmade obstacles restrict or deny maneuver within the urban area. Bridges, walls/fences, canals, streams, rivers, as well as rubble created by the effects of weapons should be thoroughly analyzed. Construction sites and commercial operations such as lumberyards, brickyards, steelyards, and railroad maintenance yards are primary sources of obstacle and barrier construction materials.

These sites can also supply engineers with materials to strengthen existing obstacles or to set up antitank hedgehogs or crib-type roadblocks.

(e) A: Avenues of Approach. Avenues to the city should support maneuver and be concealed either by terrain, darkness, smoke, or a combination of the three. Avenues that canalize or choke maneuver, due to the density of built-up area or natural terrain, should be identified and avoided. Roads, rivers, streams, and bridges provide high-speed avenues for movement. **Generally, military maps do not provide enough detail for urbanized terrain analysis.** They usually do not show the underground sewer system, subways, underground water system, mass transit routes, and power plants. Local maps of intracity road networks and subway systems and city hall/department of public works blueprints of all city buildings (if available), coupled with aerial photos, should be used.

(2) Weather. As in any military operation, weather affects equipment, terrain, and visibility, but its greatest impact is on the individual Marine. Snow, ice, dust, wind, rain, humidity, and temperature extremes reduce human efficiency. Weather extremes coupled with stress and the physical strain of urban combat can be minimized with effective small-unit leadership. Weather factors include:

(a) Precipitation. Rain or melting snow may flood basements and subway systems. This is especially true when automatic pumping facilities that normally handle rising water are deprived of power. Flooding makes storm and other sewer systems hazardous or impassable. In an NBC environment, chemical agents can be washed into underground systems by precipitation. As a result, these systems may contain chemical agent concentrations that are much higher than surface areas and thus become contaminated "hot spots." Hot spot effects become more pronounced as agents are absorbed by brick or unsealed concrete sewer walls.

(b) Fog. Many major cities experience fog, especially those located in low-lying areas and along canals or rivers. Industrial and transportation areas are the most likely to be affected by fog because of their proximity to waterways. Fog adversely affects vision and optical aids. It may also be used to help conceal friendly movement.

(c) Inversion Layers. Air inversion layers are common over cities, especially cities located in low-lying "bowls" or in river valleys. Inversion layers trap dust, smoke, chemical agents, and other pollutants, reducing visibility and often creating a greenhouse effect, which raises ground and air temperatures.

(d) Temperature. Built-up areas are often warmer than surrounding open areas during both summer and winter. This difference can be as great as 10 to 20 degrees hotter throughout the year.

(e) Wind Effects. Generally, wind chill is not as pronounced in built-up areas. However, the configuration of streets, especially in the city core and outlying high-rise areas, can cause wind tunneling. This increases the effects of the wind along streets that parallel the wind direction, while cross-streets remain relatively well protected. Wind tunneling can have a negative impact on aviation support within the urbanized area by creating dangerous wind effects down streets and between buildings.

(f) Day/Night Differences. Night and periods of reduced visibility favor surprise, which in turn, may facilitate infiltration, detailed reconnaissance, attacks across open areas, seizure of defended strongpoints, and reduction of defended obstacles. Because of the difficulties of night navigation in restrictive terrain (usually without reference points and in close proximity to the enemy) forces may consider using simple maneuver plans with easily recognizable objectives.

Rotary-wing aircraft are best utilized at night when operating on urbanized terrain. The most effective method of protecting helicopters from a successful engagement by hostile ground forces is to remain unseen. During night flights, the use of night vision goggles (NVGs) should be weighed carefully. Artificial lights can render NVGs partially ineffective and can increase the possibility of mid-air collisions by degrading the goggles to the point where nearby aircraft and aerial obstructions are not seen.

(g) Aviation Weather Considerations. Weather conditions can be a critical factor in determining the amount of aviation support the aviation combat element can provide in an urbanized area. Aviation forces should consider the following:

- Presence or absence of fog, industrial haze, low clouds, heavy rain, and other factors that limit visibility for pilots

- Illumination and moon phase/angle during night vision goggle operations

- Ice, sleet, and freezing rain that degrade aerodynamic efficiency

- Updrafts and downdrafts in the urban canyon that may present an unpredictable hazard

- High temperatures and/or high-density altitudes that degrade aircraft engine performance and lift capability

- High winds (large gust spreads) and crosswinds, to include the tunneling effect, that may create localized and unpredictable hazards to aviation

- Weather conditions that create hazards on pick-up zones and LZs, such as blowing dust, sand, or snow.

d. Troops and Support Available. The commander must be aware of all of his combat power (e.g., types of weapons systems, their numbers, and their capabilities) when performing analysis for operations in a built-up area. In the initial planning phase, force size must be evaluated in relation to urban size, enemy forces, and mission assignment. In the attack of a built-up area (population 100,000+), the GCE of a MEF would be a Marine division. However, in the initial stages of an amphibious or MPF operation, a reinforced infantry battalion or regiment may provide the initial forces. No matter what the size GCE, operations will be conducted by task-organized, reinforced battalions and their companies/platoons. Therefore, by using the infantry battalion as the basis for projecting the size of an overall force, the number of regiments or divisions required to secure a built-up area can be determined. Whatever size MAGTF is required for operations in an urbanized area, much of the fighting may be generally conducted by small teams of Marines. Consequently, our success in urban fighting may largely depend upon small unit and individual tactical skills. Specifically, leaders at all levels must analyze the following factors:

- Number and type of available units
- Task organization for urban combat (See Appendix A.)
- Availability of critical weapons systems
- State of training and discipline (training for urban warfare is imperative)
- Strength in terms of men and materiel in relation to enemy and size of built-up area
- Aviation support available
- State of maintenance and supplies
- Available combat support (based on availability and mission requirements)
- Available CSS (critical logistical and maintenance items)
- Host nation support available. (If the city's occupants are friendly to the attacker, then support in intelligence, deception, and diversion may be possible.)

e. Time Available. The following issues should be considered when analyzing the time available for an attack on urbanized terrain:

- A significant amount of time is required for clearing buildings, blocks, or axes of advance.
- Marines tire more quickly when clearing buildings because of stress and additional physical exertion.

- Adequate time should be allocated for thorough reconnaissance, planning, and rehearsals. Ultimately, this can result in higher tempo in the execution of the plan.

Once an estimate of the situation (using METT-T) has been conducted, planning for offensive operations can proceed. It is important to remember that the initial estimate of the situation must be continually updated throughout the planning process and the conduct of operations to account for changes in METT-T. (See MCWP 5-1 *Marine Corps Planning* [under development])

2103. Phases of the Attack. Attacks are categorized as either hasty or deliberate. Both hasty and deliberate attacks should take advantage of as much planning, reconnaissance, and coordination as time and the situation permit. Regardless of the size of the attacking force or of the objective to be secured, the phases of the attack (i.e. reconnoiter, isolate, secure foothold, and control the objective) remain constant.

a. Deliberate Attack. A deliberate attack *is a type of offensive action characterized by preplanned coordinated employment of firepower and maneuver to close with and destroy or capture the enemy* (Jt Pub 1-02). It is a fully coordinated operation that employs the MAGTF's combined arms team against the enemy's defense. It is used when enemy positions are well prepared, when the built-up area is large or severely congested, or when the element of surprise is lost. Given the nature of urbanized terrain, the deliberate attack of a built-up area is similar to the assault of a strongpoint (see MCWP 3-1, *Ground Combat Operations* [under development]). The deliberate attack of a built-up area is conducted in the following phases:

(a) Phase I: Reconnoiter the Objective. Intelligence gathering and reconnaissance/surveillance are critical to the planning process and success of the operation. All available intelligence resources should be used to gain vital information on the enemy. Whenever possible, a leader makes a personal reconnaissance of the objective area to collect first-hand information regarding the area to be attacked. A thorough reconnaissance yields a wealth of information to use in the continued development of the plan. Avenues of approach, observation posts (OPs), supply routes, and the emplacement positions of direct- and indirect-fire weapons systems are all examples of information that may be amassed during the reconnaissance of the objective area. Composition and structure of buildings and roadbeds, cover and concealment opportunities, and other information not apparent in a map study may have a significant impact on the plan.

(b) Phase II: Isolate the Objective. The objective can be isolated by seizing natural and man-made features that dominate the area. Isolation may also be accomplished by coordinated use of supporting arms to seal off enemy lines of communication. This phase may be conducted simultaneously with Phase III (securing a foothold). Figure 2-1 shows a task-organized, reinforced infantry battalion isolating the objective.

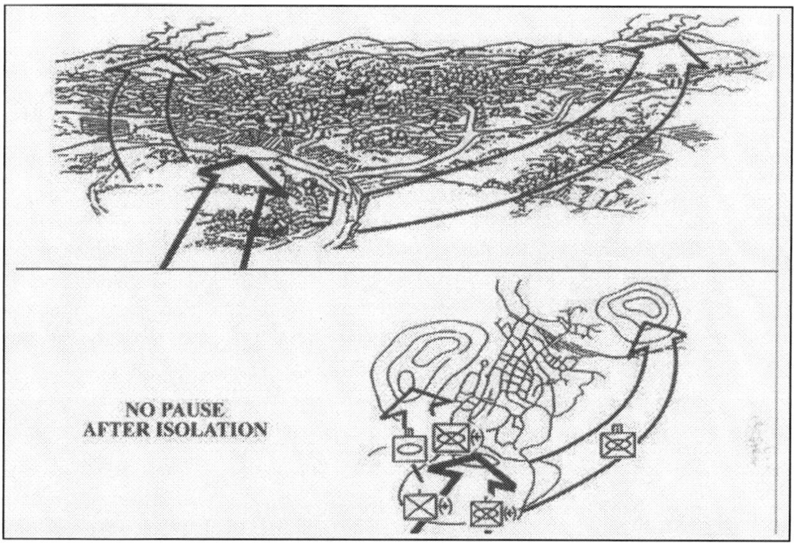

(c) Phase III: Secure a Foothold. Once the objective is isolated, a foothold should be secured as soon as possible in order to maintain tempo. The seizure of a foothold provides the attacking force with a position from which they can continue the assault through the objective area. The assault is supported by direct and indirect suppressive fires and smoke. Figure 2-2 (pg. 2-10) shows the battalion objective being isolated. One company is providing suppressive fires while another company seizes an initial foothold. Fire support assets are providing suppressive fires on the objective.

Figure 2-1. Isolation of the Objective Area

(d) Phase IV: Seizing the Objective. Once a foothold is seized and consolidated, supporting forces move to the built-up area to support the seizing of the objective area. To maintain tempo, the transition between the phases should be seamless. Once the foothold has been established, forward units continue the attack through the objective area. Supporting units assist as required. The momentum of the assault is continued until the objective area is cleared or controlled.

The assault force should establish limited objectives to ensure that the attacking forces do not get strung out along the axis of advance. Gaps may give the enemy the opportunity to infiltrate along the line of advance or make isolated friendly forces vulnerable to attack.

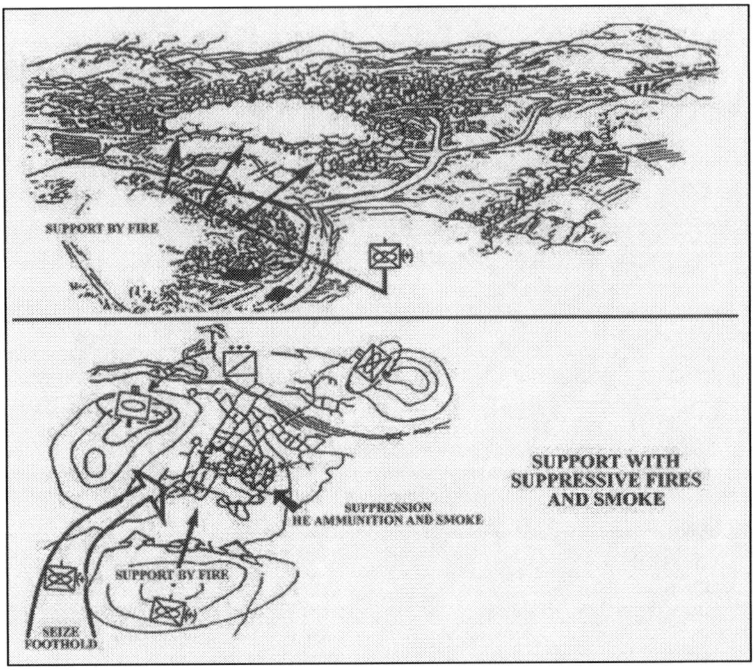

This phase also includes consolidation of the objective area and reorganization for future taskings. By setting limited objectives, the attacking forces have an opportunity to reorganize and defend against counterattacks while maintaining momentum. Once the limited objectives have been secured, the urbanized area may be sectored for detailed clearing. Clearing by sectors may help in the allocation of forces and in the prevention of fratricide. In addition to defensive preparation against counterattack, preparation should be made for follow-on missions (e.g., restoration of civilian facilities, marking and clearing minefields, etc.).

Figure 2-2. Seizing a Foothold

Seizing may consist of a systematic house-by-house, block-by-block advance through the entire zone of action, or it may be a rapid advance through a lightly defended area to seize a key objective. The difference between the two techniques is made by the commander on the basis of mission requirements and METT-T analysis. Seizing an objective in urbanized terrain requires detailed planning, coordination, decentralized command and contol, and small-unit execution. It is a continuation of the assault until all mission requirements are met.

(1) Rapid Advance. Rapid advance (Figure 2-3) may be used to maintain

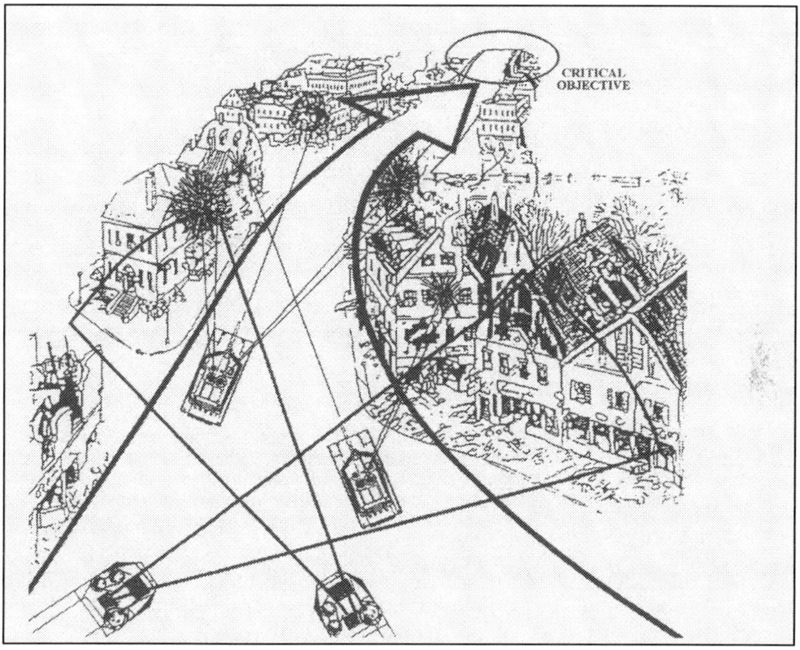

momentum through the zone of action and to seize key objectives. Some enemy positions may be isolated or bypassed in order to maintain tempo. This procedure may be used when:

- A key objective is identified (an enemy critical vulnerability or an objective that supports the friendly scheme of maneuver)
- Time is critical in reaching the objective
- Only selected buildings on the route of advance require clearing.

Figure 2-3. Rapid Advance En Route to a Key Objective

(2) Systematic Clearance. Systematic clearance (figure 2-4 on pg. 2-12) involves the deliberate reduction of all significant enemy positions throughout the objective area. It is usually used when time is not a critical factor.

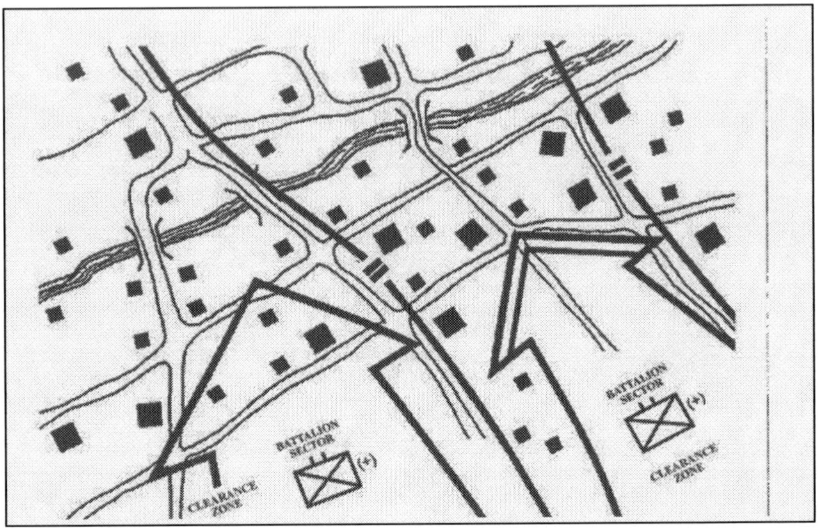

b. Hasty Attack. A hasty attack is, *in land operations, an attack in which preparation time is traded for speed in order to exploit an opportunity* (Jt Pub 1-02). Normally, hasty attacks are conducted as a result of unexpected contact during movement through the built-up area or when an opportunity for exploitation presents itself. Upon contact, the commander immediately deploys, returns fire, reports the contact and situation, develops the situation, and chooses a course of action. The hasty attack follows the same four phases as the deliberate attack; however, the time between preparation and response is compacted.

2104. Warfighting Functions. The warfighting functions (WFs) encompass all activities in the battlespace. Planners consider and integrate the warfighting functions when analyzing how best

Figure 2-4. Systematic Clearance Within Assigned Sectors

to accomplish the mission. The WFs include maneuver, intelligence, fires, logistics, command and control, and force protection.

a. Command and Control. The commander positions himself where he can best observe and influence the battle. Positioning is relative to the level of command. For example, a squad leader locates where he can observe and direct his fire teams, while the battalion commander locates where he can observe and direct his companies.

Close-quarter fighting in and around buildings makes command and control exceedingly difficult. Commanders and small-unit leaders rely on guidance and intent provided by higher

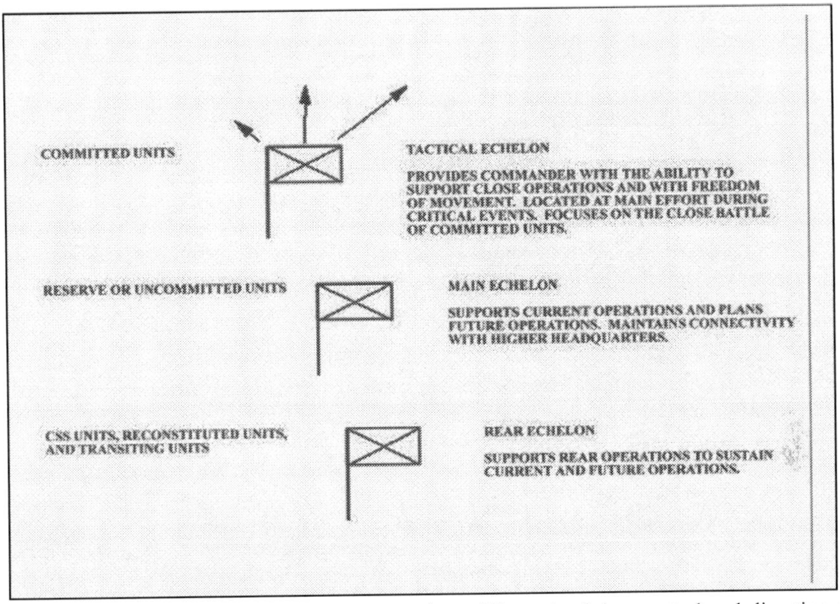

headquarters to facilitate decentralized execution while maintaining control and direction. In close-quarter fighting, commanders may have to rely upon messengers to communicate with their units and higher headquarters.

At battalion and higher levels, command and control is accomplished through three command echelons, which can be collocated or echeloned across the battlefield. The three echelons are tactical, main, and rear (see fig. 2-5). Each echelon may have a combat operations center (COC) capability. The COCs are the facilities (buildings, vehicles, or tents) used by the commander or staff at each echelon to plan, direct, control, and coordinate operations.

Figure 2-5. Tactical, Main, and Rear Echelons

(1) Tactical Echelon. The tactical echelon is normally collocated with the main echelon until the situation dictates separation. When the tactical echelon separates, designated command group members displace forward. Normally, the commander will move with the tactical echelon. The tactical echelon must have the communications and mobility necessary to allow the commander to exercise command. This echelon is usually small in size so the commander and command group can quickly displace forward to observe and influence actions. Organization of the tactical echelon should be addressed in the unit's standing operating procedures (SOPs).

(2) Main Echelon. The main echelon contains a COC, which is the nerve center for command and control of battalion through division-sized organizations. A COC consists of the command group (primary staff members with support personnel) and possesses the necessary mobility and communications for command and control of tactical operations. The main echelon COC monitors the battle, coordinates fire support, issues orders, maintains communications with subordinates and higher and adjacent commands, conducts current and future tactical plans, and assists the commander and subordinate commanders as appropriate. On urbanized terrain, the main echelon COC is normally located in buildings that provide good communications with subordinates and higher and adjacent commands. The command post (CP) is where the commander is physically located. Therefore, the CP can be located at any of the three echelons based on the commander's presence.

(3) Rear Echelon. The rear echelon's COC consists of the logistical and administrative personnel required to plan, coordinate, and execute logistics support. During urban warfare, the rear echelon COC is usually closer to combat forces than normal as it uses the city's captured buildings and facilities to provide cover and concealment for stocking supplies, conducting maintenance repair, and treating the injured.

Attack of a built-up area requires detailed planning due to restrictive terrain, close proximity to opposing forces, fire coordination requirements, and reduced communications capability. The GCE will frame a centralized plan of attack, fire support, and logistics support to conduct urban operations. However, the actual attack of a built-up area is a series of isolated, close-quarter battles carried out by small units. The execution of these attacks should be decentralized.

Communications planning procedures, as outlined in MCWP 3-40.3, *Communications and Information Systems* (under development), are applicable to the urban battle. Planners should be meticulous in their scrutiny of the environment and the effect it will have on communications. Manmade structures can create problems for single-channel radios. These structures inhibit LOS radio communications by absorbing or reflecting transmitted signals. However, the urban environment may have exploitable advantages such as the availability of electrical power, commercial telecommunications networks, and environmental control systems. Electrical power generation stations and other emergency power systems are normally found in protected structures and are probably usable. Due to their value to the enemy, these areas are more likely to be heavily defended. Enclosed areas offer excellent concealment and protection of communications and other command and control support equipment. Extensive commercial communications networks composed of miles of underground protected cable connecting central telephone exchanges are likely to be available, as well as a multitude of public service radio nets (police, fire, civil defense, taxi, etc.) complete with existing antenna systems and retransmission stations. To communicate effectively and continuously, commanders must minimize limitations imposed by the urban environment and maximize the advantage of existing civil communications.

b. Intelligence. Collection of intelligence information on urbanized terrain is difficult. Urbanized terrain provides overhead cover and concealment from observation by aircraft, reconnaissance personnel, and satellites.

(1) Information Requirements. In addition to information about the location of enemy units, nature of the terrain, trafficability of roads, etc., that all operations require, urban warfare requires detailed information on the following:

- Population density, characteristics, and culture
- Location and quality of water supply
- Width and construction of streets
- Layout of road networks
- Location, width, and load capacity of bridges
- Layout of underground networks such as subways, sewers, and utility ducts
- Types and construction of buildings
- Location of key communications and transportation facilities
- Location of canals and waterways.

Information on port facilities should include details on harbor sites, pier networks, unloading capabilities, warehouse/dockside facilities, harbor currents, and sandbars or other midstream obstructions.

(2) Collection. Accurate, detailed, and timely information is vital to the success of any urban operation. Collection operations conducted by national intelligence organizations may provide valuable information on the layout of the city, recent changes to transportation routes, political affiliations of various groups, and locations of military units. This information needs to be current and must be verified and updated as attacking forces close on the city. Reconnaissance activities such as the study of large-scale maps, aerial photographs, and background intelligence reports keep the commander's information current. The latest information on the enemy and the terrain should be continuously fed to the intelligence section using long-range reconnaissance patrols, human intelligence (HUMINT) reports, unmanned aerial vehicle (UAV) missions, and electronic intelligence. Use of reconnaissance assets should be tempered with a risk-versus-gain judgment. Reconnaissance assets should not be risked indiscriminately if the information can be collected by other means.

(3) Maps. Tactical maps should be supplemented by city street maps that contain street names and information on the location of important buildings and transportation terminals. These city street maps should be distributed to all units operating within the city. Although tactical maps are key in controlling fire support missions, simple street maps assist ground troops in maintaining their orientation within the city and in tracking buildings and areas that have been cleared.

c. Maneuver. Maneuver is the employment of forces on the urban battlespace through movement in combination with fire or fire potential to achieve a position of advantage (see MCWP 3-1). Maneuver is an essential element of combat power. Maneuver in itself cannot produce decisive results. Combined with mass, offensive action, economy of force, and surprise, maneuver provides favorable conditions for closing with the enemy. Maneuver contributes significantly to sustaining the initiative, exploiting success, preserving freedom of action, and reducing vulnerability. It is through maneuver that an inferior force can achieve decisive superiority at the necessary time and place. In many cases, maneuver is made possible only through the control of tempo and effective employment of firepower. The commander integrates supporting fires with the scheme of maneuver to create a dilemma for the enemy. Likewise, movement without fires exposes the force to effective enemy counteraction and risks losing the initiative and momentum. Maneuver that does not include violent action against the enemy will not be decisive. At all levels, successful application of this principle in the urban environment requires flexible and innovative plans and actions.

(1) Infantry/Armor Movement. Armored vehicles are vulnerable in built-up areas where streets and alleys provide ready-made fire lanes for defenders. Motorized traffic is greatly restricted, canalized, and vulnerable to ambush and close-range fire. Tanks are at a further disadvantage because their main guns cannot be depressed sufficiently to fire into basements or be elevated to fire into upper floors of buildings at close range (Figure 2-11). Whenever tanks are used in the direct-fire role, infantry must provide security against enemy ground attack. Movement of armor down narrow streets or down narrow paths through debris requires close coordination with infantry. Infantry and armor may use bounding overwatch techniques during movement (See MCWP 3-12, *Marine Tank Employment* [under development]). With this technique, infantry moves ahead of armored vehicles to clear the buildings on each side. Lead infantry units should not get so far ahead of armor that mutual support is lost. Conversely, if armor gets too far ahead of infantry, it may be quickly damaged or destroyed. Coordinated movement also contains a rear security element to protect armor from behind.

During movement down a street tanks should deploy with one tank on each side of the street focusing on the street and lower levels of buildings. *In MOUT, tanks are best deployed in pairs.* Other tanks of the tank platoon or section should move behind the infantry and fire at targets in the upper stories of buildings. When space is inadequate to deploy abreast, tanks may deploy in single file. Movement across open areas should be well coordinated to ensure mutual support. Suppression fire and smoke are used to cover force movements.

Armor can also be used in the reduction of obstacles. For example, grappling hooks and ropes attached to a tank can be used to pull down wire obstacles.

(2) Cover and Concealment. Walls and buildings can provide cover, concealment, and protection from enemy ATGM or heavy machine gun fire. Before armor enters a building, infantry should clear the building and check the ground floors to ensure that there is no basement into which an armored vehicle could fall and become trapped.

Tank and AAV generated smoke may be useful in concealing the location and movement of assaulting forces. Additionally, the thermal sights of the tank are very effective as they can use heat differentials to detect the movement or presence of concealed enemy soldiers, weapons, and vehicles.

(3) Tactical Considerations for Helicopterborne Movement. Helicopterborne operations are planned and conducted using the same combat fundamentals that apply to ground operations. However, there are considerations that apply only to helicopterborne operations. See MCWP 3-24, *Assault Support* (under development) for a detailed discussion on helicopterborne operations. These considerations include:

(a) Helicopterborne forces, once they have landed, lack vehicular mobility and heavy weapons. For this reason, it is important that the forces land on or near the objective. Such a force may be isolated for a time pending linkup with ground forces.

(b) Helicopters may be used for tactical deception. Helicopters may be used to make demonstration landings in several different zones to deceive the enemy as to the true objective of an operation.

(c) Helicopterborne assaults are normally conducted against undefended or lightly defended objectives. When attacking a well-defended objective, an LZ should be selected which provides cover and concealment for the assault force and is close to the objective.

(d) Helicopterborne forces are vulnerable to attack helicopters, fixed-wing aircraft, and surface-to-air missiles. To counter this vulnerability, friendly attack helicopters

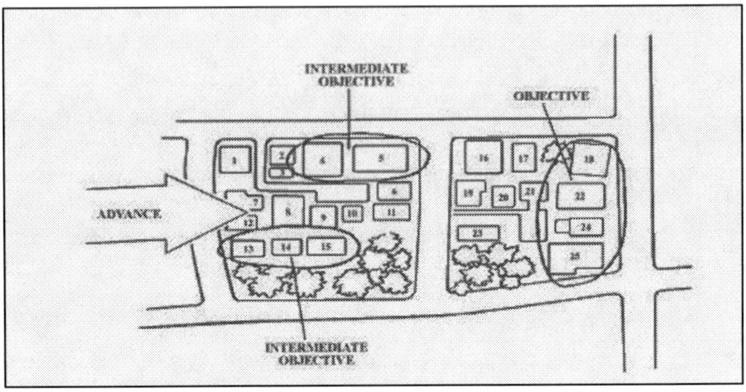

escort the helicopterborne forces during air movement, and indirect fires are used to suppress or neutralize enemy air defenses.

(4) Control Measures. The following control measures assist in the manuever of forces in MOUT:

(a) Objectives. When attacking to seize a foothold, the infantry unit may assign subordinate units the first block of buildings as their first objective. When an objective extends to a street, only the near side of the street is included. The final objective may be buildings or key terrain at the far edge of the built-up area. Key buildings, or groups

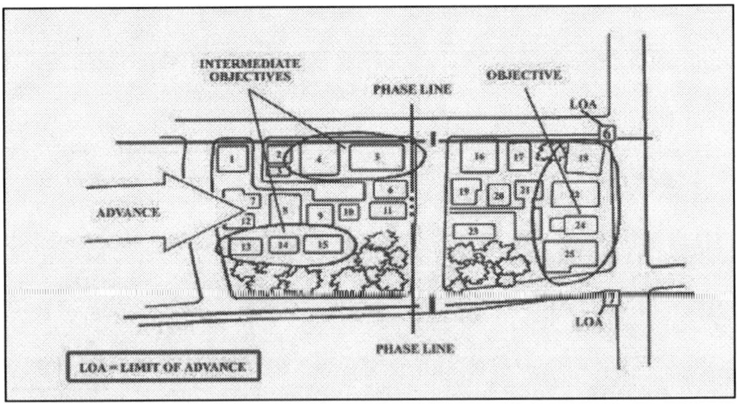

of buildings, may also be assigned as intermediate objectives. Buildings along the route of attack should be identified by numbers for clarification (Figure 2-6 on page 2-18).

During the seizing phase, bypassing buildings may increase the risk of attack from the rear or flank. It may be necessary for the unit to enter, search, and clear each building in its zone of action. A single building may be an objective for a rifle squad or, if the building is large, for a rifle platoon or company. When the commander's concept of operations is based on speed (rapid advance method) or when the force is conducting a hasty attack, a unit may be directed to bypass certain positions within its zone.

Figure 2-6. Objectives and Numbering System

(b) Phase Lines. Phase lines are control measures used to report progress or to control the advance of attacking units. Phase lines should be oriented on readily identifiable terrain features such as principal streets, rivers, and railroad lines. They should be placed on the near side of a street or open area. In an attack, a unit may have the mission to seize positions in its zone of action up to a particular phase line. (Figure 2-7).

Figure 2-7. Phase Lines

(c) Boundaries. Unit boundaries are used to define zones of action and are usually set within blocks so that a street is included in the zone. Both sides of a street should be included within the same unit's boundries.

(d) Checkpoints and Contact Points. Checkpoints aid in reporting locations and controlling movement. Contact points are used to designate specific points where units make physical contact. Checkpoints and contact points are designated by each unit as appropriate for command and control at street corners, buildings, railway crossings, bridges, or any other easily identifiable feature.

(e) Attack Position and Line of Departure (LD). A designated attack position may be occupied by forward units for last-minute preparation and coordination. The attack position is often behind or inside the last large building before crossing the LD. The LD should be located on the near side of an open area running perpendicular to the direction of attack, such as a street or rail line.

(5) Time of Attack. Ideally, the time of attack should be conducted during hours of darkness or limited visibility. Marines can exploit the poor visibility to cross open areas, to gain access to rooftops, to infiltrate enemy areas, and to gain a foothold. When attacking in unrestricted visibility, units should use smoke to conceal movement.

(6) Reserve. Reserves should be mobile and usually follow in trace of forward units. Battalion reserves normally follow one to two blocks to the rear of the lead company. If a company reserve is available, it follows within the same block so that it can immediately influence the attack. A unit with a reserve mission can be called upon to:

- Exploit an enemy weakness or friendly success
- Attack from another direction
- Attrit and/or clear bypassed enemy positions (normally requires designation of a new reserve)
- Secure the rear or a flank
- Maintain contact with adjacent units
- Counterattack

(7) Security. Each unit should ensure security of its flanks and rear (Figure 2-8 on page 2-20). Fighting in a built-up area is unique in that attacks can come from all three dimensions. Units assigned a mission to isolate can also provide security. Small units conduct security patrols and man OPs to supplement higher level reconnaissance and security operations.

Figure 2-8. Enemy Firing From Flank

(8) Engineers. Leading units should have engineers attached for immediate support. Tasks commonly assigned to engineers include:

- Preparing and using explosives to breach walls and obstacles
- Finding and exploding mines in place or helping to remove them

- Cratering roads and performing other countermobility measures
- Clearing rubble and obstacles

Requirements for engineer support in MOUT normally exceed capabilities. To compensate, all Marines should receive basic demolition training prior to conducting urban operations.

d. Fires. The attack of a built-up area may require extensive air and artillery preparation prior to the ground attack. Supporting fires suppress the defender's fire, restricts his movement, and possibly destroys his position. Consideration should be given to the rubbling effect produced by aerial and artillery bombardment. The assault should closely follow air and artillery fire to exploit its immediate effect on the defender. Maneuver units move near the final coordination line while the enemy is engaged by supporting fires. As the attacking force assaults, supporting fires lift and/or shift to block enemy withdrawal or to prevent the enemy from reinforcing their position. (See Chapter 4 for detailed discussion on air and artillery support during urban warfare.) Fire can be categorized into indirect and direct fires.

(1) Indirect Fire

(a) Artillery. Indirect artillery fire is planned to isolate objectives, to prevent reinforcement and resupply, to neutralize known and suspected CPs and OPs, and to suppress enemy defenders. Urbanized terrain requires that most indirect artillery fire use a high-angle trajectory.

(b) Mortars. Mortars are the most responsive indirect-fire weapon for the urban environment. They can be employed to hit targets of opportunity at the close ranges that are typical of combat in built-up areas.

(c) Other. Urban conflict is conducive to innovation. Many direct fire weapons can be employed in the indirect fire mode. Indirect fire weapons may also be effective at clearing the tops of buildings using variable-timed fuses.

(2) Direct Fire.

(a) Armor. Tanks are normally task organized with mechanized units for protection and security operations. The commander can employ tanks to take advantage of their long-range lethality, high-speed mobility, and survivability. They can be used outside the built-up area to cover high-speed armor avenues of approach. Tanks are normally employed in this role during the isolation phase. Tanks also support by fire the assault to seize a foothold and may support attacks to seize objectives within the built-up area. In MOUT, tanks can also be organized into special assault teams. The tank cannon's direct fire is very effective against structures. (Figure 2-9 on page 2-22 and Appendix B)

(b) Howitzers. Artillery rounds delivered by direct fire can be very effective for destroying targets in buildings. When used in this manner, the mobility, flexibility, responsiveness, and survivability of gun crews may be reduced. Artillery used in the direct fire role normally use HE shells.

(c) Infantry. Direct-fire systems, mainly machine guns, antitank guided missiles (ATGMs), shoulder- launched multipurpose assault weapons (SMAWs), and AT4s—are initially employed to support the seizure of a foothold. The positioning of antitank weapons in buildings must allow enough space for the effects of backblasts. Antitank weapons have limited as effectiveness when neutralizing targets behind walls. Designed primarily as armor penetrators, they neutralize a target only if that

Figure 2-9. Tank Direct-Fire Support Within the Built-Up Area

target is located directly behind the point of impact. Consideration must be given to weapons with minimum arming distances. Many of these factors can combine to limit the effects of weapons in the urban environment.

(d) Aviation. The vertical nature of large urban areas may block the pilot's line of sight with the target unless careful consideration is given to inbound headings and altitudes. Plan attack headings to take advantage of large open areas and street axis. Friendly ground forces must be clearly distinguishable from targets. The use of smoke, visible or infrared strobes, or other visual devices must be effectively employed by ground forces and clearly understood by friendly aircraft.

e. Logistics. Logistics support in urban operations should be timely, flexible, and responsive. Logistics planners, both in the supported units and the CSSE, should be fully integrated in the planning process. The final logistics plan must be thoroughly coordinated with the scheme of maneuver. Units may need special equipment such as ropes, grappling hooks, ladders, sandbags, and wire; medical supplies must be readied in anticipation to support increased casualties; and additional supplies must be prestocked. (See Chapter 5 for a detailed discussion on logistics and urban warfare, and MCWP 4-1, *Tactical Logistics*, for additional planning considerations.)

f. Force Protection. Force protection encompasses all those measures that preserve the fighting potential of the command. It is integrated throughout all the warfighting functions. It does not entail security operations alone. It involves the continuous implementation of active and passive measures by all units during all phases of an operation.

Some particular active and passive measures in MOUT include:

-Manmade structures in the urban environment provide cover and concealment for both the attacker and defender.

-Lines of communication can be more easily interrupted due to multiple avenues of approach.

-All units must be prepared to implement force protection in a 3-dimensional environment.

-Proximity of civilian population makes identification of non-combatants more difficult.

-ROE may be more restrictive

-Snipers and man-portable shoulder lauched air defense missiles are usually more difficult to detect.

Section II

Infantry Battalion Operations

2201. Overview. This section addresses some of the operations that a task-organized and/or reinforced infantry battalion could conduct in MOUT. These operations are seldom conducted independently and are usually part of a larger joint operation. In the example given below, the battalion has seized a foothold and is operating in its sector of a town of 100,000 people. Concurrent operations are on-going through other parts of the town. The threat in the battalion sector is a company (-) with a capability to reinforce.

2202. Seizure of a Key Objective. The tactical situation may require the immediate seizure of a building, industrial complex, or transportation site that is deemed decisive terrain. Examples of such objectives may include key bridges, intersections, railheads, or facilities that constitue the center of gravity for the enemy. As an example, a battalion may be assigned the mission of seizing the only bridge that is capable of supporting armored vehicles and resupply by heavy transports along a river bisecting a town. The intent is to isolate enemy forces and secure the bridge for follow-on forces. Intelligence indicates that the bridge is lightly defended at this time. In this case, speed of attack and surprise is critical to capture the objective. Some key intelligence information is required prior to the attack. This would include the number of enemy defending the bridge, their weapons, disposition, their ability to reinforce their defense, location of LZs near the bridge, whether the bridge has been set for demolition, and the location of enemy indirect fire assets that can range the objective.

The commander decides to conduct a penetration using a combination of helicopterborne attack to rapidly seize the bridge followed by a mechanized ground attack to link-up with the helicopterborne forces (Figure 2-10). The helicopterborne force can achieve surprise and speed of movement required while the mechanized force can thrust quickly along high-speed avenues to minimize the time the helicopterborne force must hold the objective against enemy counterattack.

For the ground attack, the battalion commander makes a determination that speed of movement outweighs the risks of rapidly extending his lines of communication. He directs, on the basis of his knowledge of the enemy situation and his mission, that his maneuver forces avoid decisive engagement with the enemy along the route to minimize the time it takes to link up with the helicopterborne force. The commander provides guidance on those enemy forces that should be bypassed and those that should be isolated and reduced.

The commander task organizes his battalion for rapid movement on two axes to allow flexibility in reacting to unexpected enemy contact. Lead and rear security is conducted throughout the operation. As lead units locate enemy positions, they fix them by fire, and quickly reduce or bypass them. They also report the contact so that follow-on forces can then deal with the threat.

As the ground attack force reaches the edge of the built-up area, Marines remain mounted so as not to slow the advance. All vehicle machine guns fire to flanks and into doors, windows, and

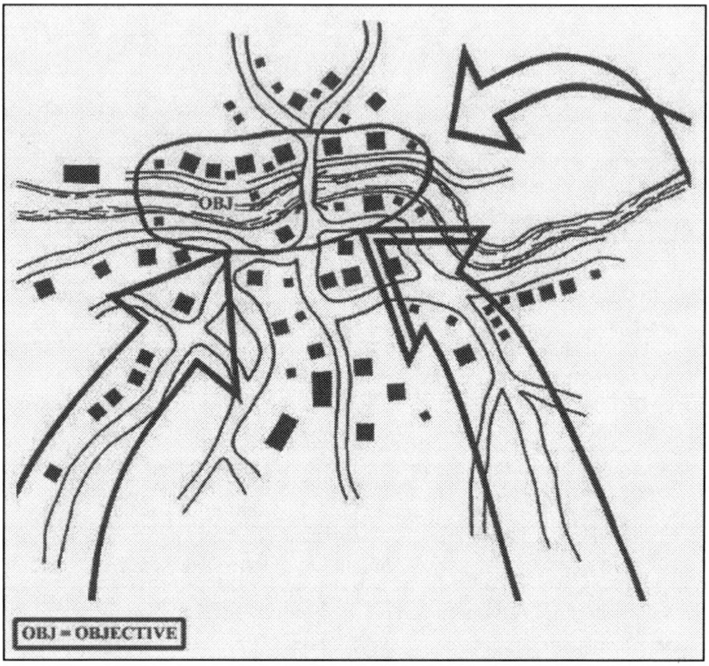

Figure 2-10. Seizure of a Key Objective

alleys to provide suppressing fires (Figure 2-11 on page 2-26). If heavy resistance is met and the column is halted, Marines dismount and the enemy and/or obstacles are cleared as quickly as possible. Once resistance and/or obstacles are cleared, the attack is immediately resumed. If necessary, units are dropped off to assume blocking positions and await follow-on clearing forces.

Once the objective is seized or linkup is accomplished, the battalion establishes a perimeter defense. The companies seize buildings around the objective and expand the size of the perimeter until it is large enough to secure the bridge. Attached engineers can then examine the bridge and clear it of any explosives.

2203. Infiltration. Infiltration is *"the movement through or into an area or territory occupied by either friendly or enemy troops or organizations. The movement is made, either by small groups or by individuals, at extended or irregular intervals. When the term is used in connection with the enemy, it infers that contact is avoided"* (Joint Pub 1-02).

Figure 2-11. Rapid Advance Movement

.3

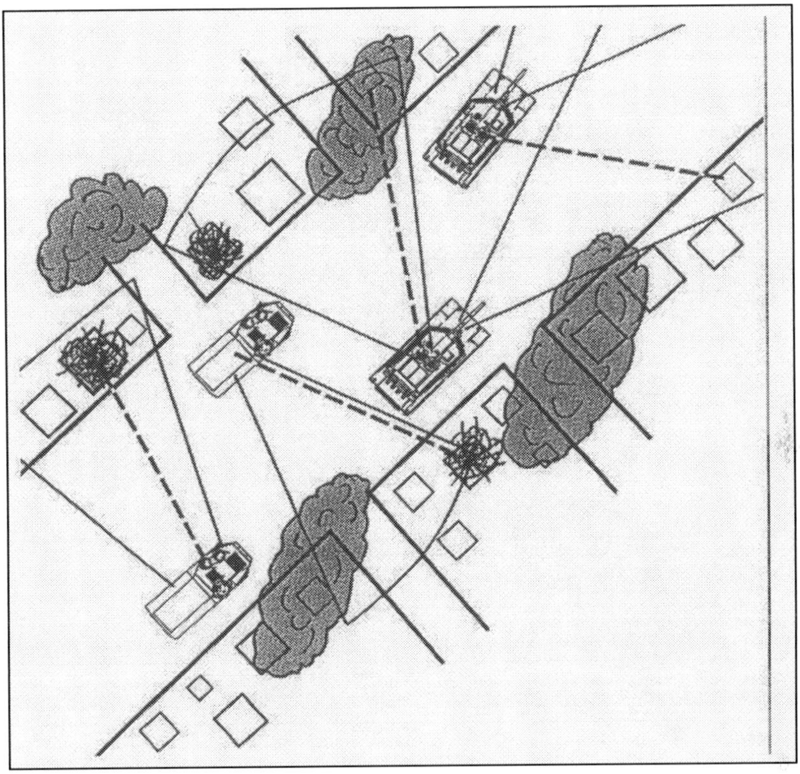

The outskirts of a town may not be strongly defended. The town's defenders may have only a series of antiarmor positions, security elements on the principal approach, or positions blocking the approaches to key features in the town. The enemy strongpoints and reserves may be positioned deeper in the city. In such a situation, an infantry battalion may be able to seize a part of the town by infiltrating companies and platoons between the enemy positions on the outskirts (Figure 2-12).

Moving by stealth on secondary streets, using the cover and concealment of alleys and buildings, the battalion may be able to seize key street junctions or terrain features, isolate enemy positions, and help following units move into the built-up area. Such an infiltration should be performed when visibility is poor.

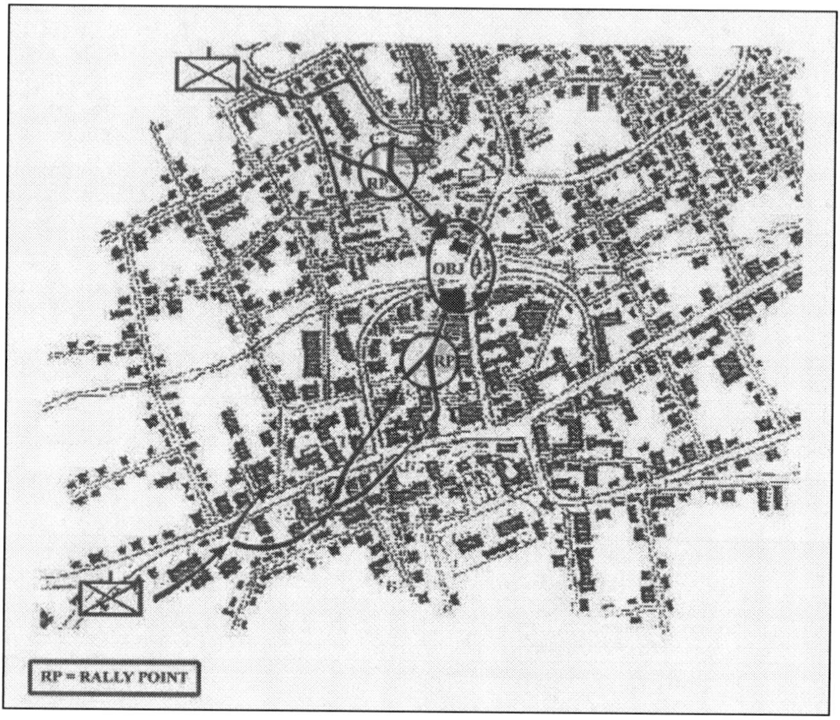

Figure 2-12. Infiltration

The infantry battalion is organized into infiltration companies with engineers attached and a designated reserve. Each company is assigned an infiltration lane that allows for infiltration by company-sized or smaller units. Depending on the construction of the built-up areas and streets, the infiltration lane may be 500 to 1,500 meters wide.

Mortar and artillery fire can be used to divert the enemy's attention and cover the sound of infiltrating forces. Armored vehicles and antitank weapons should be positioned to cover likely avenues of approach for enemy armored vehicles. Designated unit(s) provide isolation and security for infiltrating rifle companies. As the companies move into the built-up area, they secure their own flanks. Security elements may be dropped off along the route to warn of a flank attack. Engineers assist in breaching or bypassing minefields or other obstacles. Enemy positions are avoided but reported. The infiltrating companies proceed until they reach their rally points. At that time, they reorganize and make final preparations for attack of the objective.

.3

2204. Route Security. In MOUT, a battalion may be tasked to provide route security. The process of securing the route may consist of a series of small-unit actions to clear buildings along the route. Fires should be planned to isolate the route as well. It may also require the reduction of obstacles and the emplacement of flank security along the entire route. In an outlying area, the forward units may proceed by bounds from road junction to road junction. Other units could augment flank security by moving down parallel streets and by probing the flanks.

The speed of movement required and the enemy situation determine whether the unit should be mounted or dismounted. For mechanized forces, narrow streets should be avoided if possible to avoid being trapped. Designated armored vehicles overwatch the squad to its front, keeping watch on the opposite side of the street. While armored vehicle movements and positioning provide mutual support, armored vehicles should be protected by dismounted infantry.

When contact is made with the enemy, armored vehicles seek covered positions and provide direct-fire support. Supporting fires fix and isolate enemy positions. Infantry dismount then maneuver to attack. Phase lines are used to coordinate movement along the route. For example, at each phase line, the forward companies might reestablish contact, reorganize, and continue clearing (Figure 2-13).

2205. Battalion Considerations. Some of the concerns for the commander to consider include:

a. Command and Control. CP location can be critical due to line of sight restrictions in the urban environment. Consideration should be given to maintaining a highly mobile CP with links to airborne retrans stations.

b. Intelligence. Organic collection capability resides primarily from units in contact with the enemy. The Scout Sniper Platoon also provides information but they may have obstructed visibility in the urban environment. Commanders may chose to employ them more often in a sniper role. Links to higher command intelligence sources may be key to maintaining current information on the enemy. The same line of sight restrictions encountered with command and contol will also restrict intelligence dissemination. Commanders may have to rely on messengers.

c. Manuever. Maneuver of mechanized forces in an urban environment may be restricted to the size, width, and load capabilities of the streets. Consideration must be given to plan for alternate routes for any attack. Significant manmade obstacles can be encountered throughout the battlespace. Additionally, the reserve is normally kept closer to the assault force. Movement of forces by aviation is less restrictive, but may also be vulnerable to surface to air missiles and small arms threat. Availability of LZs may also be a limiting factor.

d. Fires. A key consideration on the use of fires will be the ROE and the collateral damage that can result by their use. Rubbling of buildings can complicate an already restricted maneuver capability. Weapons effects can be more pronounced and can create a hazard for friendly forces. IPB determines potential firing positions for indirect fire assets that accounts

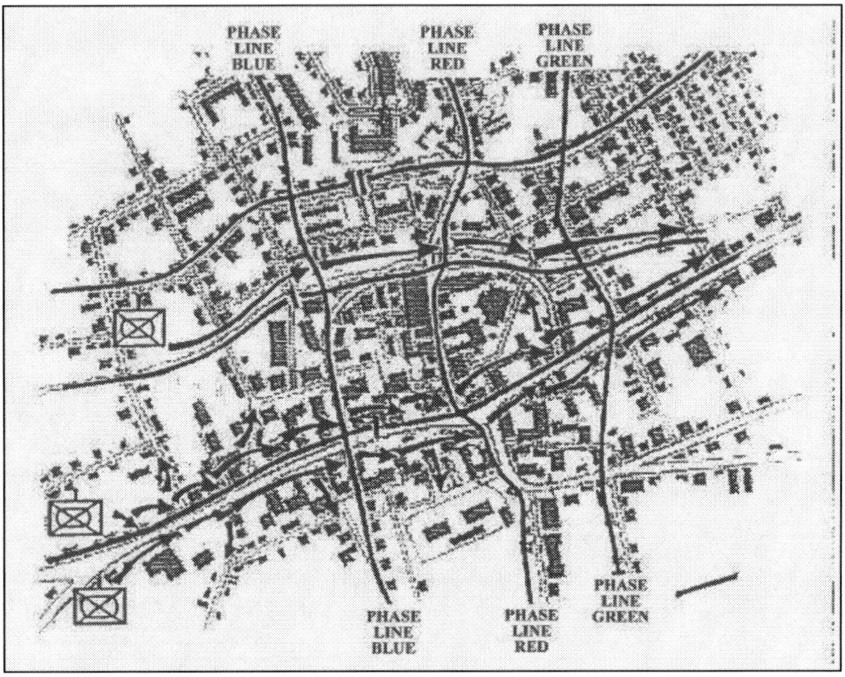

Figure 2-13. Clearing Along a Route

for the overhead masking by buildings and the ability to seat baseplates. Traditional indirect fire assets, such as artillery, may be used in the direct fire role.

e. Logistics. Historically in MOUT, the utilization of class I-V and VIII supplies is much greater. More vehicular assets should be allocated to the small unit level to transport ammunition and support the increased casualty evacuation requirements.

f. Force Protection. Lines of communication require extensive flank security in the urban environment. The commander must determine where his critical security requirements are and concentrate his force protection efforts in those areas while economizing in other areas. Manpower may further be depleted by security requirements for refugees and EPWs.

Section III

Rifle Company Operations

2301. Overview. This section addresses some of the operations the infantry battalion could assign to the rifle company in MOUT. For our focus, the rifle company is a task-organized and/or reinforced unit. Company operations will normally be conducted as part of a larger battalion operation.

2302. Attacking within a Built-up Area. Attacking within a built-up area is a task the rifle company may be assigned (Figure 2-14). This task involves the isolation, attack, and clearing of a specified block of structures. It is usually conducted in conjunction with adjacent friendly attacks or security operations. (See Appendix A.)

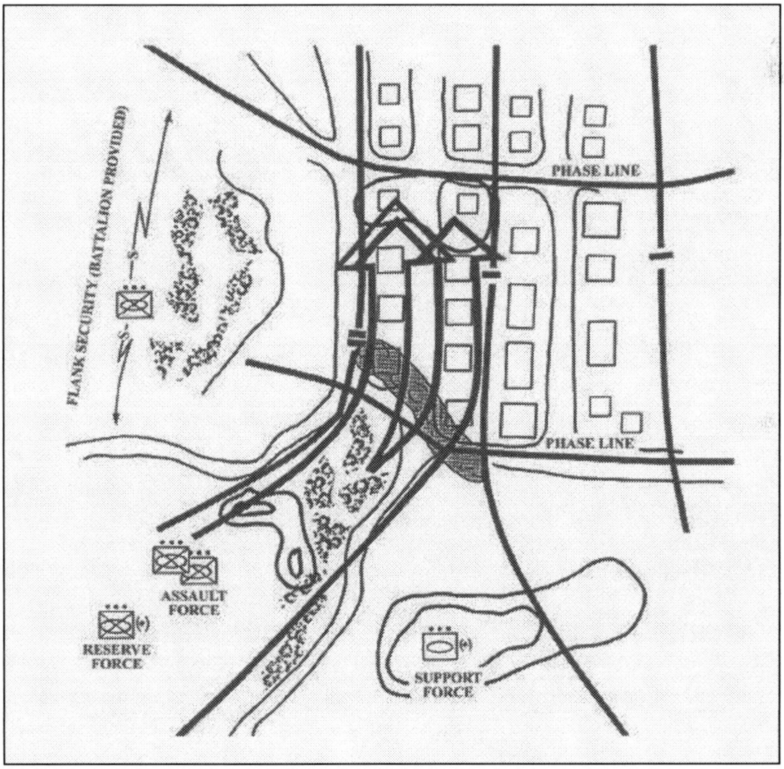

Figure 2-14. Company Attack of a Block

Before an assault, the company commander employs smoke to conceal the movements of assaulting platoons. He provides suppression fires to their flanks with direct-fire weapons. The battalion may provide additional security forces for flank protection.

(1) Concealed by smoke and supported by direct-fire weapons, an assaulting platoon attacks the first isolated building. The platoon must close on the building quickly while the enemy is still stunned by supporting fire. The company commander must closely coordinate the assault with supporting fire so that the fire is shifted at the last possible moment.

(2) The platoons continue to seize each building until the block is cleared. (See Appendix C for a detailed discussion on clearing of buildings.) After seizing the block, the company consolidates and reorganizes to repel a counterattack or to continue the attack.

The company commander directs platoon attacks on a designated front supported by direct and indirect fires. Success depends on isolating the enemy positions (which often become platoon objectives), suppressing enemy weapons, securing a foothold in the block, and systematically clearing the block's buildings room by room.

During the attack of a block, the company is usually task organized usually into four elements: *assault*, *support*, *security,* and *reserve*. The assault force is organized to conduct the actual assault of the objective. The support force provides direct- and/or indirect-fire support. The security force isolates the objective area and provides security. The reserve is organized, equipped, and positioned to exploit success. For example:

(1) The rifle company fighting in the outskirts of a city might organize as follows:

- Assault force: two platoons reinforced with the assault section
- Support force: armored vehicles, mechanized squad, and other direct- and indirect-fire weapons
- Security force: one reinforced squad.
- Reserve: mechanized platoon (-) reinforced.

(2) In a core or core periphery area (See Chapter 1), that same company might be organized as follows:

- Assault force: one rifle platoon, reinforced with engineers, SMAWs, Dragons, and a tank section (two tanks) per platoon
- Support force: One squad with direct- and indirect-fire weapons

.3

- Security force: one rifle platoon (-)

- Reserve: one rifle platoon, mechanized.

c. Armored vehicles, machine guns, and other direct-fire weapons under the company commander's control are assigned to isolate the objective and provide suppressive fires from covered positions. These weapons should not be fired for prolonged periods from one position. The gunners should use a series of positions and displace from one to another to gain better fields of fire and to avoid being targeted by the enemy. Direct-fire support tasks are assigned as follows:

- Machine guns fire along streets and into windows, doors, alleys, etc.; however, innovative use of direct fire weapons can be key in MOUT

- Tanks, ATGMs, and AT4s fire at enemy tanks and other armored vehicles.

- SMAWs can be used to make entrances in buildings and destroy identified bunker/fighting positions.

- Tanks can fire at targets protected by walls, make entrances in buildings, and engage enemy armored vehicles with their main gun and other targets with their machine guns.

- Riflemen engage targets of opportunity.

2303. Attack of an Enemy Outpost. A motorized or mechanized force may encounter enemy outposts, checkpoints, or strongpoints. In a hasty attack on an enemy outpost, the company commander makes a quick assessment based on the mission and situation, and then reacts in accordance with the commander's intent. If not able to bypass an enemy outpost, he may have to attack and reduce it.

The company occupies positions of advantage. The tanks or other armored vehicles support the dismounted infantry in the attack by assuming support-by-fire positions from which they can fire on the outpost, prevent the enemy from escaping, and destroy any reinforcements.

The infantry platoons dismount and move to engage the enemy (Figure 2-15). They should attack from a covered route to hit the outpost at a vulnerable point. As the platoons approach the outpost, smoke may be employed to screen their movement, and supporting fires are shifted. After platoons close on the outpost, they clear the buildings quickly and consolidate. The company is then ready to continue operations.

2304. Seizure of a Traffic Circle or Major Intersection. A company may have to seize a traffic circle or major road or traffic intersection (Figure 2-16 on page 2-34). This operation consists of seizing the buildings adjacent to the traffic circle or intersection while bringing direct-fire weapons into position to fire on likely avenues of approach.

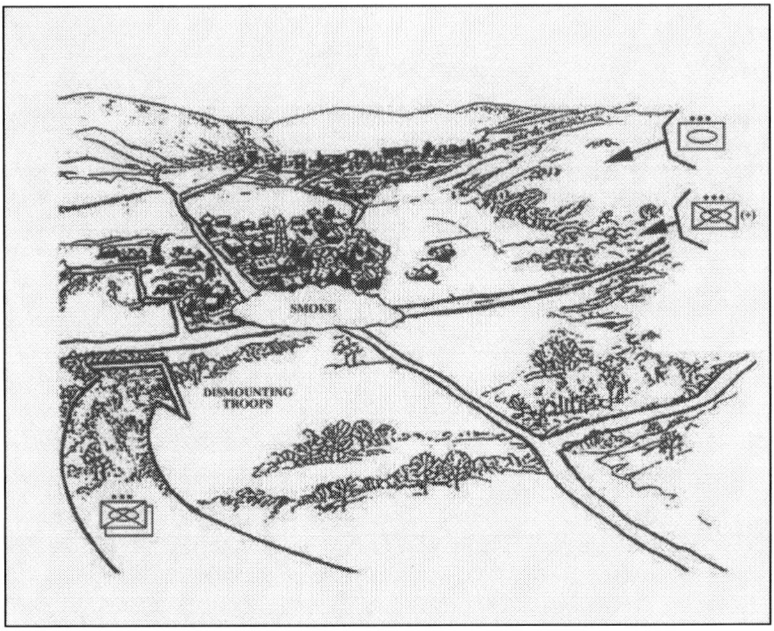

Figure 2-15. Hasty Attack of an Enemy Outpost

The company commander takes the following actions to seize a traffic circle or major intersection:

- Isolate the objective area.
- Seize and clear the buildings overlooking the traffic circle or major intersection under cover of direct-fire weapons.
- Secure the flanks against enemy attack.
- Consolidate and prepare for counterattack.

2305. Seizure of Key Terrain. The seizure of certain key terrain such as a bridge, intersection, facility, or structure may be critical to the success of an operation. Seizing key terrain for friendly use is a possible mission for a rifle company.

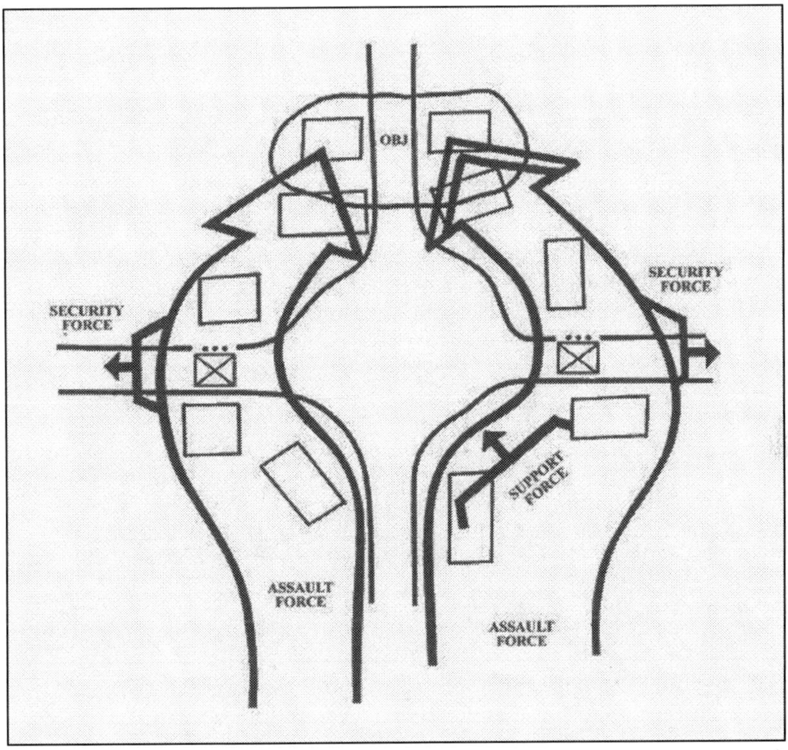

Figure 2-16. Seizure of a Traffic Circle or Major Intersection

To seize a bridge, a company should:

- Clear the buildings on the *near bank* that provide clear fields of fire for supporting weapons
- Suppress enemy weapons on the *far bank*
- Use screening smoke to obscure enemy observation
- Clear the buildings on the *far bank* that dominate the bridge
- Secure a perimeter around the bridge so that the engineers can clear obstacles or remove demolitions from the bridge.

The commander determines which buildings dominate the approaches to the bridge. Buildings that permit him to employ AT4s, ATGMs, machine guns, and riflemen are cleared. Supporting fire prevents the enemy from reinforcing units on the far bank. Enemy positions that can fire directly onto the bridge should be destroyed first. The company's forward observer (FO) should plan artillery and mortar fires to suppress the enemy's positions on the far bank.

The objectives of the assaulting platoons are buildings that dominate the far side of the bridge. One or two platoons are supported by the remainder of the company and assault across the bridge. Supporting fires are shifted and the infantry start clearing buildings. The assault continues until all the buildings in the objective area are cleared.

At this point, the engineers clear the bridge and its approaches of all mines, demolitions, and obstacles. The company commander may expand his perimeter to prepare for counterattack. Once the bridge is cleared, the armored vehicles are brought across to the far bank (Figure 2-17).

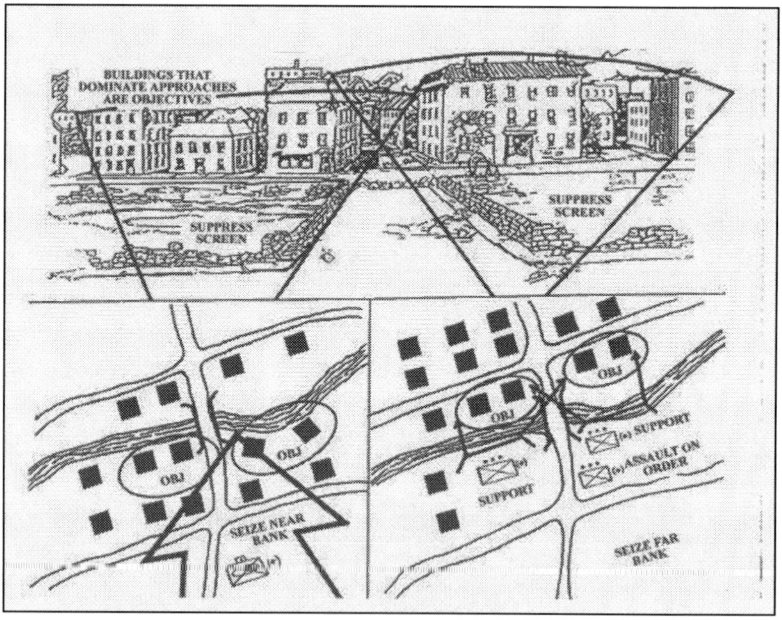

Figure 2-17. Seizure of a Bridge

.3

2306. Movement to Contact. A rifle company may conduct a movement to contact along a road. Similarly, the company may have to reconnoiter such a route to prepare for a battalion attack. This type of mission can best be accomplished by a mechanized company with appropriate attachments. (Figure 2-18 on page 2-36).

Figure 2-18. Movement to Contact Along a Highway (Commercial Ribbon)

The speed of movement depends on the mission, terrain, and enemy situation. In open areas where rapid movement is possible, tanks should lead. In close terrain, the infantry may lead while being overwatched by the tanks. Artillery fire should be planned along the route. Engineers may accompany the lead platoon on the main route to help clear obstacles and mines as they may be encountered. Between key terrain, when contact is not likely, the company remains on their vehicles. At key terrain or when enemy contact is likely, the company moves on foot to clear enemy positions or to secure the key terrain. Armored vehicles support the infantry.

Section IV

Rifle Platoon Operations

2401. Overview. This section discusses some of the various tactical missions that might be assigned to a rifle platoon.

2402. Attack of a Building. One offensive mission assigned to a rifle platoon in MOUT is the attack of a building. The attack requires isolating the building and entering the building at the least-defended point or through a breached hole (Figure 2-19 on page 2-38). When possible, clear the building from top to bottom. Close coordination between the assault and support forces is required.

When attacking a building, a platoon should be organized into an assault force, a support force, and a security force. The assault force provides the forces that actually conduct the assault. A support force provides supporting fires and efforts that assist the assault force. A security force covers the platoon's flanks and rear. (See Appendix A.) The platoon may be augmented by armored vehicles and from the company's weapons platoon or the battalions weapons company. If a reserve is required it should be provided by the company.

The assault consists of three steps:

(1) Isolate the building and deny the enemy the ability to maneuver or reinforce.

(2) Enter the building to secure a foothold for further action.

(3) Clear the building systematically.

Clearing is performed by individual squads and teams, which leapfrog as rooms and floors are secured. Platoons tasked to clear buildings may be reinforced with combat engineers.

2403. Movement Down a Street. When moving in urban terrain, a rifle platoon normally follows the same principles of movement as in other environments. However, some movement techniques must be modified to adjust to the unique dangers encountered in an urban environment. (See Appendix A, Section VII.) This section describes the movement down the street of the lead platoon of a company.

The speed of movement depends on the enemy situation. In outlying or lightly defended areas, a mechanized platoon may proceed along the street mounted but send dismounted infantry forward to reconnoiter danger areas. In the center of a built-up area the platoon dismounts to move on foot with two squads leading—one on each side of the road, using all available cover. They move through the buildings, if feasible, to avoid exposure on the streets. The squads coordinate to provide each other with all-around observation and mutual support.

Figure 2-19. Attack of a Building

The platoon should organize and practice to move as two forces: a *maneuver* force and an overwatch force. The maneuver force (normally of one squad on narrow streets or two squads on wide streets) moves forward, scouts danger areas, and closes with the enemy. An *overwatch* force (the remainder of the platoon and its attached or supporting crew-served weapons), moves behind the maneuver force, secures the platoon's flanks and rear, and provides direct-fire support when needed. These two forces, can exchange roles from time to time to avoid overfatigue of the maneuver force (Figure 2-20).

2404. Utilization of the Reserve. In the offense, a rifle platoon may be assigned as a reserve. The reserve can be used to reinforce and attack or exploit success. Reserves are normally held at the company level.

Figure 2-20. Movement Down a Street

In outlying areas, where the terrain is relatively open, a tank platoon reinforced with mechanized assets (tank-heavy task force) may be tasked as the reserve. However, if inside the urban area, a mechanized platoon reinforced with tanks (mech-heavy task force) would most probably be directed to provide the reserve. Regardless of what unit is designated as the reserve, the following items must be considered for their employment in an urban environment:

- The reserve should plan to use a different axis of advance
- Route selection for the reserve may be more constructive since it may have to use building, side streets, sewers, etc.
- Reserves are kept closer to assault forces in MOUT than on open terrain
- Engineers should normally be attached
- Reserves should maintain a high degree of mobility

Chapter 3

Defensive Operations

The essence of defensive tactics is to place the enemy into a position that permits his destruction through the intelligent use of terrain and firepower.

3001. Introduction. In urban combat, the defender possesses key advantages over the attacker. The defender can shape the battlespace to his advantage by maximizing the natural restrictions and obstacles found in the urban environment. ROE for the attacker can add to the defender's advantage by placing restrictions on the application of force by the attacker. Knowledge of the terrain and time available for preparing defensive positions are advantages which may enable the defender to successfully resist a numerically superior force. A spirited and stubborn defense may persuade an attacker to abandon the attack. In some battles, urban defenders can be credited with repelling or decisively delaying an attacker which ultimately influenced the course of the war in their favor.

Marines may be called upon to defend a port city, an embassy located on or near a coastline, a transportation link, or the capital of a foreign government. They may also be tasked to defend from an urban area in an economy-of-force role in order to support operations elsewhere. In combat, offensive and defensive operations are inseparable. Marines need to be capable of conducting both offensive and defensive operations in an urban environment.

The fundamentals of defense do not change in an urban environment. The defenders of a city, however, usually have detailed knowledge of the terrain. This will allow them to establish an extensive defensive network that is designed to force an attacker to expend exorbitant amounts of time, supplies, equipment, and manpower. Commanders need to recognize both the advantages and disadvantages of defensive operations in an urban environment.

3002. Decision to Defend

a. Reasons for Defending Built-Up Areas. Historically, the following are some of the reasons for defending built-up areas:

(1) Denial of Important Strategy/Political Objectives. Capitals and cultural centers can be defended strictly for psychological or national morale purposes even if they do not offer a tactical advantage to the defender. Defending a city can cause an attacker to commit a significant amount of his forces which reduces his capability to attack elsewhere. The defense of a city can delay the overall offensive capability of the attacker.

(2) Retention of Key Economic Centers. In many countries, the entire nation's economic well-being may be tied to a few key cities. These key cities usually contain the country's primary industrial, transportation, and communications base. The capture of these key centers could result in the overthrow of the current government, or deny that

.3

government the ability to adequately support combat operations against enemy regular or insurgent forces.

(3) Control of Avenues of Approach. Most avenues of approach to large cities are straddled by small towns every few kilometers which must be controlled by defending forces. These areas can be used as battle positions or strongpoints to control the avenues of approach. For the attacker to utilize these avenues of approach he must sacrifice speed and expend resources. The defense of these cities or towns may constitute a large part of the overall defense of a city.

(4) Economy of Force. The tactical advantages provided by urbanized terrain allows the defender to engage a numerically superior force. The defender can conserve the bulk of his combat power so that it is available for use in other operations.

(5) Concealment of Forces. Reconnaissance and combat identification is more difficult in an urban environment. CPs, reserves, CSS complexes, and combat forces emplaced in built-up areas are much harder to detect.

3003. Reasons for Not Defending Built-Up Areas. The commander considers the following reasons for deciding not to defend built-up areas.

a. Unnecessary to the Defensive or Offensive Plan. If the built-up area is too far forward or back in a unit's defensive sector, is isolated, or is not astride an enemy's expected avenue of approach, the commander may choose not to defend it.

b. Bypassable. If the nature of nearby terrain allows the enemy to bypass the city, then it will not make a good blocking position. Some built-up areas, mainly smaller ones, are easily bypassed by existing main road and highway systems. A built-up area that can be easily bypassed normally will be, thereby effectively isolating that area's defenders from the remainder of their forces.

c. Inadequate Structures for Defense. Extensive areas of lightly built or flammable structures offer little protection to the defender. Built-up areas near flammable or hazardous industrial areas, such as refineries or chemical plants, also may not lend themselves to successful defense.

d. Adjacent Dominating Terrain. If the built-up area is small and dominated by close, prominent terrain, the commander may choose to defend on that terrain rather than in the built-up area.

e. Better Fields of Fire Elsewhere. The commander may choose to base all or part of his defense on the better fields of fire that exist outside built-up areas. An example would be an armor-heavy force defending in sectors from multiple, small, built-up areas surrounded by open or farm-type areas.

f. Open City. See chapter 2.

Section I

Defensive Planning

3101. Commander's Estimate. As in any other defensive operation, planning begins with the commander's estimate of the situation. The complxities involved with the defense of a built-up area require detailed and centralized planning. This section details the general procedures and principles for planning and organizing the defense of an urban area.

3102. METT-T

a. Mission. Mission analysis is the first step in conducting the estimate of the situation. When conducting mission analysis commanders and their staffs should analyze their assigned tasks and commander's intent to determine the scope of the defense. For example:

- Do I need to control every building or can I control the frontage from key buildings?

- Are certain parts of the urban area more significant? (i.e. main transportation centers, communication nodes government buildings)

b. Enemy. The commander must focus on the enemy and build his defense to concentrate his strength against enemy weakness. The IPB process is used to analyze the terrain and enemy capabilities. (See FM 34-130.) For example:

- If the attacker relies heavily on aviation for mobility, the defender may defend potential LZs and emplace manpad teams to counter the air threat

- If the enemy relies heavily on mechanized movement, the defender may employ additional countermobility assets.

c. Terrain and Weather

(1) Terrain. Analysis of all manmade and natural terrain features is critical for planning a defense in built-up terrain. A defending commander has the advantage of occupying the terrain on which the fight will occur, and therefore usually knows it better than the attacker.

The defender must make best use of the battlespace that he controls: ground level (streets and parks), above ground (buildings), and below ground (subways and sewers). Defenders must be aware that built-up areas are normally interdependent. A reservoir supplying water to a city may be many miles away, electricity can come from remote power stations, and the control points for these services and others, such as communications lines, may be vulnerable.

.3

(a) K: Key Terrain. Primary examples of key terrain in the urban environment are ports, airfields, power grids, communication nodes, bridges, government building complexes, or parks. We identify key terrain so that we can better select our defensive positions and also better determine the enemy's objectives.

1 **Villages.** Villages often dominate high-speed avenues of approach. If the buildings are constructed of stone, brick, or concrete, providing good protection against both direct and indirect fires, a formidable defense can be constructed (Figure 3-1). Company-sized battle positions can be established in these small villages to block approaches into the larger urban area.

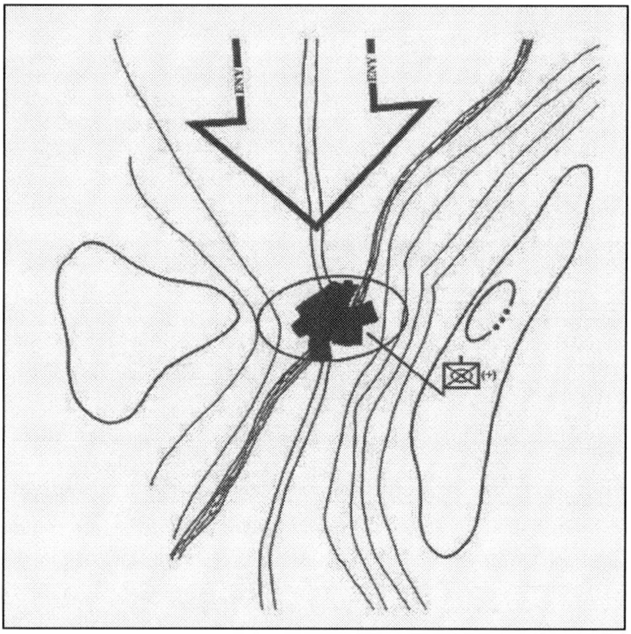

Figure 3-1. Village Defense (Chokepoint)

2 **Strip Areas.** Strip areas are built along roads or down valleys between towns and villages. They afford the defender the same advantages as villages. If effective fields of fire are available, a unit can occupy a few key positions within the strip to create an effective defense. (Figures 3-2A and 3-2B).

3 **Towns and Cities.** When a small town or city is positioned on a key enemy avenue of approach, a force can defend from the urban area while other forces

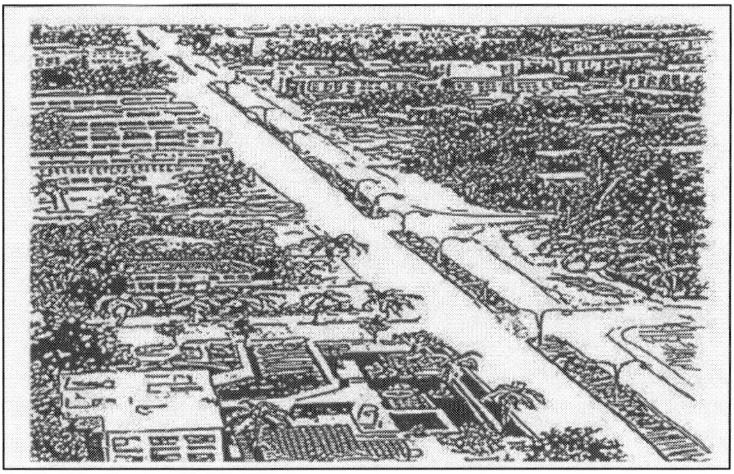

deny the enemy the ability to bypass. The town or city can be made into a strongpoint defense. (Figures 3-3A and 3-3B on page 3-6).

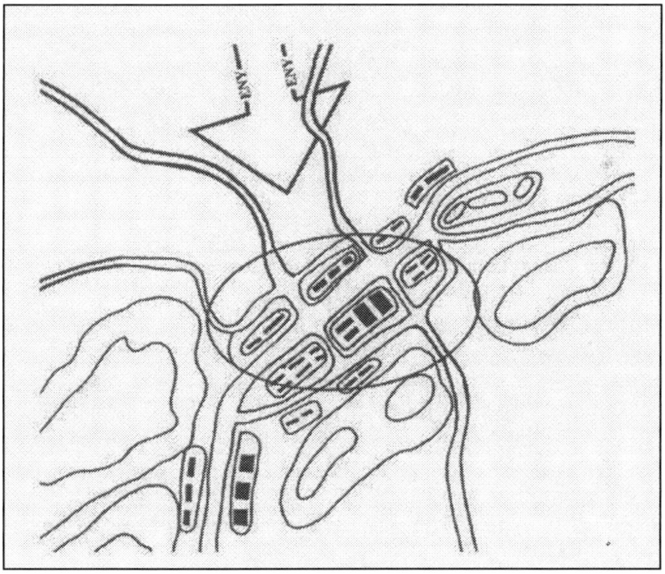

Figure 3-2A. Strip Area

.3

Figure 3-2B. Strip Area Defense

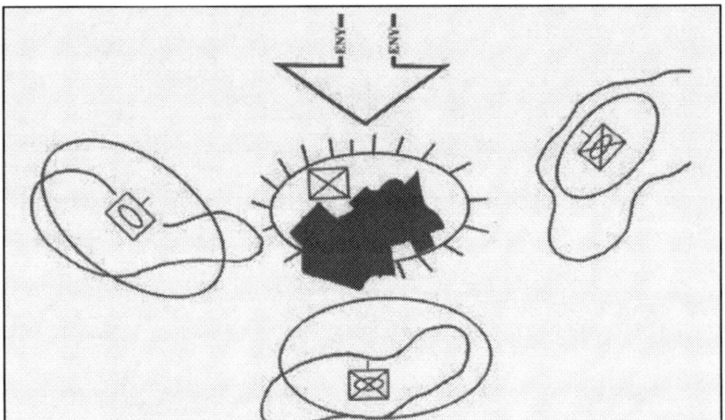

Figure 3-3A. Towns and Cities

Figure 3-3B. Defense of a Town or City

4 **Large Built-Up Areas.** Large buildings that are close together may require more forces and smaller defensive sectors than natural open terrain. The density of buildings, rubble, and street patterns will dictate the frontage of the unit.

(b) **O: Observation and Fields of Fire.** Attackers must generally advance by crossing streets and open areas between buildings where they are exposed to fires from concealed positions.

1 **Weapons and Range.** The defender must position weapons to obtain maximum effect and mutual support. FACs and FOs should be placed well above street level for increased observation. Fires and final protective fires should be preregistered on the most likely approaches.

2 **Limited Visibility.** The commander can expect the attacker to use periods of limited visibility to sustain or gain momentum. (See Appendix H for more detailed discussion.) The commander considers the following during periods of limited visibility:

- Unoccupied areas that can be observed and covered by fire during daylight may have to be occupied or patrolled at night.

- Remote sensors should be emplaced in dead space.

- Mines, noise-making devices, tanglefoot, and OPs should be positioned on avenues of approach to provide early warning.

- Artificial illumination to include street lamps, stadium lights, and flares should be integrated into the overall defense.

- When defenses are probed, indirect-fire weapons, grenade launchers, and hand grenades can be used to avoid disclosure of defensive position locations.

(c) **C: Cover and Concealment.** Although walls, floors, and ceilings can be used as protective cover, Marines should always improve these positions by using sandbags, rubble, etc. A defender can reduce his exposure by establishing routes between positions using:

- prepared breaches through buildings

- reconnoitered and marked underground systems
- trenches and sewage systems
- concealment offered by smoke and darkness to cross open areas.

(d) O: Obstacles. A city is an obstacle because it canalizes and impedes an attack. Avenues of approach should be blocked by obstacles and covered by fire. (See Paragraphs 3404.d., 3404.f., 3405.e., and Appendix F.)

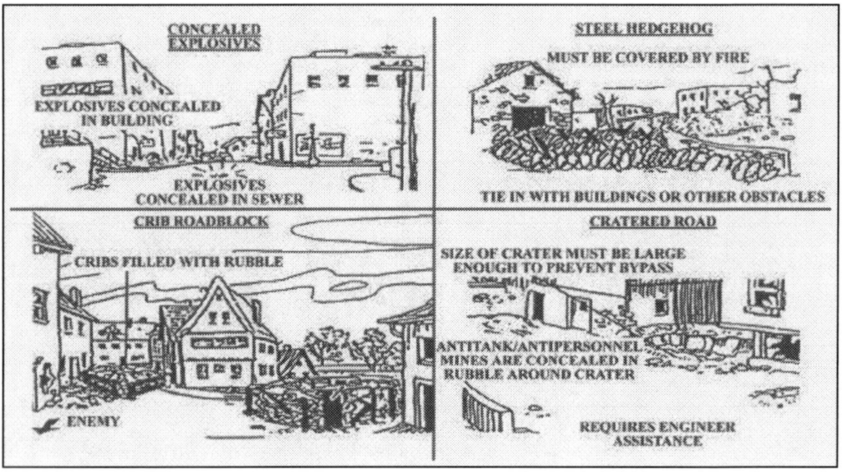

(e) A: Avenues of Approach. Avenues of approach (surface and subsurface) should be denied to the enemy in keeping with the overall defensive plan (Figure 3-4).

(2) Weather. The effects of weather as discussed in Chapter 2, Section II, pertain equally to defensive operations.

D. Troops and Support Available. The defender of a built-up area has a terrain advantage and can resist the attacker with much smaller forces. The defender stands a greater chance of success when he fights with combined arms and task organizes his forces. The discussion of troops and support available in Chapter 2, Section II, also pertains to defensive operations.

Figure 3-4. Denial of Avenues of Approach

e. Time Available. The commander establishes priorities of work, which depend on the time available. Many tasks can be accomplished at the same time, but priorities for work are accomplished according to the commander's order. An example priority-of-work sequence follows:

- Establish security and communications
- Assign sectors of responsibility, final protective fires
- Clear fields of fire
- Select and prepare initial fighting positions
- Establish and mark routes between positions (including alternate and supplementary routes)
- Emplace obstacles and mines
- Improve fighting positions.

3103. Organizing for a Defense

a. Types of Defense. The defense of a built-up area should be organized around key terrain features that preserve the integrity of the defense and provide the defender ease of movement. There are two types of defense: position and mobile. (See MCWP 3-1, *Ground Combat Operations* [under development] for details on defensive operations.) Most defenses will include a combination of position and mobile defense. The type of defense chosen is predicated on commander's intent and METT-T.

(1) Position Defense. The type of defense in which the bulk of the defending force is disposed in selected tactical localities where the decisive battle is to be fought. Principal reliance is placed on the ability of the forces in the defended localities to maintain their positions and to control the terrain between them. The reserve is used to add depth, to block, or restore the battle position by counterattack. (Joint Pub 1-02.) The position defense is usually characterized by defending key terrain. In the urban environment, this equates to the physical occupation of key public buildings, avenues of approach, transportation centers, industrial parks, etc. This defense focuses combat power to repel the attacker's advance and/or penetration while retaining the terrain. This type of defense is common for battalion-sized units and below.

The commander conducting a position defense organizes forces utilizes the same fundamentals that apply to any defense. Forces can be assigned sectors to defend and/or battle positions or strongpoints based on the commander's analysis of METT-T. Reserves are identified and located to respond quickly to exploit success.

(2) Mobile Defense. Defense of an area or position in which maneuver is used with organization of fire and utilization of terrain to seize the initiative from the enemy. (Joint Pub 1-02.) The mobile defense is characterized by its focus on the enemy, not on terrain. Terrain is still important. However, this defense is organized with a mobility capability to exploit enemy weakness. In order to conduct the mobile defense, the defender must have enough forces to defend the built-up area, and to rapidly conduct an enveloping or flanking attack.

As in the position defense, the sector, battle position, strongpoint, or any combinations thereof are also used in shaping the battlespace by canalizing the enemy into a designated location where he can be destroyed.

b. Defensive Options

(1) Defense Outside the City. The commander may defend from outside the city. This option is often selected when the terrain surrounding an urban area offers an advantage to a defender and when his defending force is of sufficient strength (See Cherbourg paragraph in Chapter 1). The presence of a large, friendly, civilian population, or buildings of historic significance or specific cultural value, would also contribute to deciding on an outside defense. A situation in which the built-up area has a preponderance of building structures that offer little protection to defenders or terrain that favors the attacker.

A commander organizes his defenses around strongpoints selected on the outer edges of the urban area when these may be suitable structures for siting weapons to gain maximum firepower range. The organization of an outer-city defense is also heavily influenced by the number and type of weapons with which the hostile force is armed. A defense against a predominantly infantry threat would differ from that planned to defeat an armored threat. The decision to defend or concentrate assets solely on the outer edge of a city are weighed against the enemy's ability to mass fires, infiltrate, or bypass strongpoints.

(2) Defense Inside the City. The commander may decide that his best defense is to defend within the city. This option is usually chosen when hostile forces possess strong, accurate, long-range fires and the defender wishes to minimize their effectiveness (See Quang Tri City example in Chapter 1). The strength of the defending force relative to the attacker may also lead to this option as an economy-of-force measure.

Structures on the outer edge of an urban area that obstruct fields of fire or that are likely to be used by an attacker should be prepared for demolition or demolished ahead of time. All defensive positions should be supplemented or covered by direct-fire weapons. Indirect fires should be planned for most likely avenues of approach. Additionally, close air support should be planned and integrated with the defense.

Security forces are placed outside the defended area to gain early enemy contact, inflict maximum casualties, and attempt to deceive the enemy as to the location of the friendly positions.

(3) Defense of a Key Sector. This option is usually employed when only a portion of an urban area be held or when strength is insufficient to defend an entire area (See the Japanese defense of Manila in Chapter 1). This option is often employed in situations where the mission requires clearing and holding a key sector, or sectors, in order to facilitate other operations.

(4) Entrapment and Ambush. This option is preferred when the defender lacks requisite strength to man organized, in-depth, defensive positions (See the highly successful Egyptian defense of Suez City example in Chapter 1). It requires the preparation of a series of blocking and/or ambush positions along major avenues of approach through an urban area. The defending force uses mines and obstacles to block alternate or possible detour routes in order to canalize the attack force into prepared ambush sites. The defender employs security forces forward of the main battle area and establishes inner or outer defensive strongpoints with extensive obstacles designed to assist in canalize the enemy into kill zones.

The entrapment and ambush option requires detailed planning and rehearsal by all echelons. Alternate plans should be prepared in the event the ambush is only partially successful or if the enemy fails to enter prepared ambush routes. Alternate ambush positions or counterattacks are planned.

Ambush planning in an urban environment does not differ significantly from planning in other environments. However, the urban environment offers an abundance of kill zones that afford the enemy little or no opportunity to escape. Ambush positions are established in selected buildings in the kill zone which offer optimum fields of fire for all weapons. Indirect fires and close air support should be planned to obtain maximum effect in the kill zone. The kill zone should also be covered with targets for indirect fire and CAS. Entrances into buildings not used as firing points should be blocked to entrap hostile forces in the kill zones.

(5) Defense in Depth. Defense in depth should be incorporated into any defensive plan. Units should be given enough terrain to allow for primary, supplementary, and alternate fighting positions as well as successive battle positions or strongpoints. This allows enough room for multiple ambush/entrapment kill zones and the positioning of a reserve.

c. Defensive Organization. The defensive battlespace is divided into three areas: the *security area*, *main battle area*, and *rear area*. Depending on the mission of the unit, forces defending a built-up area may have missions in any one or more of these areas. (Figure 3-5 on page 3-12).

.3

(1) **Security Area.** The defensive battle begins with a combined-arms force conducting security operations well forward of the main battle area. Security area missions can include screen , guard, and cover. Based on their assigned mission, security forces use all

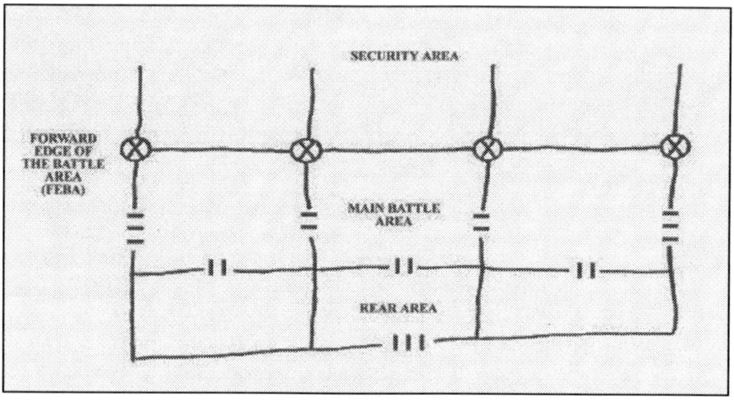

available combat power to gain time, destroy the enemy, or slow his momentum. The initial battle is fought using aviation assets and indirect fire weapons.

(a) **Value of Security Area Operations.** Security area operations complement to the defense by:

- Alerting the defense to the strength, location, and general direction of the enemy's main and supporting attacks
- Delaying enemy first-echelon detachments
- Initiating early engagement of enemy forces
- Deceiving the enemy as to the true location of the defender's main forces and main effort.

Figure 3-5. Organization of the Battlefield

(b) **Withdrawal of Security Area Forces.** The urban environment may complicate battle handover from the security force to the main battle area force. However, this transition must be accomplished smoothly to prevent the enemy from gaining momentum. The withdrawal of the security area forces must not result in an easing of pressure on the enemy.

(2) Main Battle Area. The decisive battle is fought in the main battle area. However, the

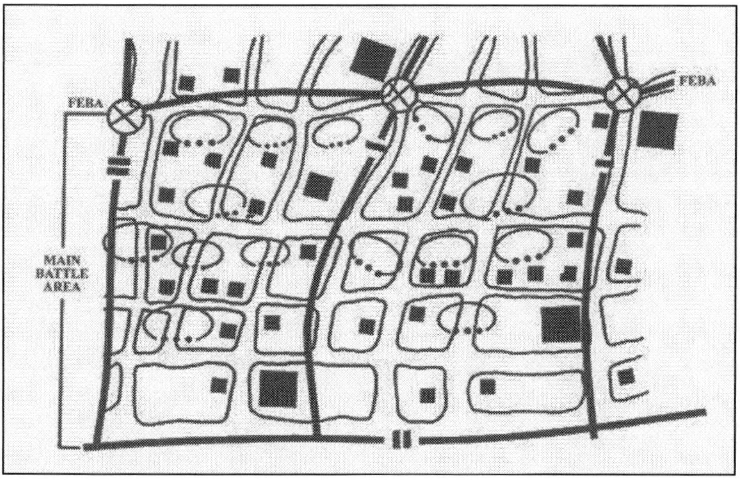

commander may deploy units on the forward edges of the city or in battle positions in depth. The defensive scheme should include forces to defend along the flanks. The commander normally employs security forces to the front to provide early warning and to deny the enemy intelligence on the unit's defensive dispositions (counterreconnaissance) (Figure 3-6).

(a) The size and location of battle positions depend on METT-T and the type of defense prepared. A sophisticated defense will often include a combination of both position and mobile defense but will always be planned in depth.

(b) A unit assigned battle positions on the forward edge of a city or town should:

- Provide early warning of the enemy's advance
- Engage the enemy at long range
- Deceive the enemy as to the true location of the defense.

Figure 3-6. Main Battle Area

(c) The defender employs all available fires to destroy and suppress the direct-fire weapons that support the ground attack. Depending on the concept of operations, weapons may engage targets at their maximum effective range or engage simultaneously for an ambush.

(d) As the attack develops, the actions of small-unit leaders become increasingly important. It is imperitive that all leaders understand their commander's intent.

(3) Rear Area. The rear area is located behind the main battle area and it contains the CSS and administrative support. Protection of these units is vital. Fortunately, the urban environment usually provides sufficient cover and concealment for CSS assets as well as covered and protected routes to the forward areas. Rear area operations are covered in detail in MCWP 3-41.1, *Rear Area Operations*.

3104. Warfighting Functions. The defensive planning sequence remains the same as in any other defensive operation. Defense planning considerations, based on METT-T, must take into account all the activities in each of the warfighting functions. The complexities involved in defense of a built-up area require detailed and centralized planning. Commanders and subordinate leaders must incorporate the following planning considerations for an urban environment when conducting a defensive operation.

a. Command and Control. The commander positions himself so that he can control the action. In an urban environment, this can be more difficult because of the close nature of fighting, prolific obstacles, poor visibility, and difficulty in communication.

Graphic control measures oriented on prominent terrain features are also used in planning and conducting combat in built-up areas. Prominent streets are ideal for use as phase lines and boundries. These and other control measures assist the commander in controlling manuever and fires throughout the battlespace.

COCs should be located in secure locations. (These can be inside or underground when possible.) (See Chapter 2, Section IV, Paragraph 2402.)

Commanders consider the effects of built-up areas on communications when they allocate time to establish communications. LOS limitations affect both visual and radio communications. Wire laid at street level is easily damaged by rubble and vehicle traffic. Also, the noise of urban combat is much louder than in other areas, making sound signals difficult to hear. Therefore, the time needed to establish an effective communications system may be greater than in more conventional terrain.

Wire is the primary means of communication for controlling the defense of a city and for enforcing security. However, wire can be compromised if interdicted by the enemy. Radio communication in built-up areas is normally degraded by structures and a high concentration of electrical power lines. Nonetheless, radio is an alternate means of communication. Messengers can be used as another means. Visual signals may also be used but are often not effective because of the screening effects of buildings, walls, and so forth. Signals must be planned, widely disseminated, and understood by all assigned and attached units. Commanders consider the following techniques when planning for communications:

(1) If possible, lay wire through buildings for maximum protection.

(2) Use existing telephone systems. (Remember, telephones are not secure.)

(3) Emplace radios and retransmission sites in buildings on top floors or atop buildings that are higher than surrounding structures. Ensure that antennas are placed on the side of buildings where the radio signal can be directed at friendly units. This will enhance LOS radio communications.

(4) Use messengers at all levels because they are the most reliable and secure means of communications.

b. Intelligence. The discussion in Chapter 2 pertains equally to a defensive operation. Intelligence gathering for defensive operations is not limited to studying only the enemy. Commanders must emphasize obtaining and using all intelligence. Commanders must emphasize the utility of the urban infrastructure to enhance his intelligence gathering and dissemination effort. The defender usually has the advantage of being familiar with the intricacies of the urban terrain that he is defending. For example, the:

- The sanitation or public works department will have knowledge of the subterranean sewer lines
- Civil engineers will have extensive knowledge of water and power distribution nodes
- Law enforcement representatives may have access to lists of people sympathetic to the attacker
- Local journalists may agree to use local sources to assist in intelligence gathering effort.

c. Manuever

(1) Positioning of Forces. Finding fighting positions in urban areas that provide both good fields of fire and cover is often difficult. The forward edges of a town usually offer the best fields of fire but can be easily targeted by enemy observation and supporting fire. These areas often contain residential buildings constructed of light material. Factories, civic buildings, and other heavy structures that provide adequate cover and are more suitable for a defense are more likely to be found deeper within the town, but they have limited fields of fire.

Because the forward edge of a town is the obvious position for the defender, it should generally be avoided. However, the defender may choose to use the edge of the town if the terrain limits the enemy's ability for engagement or if strongly constructed buildings that give defenders protection are available. The mission of such a force is to provide early warning of the enemy's advance, to engage the enemy at long range, and to deceive the enemy as to the true location of the defense. This force should withdraw in time to avoid decisive engagement. If there is limited observation from the forward edge, a force should be positioned on more favorable terrain to gain better observation and to engage the enemy at long range.

Reconnaissance by all defending elements should help select routes for use by defenders moving between positions. Movement is crucial in fighting in built-up areas. Early selection and marking of routes adds to the defender's advantages.

The commander covers probable landing and drop zones such as parks, stadiums, or large rooftops and heliports with obstacles or fire to prevent air assault. (See Appendix F.) In a large built-up area, a unit is given a sector to defend and normally establishes a series of defensive positions. Although mutual support between positions should be maintained, built-up terrain often allows the enemy to infiltrate between positions. Therefore, the defender must identify the following:

- Positions that allow suppressive fires on infiltrating routes
- Covered and concealed routes, such as subways and sewers, for friendly forces to use to move between positions
- Structures that dominate large areas
- Areas such as parks, boulevards, rivers, highways, and railroads where antiarmor weapons have longer fields of fire
- Firing positions for mortars
- Command locations that offer cover, concealment, and ease of command and control
- Protected storage areas for supplies.

Buildings that enhance the general plan of defense are selected. Mutual support between these positions is vital to prevent the attacker from maneuvering and outflanking the defensive position. Buildings chosen for occupation as defensive positions should:

- Offer good protection
- Have strong floors to keep the structure from collapsing under the weight of debris

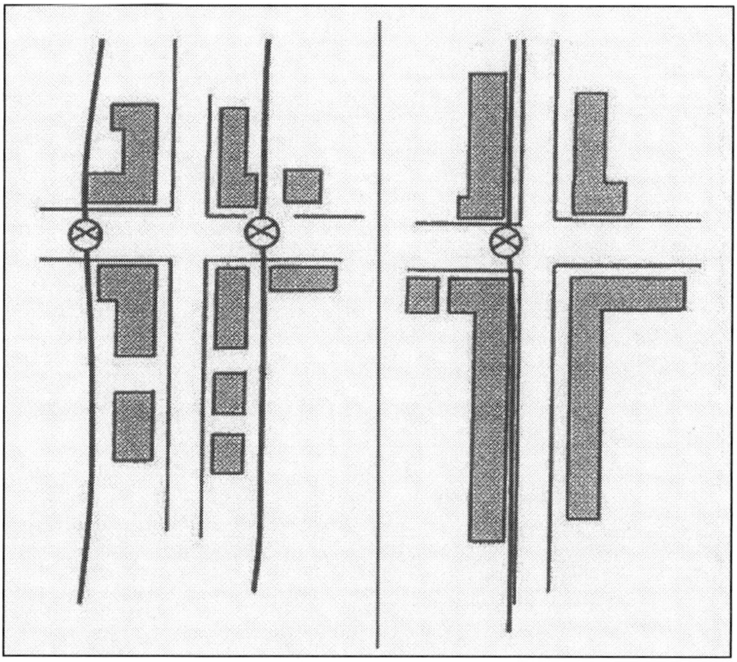

- Have thick walls
- Be constructed of nonflammable materials
- Be strategically located (corner buildings and prominent structures)
- Be adjacent to streets, alleys, vacant lots, and park sites (these buildings usually provide better fields of fire and are more easily tied in with other buildings)
- Provide for 360-degree protection and fields of fire
- Be stocked with adequate supplies as positions may become isolated and cut off for extended periods.

(2) Assign Sectors of Responsibility. Boundaries define sectors of responsibility. Sectors include areas where units may fire and maneuver without interference or coordination with other units. Responsibility for primary avenues of approach should never be split among units. In areas of semi-detached construction, where observation and movement are less restricted, boundaries should be established parallel to alleys or streets so that both sides of a street will be in a single sector. Where buildings present a solid front along streets, a boundary may have to be extended to one side of the street (Figure 3-7).

Figure 3-7. Sector Boundaries in Areas With Semi-detached and Solid-Front Construction

(3) Select and Prepare Initial Fighting Positions. Commanders select positions in depth. The unit should prepare positions as soon as they arrive and continue improving positions as long as they are occupied. Enemy infiltration or movement sometimes occurs between and behind friendly positions. Therefore, each position must be organized for all-around defense. The defender should also:

(a) Make minimum changes to the outside appearance of buildings where positions are located.

(b) Screen or block windows and other openings to keep the enemy from seeing in and tossing in hand grenades.

(c) Remove combustible material to limit the danger of fire. Fires are dangerous to defenders and create smoke that could conceal attacking forces. Defenders should store firefighting equipment (water, sand, etc.). The danger of fire also influences the type of ammunition used in the defense. Tracers or incendiary rounds should be avoided if the threat of fire exists.

(d) Turn off electricity and gas. Both propane and natural gas are explosive. Natural gas is also poisonous and is not filtered out by a protective mask. Propane gas, although not poisonous, is heavier than air. If it leaks into an enclosed area, it displaces the oxygen and can cause suffocation. Gas mains and electricity should be shut off at the main distribution nodes.

(e) Battle positions should be located so as not to establish a pattern. The unit should avoid obvious firing locations like church steeples.

(f) The unit should rehearse movements between positions.

(g) Camouflage positions.

(h) Reinforce positions with available materials, such as beds, furniture, etc.

(i) Block stairwells and doors with wire or other material to prevent or delay enemy movement. Create holes between floors and rooms to allow for covered movement within a building.

(j) Prepare range cards, fire plans, and sector sketches.

(k) Emplace machine guns in basements windows, where the guns can provide grazing fires. When basements are not used, access to them should be sealed to prevent enemy entry.

(l) Stockpile V (ammunition) and class VIII (medical supplies) items.

(4) Clear Fields of Fire. Commanders may need to rubble certain buildings and structures to provide greater protection and fields of fire. If the ceiling of a lower story room can support the weight of the rubble, collapsing the top floor of a building before the battle starts may afford better protection from indirect fires. Rubbling an entire building can increase the fields of fire and create an obstacle to enemy movement. Rubbling buildings too soon (or rubbling too many) may give away defensive locations and may destroy cover from direct fire. Rubbled buildings may also interfere with planned routes of withdrawal or counterattack.

(5) Improve Fighting Positions. When time permits, all positions, including supplementary and alternate positions, should be reinforced with sandbags and provided

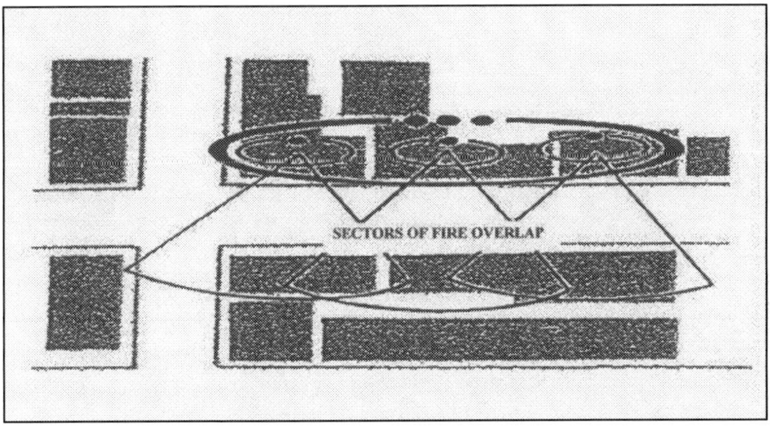

with overhead cover. Obtain support from attached engineers. (See Appendix D).

(6) Employment of Infantry

(a) The infantry battalion is normally assigned a sector to defend. However, depending on METT-T, it may be assigned to defend from a battle position or strongpoint. The battalion will usually assign its companies to either a battle position, a strongpoint, or a sector. These battle positions, strongpoints, or sectors are placed along avenues of approach to block or restrict the enemy's movement.

Depending on the type of built-up area, a company may be employed on the forward edge of the flanks of the area. This forces the enemy to deploy early without decisive engagement because it deceives the enemy as to the true location of the main defense. Other companies may then be employed in a series of battle positions and/or strongpoints in the center of the city or town. In all cases, mutual support between positions is vital. Companies and platoons should have designated alternate and supplementary positions.

(b) Once the rifle platoon commander has received his battle position or sector, he then selects the positions for his squads and crew-served weapons. Squad positions within the built-up area may be separated by rooms within buildings or be deployed in different buildings. Squad positions must be mutually supporting and allow for overlapping sectors of fire (Figure 3-8).

Figure 3-8. Sectors of Fire

(7) Employment of the Reserve. The commander's defensive plan always considers the employment of a reserve. The reserve force is organized and prepared to exploit success, to counterattack to regain key positions, to block enemy penetrations, to protect the flanks, or to assist by fire in the disengagement and withdrawal from endangered positions. For combat in a built-up area, a reserve force:

- Must be as mobile as possible
- May be a mechanized platoon, a platoon (-), or a squad at mechanized company level or a company, a company (-), or a platoon at battalion level
- May be supported by tanks or other armored vehicles
- Should be centrally located to facilitate manuever to threatened areas.

(8) Employment of Armored Vehicles. The commander plans to employ armored vehicles to take advantage of their long-range fires, heavy machine gun fires, self-generating smoke, and mobility. However, built-up areas restrict the mobility of armored vehicles and make them vulnerable to the antiarmor weapons of the enemy. Both the LAV and AAV are lightly armored and can be penetrated by heavy machine gun and antiarmor fires. Armored vehicles provide the commander with a mobile force to respond quickly to enemy threats on different avenues of approach.

(a) When armored vehicles are employed in the defense of a city, infantry should be positioned to provide them with security against close antitank fires and to detect targets for the armored vehicles. Armored vehicles should be assigned *primary*, *alternate*, and *supplementary* positions, as well as *primary* and *alternate* sectors.

(b) Armored vehicles should be located on likely avenues of approach to take advantage of their long-range fires. They may be:

- Positioned on the edge of the city in mutually supporting positions
- Positioned on key terrain on the flanks of towns and villages
- Used to cover barricades and obstacles by fire
- Placed with the reserve.

(c) Armored vehicles are normally employed as a platoon. However, sections of armored vehicles may be employed with rifle platoons or squads. This provides the armored vehicles with the close security of the infantry.

(9) Employment of Helicopterborne Forces

(a) Helicopterborne Task Force in the Defense. A helicopterborne force can defend against an infantry-heavy threat by utilizing its mobility to achieve a maneuver advantage. This force can be key in the execution of a mobile defense in both the security area and main battle area.

(b) Security Area. The helicopterborne force may be able to conduct security force operations for a larger force. Helicopters can position combat power and combat support quickly during rapidly changing situations.

(c) Main Battle Area. The mobility advantage that the helicopterborne force has over enemy infantry-heavy units may allow it to defend in greater depth in a large city. The helicopterborne force focuses on the destruction of advancing enemy forces and fights a series of battles in depth, it can attack the enemy from the front, flanks, and rear. Battle positions are selected and prepared throughout the main battle area along likely avenues of approach. Primary and alternate LZs and pick-up zones should be selected.

d. Fires. Fire support in the defense is as important as it is in the offense and has many of the same characteristics. The proximity of buildings to targets, minimum range restrictions, and repositioning requirements are all factors in the utilization of fire support in the defense. Additionally indirect fires are planned on top of and immediately around defensive positions for close support. Some of the indirect fire support considerations for the defense include:

(1) Historically, artillery fire has been used in some unique and innovative ways in an urban environment. Artillery fire support may be used in the direct- or indirect-fire mode. Artillery fire should be used to:

- Suppress and blind enemy overwatch elements

.3

- Disrupt or destroy an assault
- Provide counterfire
- Support counterattacks
- Provide direct fire when necessary
- Provide illumination in hours of darkness.

(2) Mortars at the infantry battalion and rifle company level are employed to maximize the effect of their high-angle fires. Mortars are one of the most effective indirect fire weapons in the urban environment. They should be used to engage:

- Enemy overwatch positions
- Enemy infantry before they seize a foothold
- Targets on rooftops
- Enemy reinforcements within range.

(3) Final protective fires are planned to stop dismounted assaults in front of the defensive positions. Fires within the city are planned along likely routes of advance to delay, disrupt, and destroy the enemy as he attempts to deepen a penetration.

(4) Commanders establish priorities of fire on the basis of enemy avenues of approach and threat systems that present the greatest danger to the defense. For example, during the attacker's initial advance, tanks, IFVs, and overwatching elements are the greatest threat to the defense. ATGMs should concentrate on destroying tanks first, then IFVs. Indirect fires should suppress and destroy enemy ATGMs and overwatch positions and/or elements. If enemy formations secure a foothold, priority is shifted to the destruction of enemy forces within the penetration.

(5) As the enemy attack progresses into the city, fires are increased to separate enemy infantry from its supporting armored vehicles. Friendly artillery concentrates on attacking infantry, conducting counterfire missions, and destroying reinforcements.

(6) Ample consideration must also be given to the rubbling effects of supporting arms. These effects often create massive rubbling of structures and buildings, making obstacles to both friendly and enemy movement.

(7) Mortars may be initially positioned forward in support of security forces. After their withdrawal, mortars are repositioned where they can support the in the main battle area.

(8) The commander assigns final protective fires and machine gun final protective lines (FPLs). Machine gun positions provide interlocking grazing fire and mutual support between adjacent units. FPLs are fired on planned signals. Proposed FPLs must be "walked out" to determine the extent of grazing fire available and to locate dead space. Dead space can be covered by:

- Rifle and light machine gun fire
- Grenade launchers
- Mines and boobytraps
- Indirect fires.

(9) Long-range antiarmor weapons are employed in the forward areas of the main battle area and the security area. Antiarmor weapons must be carefully positioned to prevent terrain and obstacles from reducing their range. Dismounted and manpacked antiarmor weapons should be positioned in buildings and along routes where engagement distances are reduced but where surprise and ambush are ideal.

e. Logistics. Just as in the offense, the defense utilizes a greater amount of class V and class VIII supplies in the urban environment. The defender needs to plan for multiple routes and means of supply. These can include:

- Using the city's existing transportation infrastructure to move supplies
- Prepositioning supplies in the defense in depth
- Using helicopter assets for rapid resupply
- Establishing and protecting supply routes
- Plan for utilization of existing repair and maintenance facilities
- Identify potable water sources

f. Force Protection

(1) Employment of Engineers. Engineers are employed at the battalion level or attached to companies and integrated down to squad level. Commanders must consider and assign the priority of work for engineer tasks that enhance survivability, mobility, and countermobility. Tasks that engineers can accomplish in the defense of a built-up area include:

- Constructing obstacles and rubbling for countermobility

- Clearing fields of fire
- Laying mines
- Preparing routes to the rear
- Preparing fighting positions.

(a) Employment of Obstacles. Obstacles are used to shape the battlespace. (See FM 5-102, *Countermobility*, and FM 20-32, *Mine/Countermine Operations*, for details.) The four types of obstacles are:

- **Disrupting.** Disrupting obstacles are used to disrupt the enemy's timetable and force him into conducting a time-costly bypass or breach, or cause his march columns to become staggered or "telescoped."
- **Turning.** Turning obstacles are used to canalize the enemy into engagement areas or kill zones where combined-arms fires can be concentrated and flanking shots are increased.
- **Fixing.** Fixing obstacles increase the time the enemy can be fired on in an engagement or target area and to cause increased confusion.
- **Blocking.** Blocking obstacles prevent or delay the enemy from moving past a certain point.

(b) Obstacle Belts. In defensive obstacle planning, obstacles may be emplaced in a series of belts. This enhances the overall effectiveness of the obstacle plan with successive belts designed for specific purposes. The following example lays out a three-belt obstacle plan.

1 **First Obstacle Belt.** The first obstacle belt is enplaced at the nearest buildings across from and parallel to the main defensive position. This belt consists of wire and improvised barriers (including inside buildings, in subterranean avenues of approach, and outside in open areas), danger areas, and dead space. These barriers and obstacles should be heavily boobytrapped and covered by fires. This belt is designed to impede enemy movement, breaks up and disorganize attack formations, and inflicts casualties.

2 **Second Obstacle Belt.** The second obstacle belt is placed between the first belt and the main defensive position buildings, but out of hand grenade range. It is designed to impede movement, canalize the enemy into the best fields of fire, break up attack formations, and inflict casualties. This belt is not meant to stop the enemy permanently. It consists mainly of wire obstacles, improvised barriers, road

craters, and minefields. It should be heavily boobytrapped. Triple-strand concertina wire is placed along the machine gun FPL to delay the enemy in the kill zone.

3 **Third Obstacle Belt.** The third obstacle belt is the defensive position's denial belt. It consists of wire obstacles placed around and through the defensive buildings, close-in minefields, and subterranean accesses. It is designed to impede the enemy's ability to gain a foothold in the defensive area. It should be heavily boobytrapped, and integrated with mines.

(c) Emplace Obstacles and Mines. While principles for employing mines and obstacles do not change in the defense of a urban area, techniques for their employment change. For example, burying and concealing mines in streets is usually not done. To save time and resources in preparing the defense, commanders must emphasize using all available materials (automobiles, railcars, rubble) to create obstacles. Civilian construction equipment and materials should be utilized. Obstacles must be integrated into buildings and rubble areas to increase their effectiveness and to canalize the enemy. The family of scatterable mines (FASCAM) may be effective on the outskirts of a city or in parks and open areas. (See Appendix F for further discussion on obstacles, mines, and demolitions.)

(2) Air Defense. Air defense (Stinger) weapons are positioned on rooftops and parking garages in order to obtain LOS coverage for the units they are protecting. Stinger teams can be assigned the mission of protecting specific positions or placed in general support (GS) of the battalion.

(3) Fire Hazards. All cities are vulnerable to destruction by fire, especially those with many wooden structures. The defender's detailed knowledge of the terrain permits him to avoid areas that are likely to be fire hazards. The defender can deliberately set fires:

- To disrupt and disorganize the attackers
- To canalize the attackers into more favorable engagement areas
- To obscure the attackers' observation.

Section II

Infantry Battalion Defense of a Built-Up Area

This section addresses some of the missions a task-organized and/or reinforced infantry battalion could be assigned. These missions are normally conducted as part of a larger joint operation.

3201. Defense in Sector. An infantry battalion is normally given the mission of defending a sector in a city (Figure 3-9). The interior buildings in the battalion sector form battle positions for the battalion's fixed defense. The size of a sector varies, but may cover the outlying structures and can be used to provide early warning and to delay the enemy.

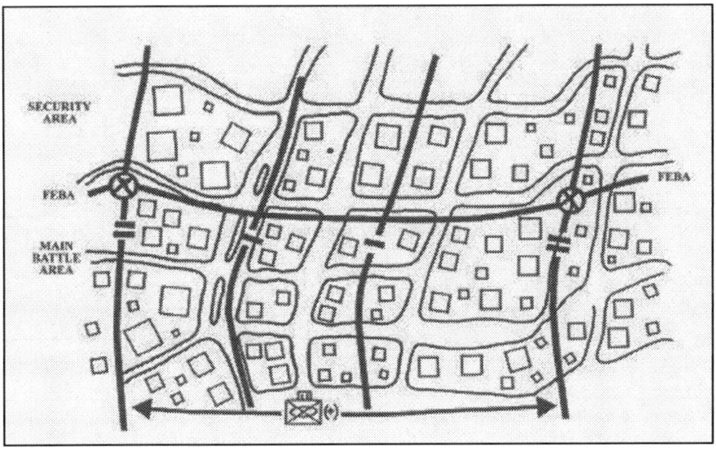

Figure 3-9. Defense of a Built-Up Sector

a. Defensive operations begin with the reconnoitering of the assigned sector to learn the terrain and to orient toward the enemy's expected advance. The main defensive positions should be formed on key terrain features, such as prominent buildings, road junctions, and positions which offer good fields of fire. Long-range antiarmor weapon systems should be deployed to engage the enemy as early as possible.

b. Depending on the situation, a reinforced rifle company may be used as a security force. The security force would operate forward of the FEBA and, depending on the mission, cause the enemy to deploy without without becoming decisively engaged. This can be done through maximum use of ambushes and obstacles while using covered and concealed routes. These routes are used for disengagement. Covered and concealed routes can be artificially made by making holes through buildings. The security force also attempts to detect the location of the enemy's main attack. Once the security force withdraws to the main battle area, it can be used as a reserve to reinforce or counterattack.

c. Based on METT-T considerations, the defense in sector (behind the FEBA in the main battle area) may consist of either company sectors, strongpoints, or battle positions. Strongpoints located on or covering decisive terrain are extremely effective in the defense.

d. Antiarmor vehicles and weapons should be used to engage enemy armored vehicles and/or cover obstacles with fire on likely avenues of approach.

e. Tanks are best used to engage enemy tanks, cover obstacles by fire, and engage in counterattacks. They should be employed in platoons where possible. However, in congested areas they may be employed in sections.

f. Artillery and mortar fire should be used to suppress and blind enemy overwatch elements, to engage enemy infantry on street approaches, provide counterbattery fire, and support counterattacks with direct and indirect fire.

g. Engineers are integrated throughout the defense. They provide support to forces in strongpoints to help prepare fighting positions. Engineers may also be attached to security forces to help in laying mines and constructing obstacles, clearing fields of fire, and preparing routes.

3202. Delay in a Built-Up Area. A delay is an operation in which a force under pressure trades space for time by slowing down the enemy's momentum and inflicting maximum damage on the enemy without becoming decisively engaged (MCWP 3-1). The delay can be oriented either on the enemy or on specified terrain such as a prominent building or industrial complex.

a. A delay in a built-up area may consist of a combination of ambushes and alternating or successive battle positions (Figures 3-10A and 3-10B on page 3-28).

(1) Ambushes are normally planned on observed obstacles. They require decentralized control and execution. Ambushes can be combined with limited-objective attacks on the enemy's flanks. They are most effective at the edge of open spaces, parks, wide streets, and so on. These attacks may be executed by dismounted infantry supported by tanks, LAVs/AAVs, or HMMWV-mounted machine guns and antitank vehicles.

(2) Battle positions should be placed where heavy weapons, antiarmor weapons, and machine guns will have the best fields of fire. Such locations are normally found at major street intersections, parks, and the edge of open residential areas. Battle positions should be carefully and deliberately prepared, reinforced by obstacles, and supported by indirect fire weapons. Battle positions are designed to inflict maximum casualties on the enemy and cause him to deploy for a deliberate attack.

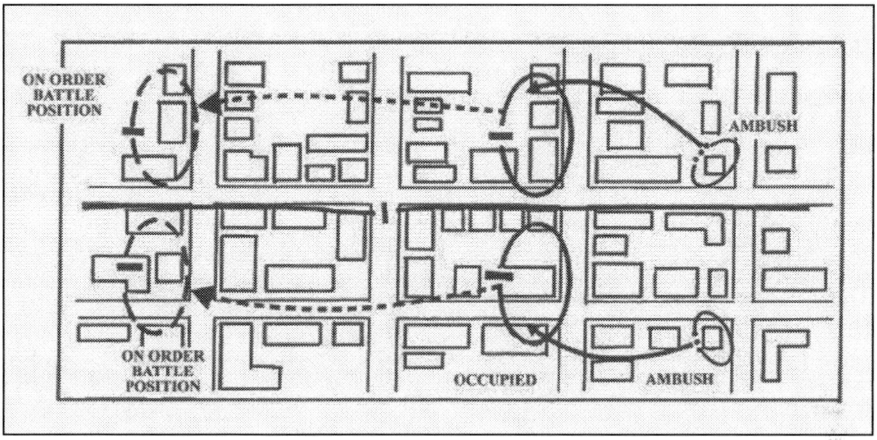

b. Tanks, LAVs/AAVs, and antiarmor weapons should prepare primary and alternate positions to improve their surivability and flexibility.

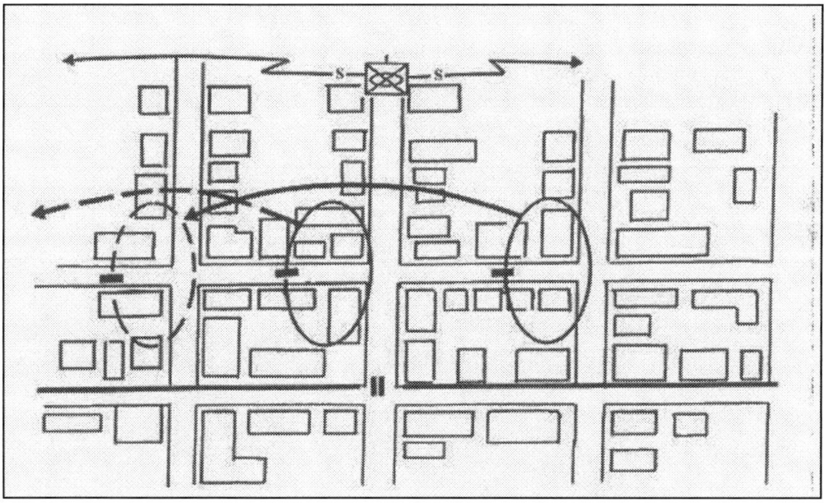

Figure 3-10A. Delay in a Built-Up Area (Successive Bounds)

Figure 3-10B. Delay in a Built-Up Area (Alternate Bounds)

c. The infantry battalion is most effective when deployed in successive delaying actions. Forces alternately engage the enemy to the maximum extent possible without becoming decisively engaged, then move on order in successive or alternate bounds. Smoke and demolitions are used to aid disengagement and cover the movement of the withdrawing force. Security units on the flank can be employed to prevent the enemy from outflanking the delaying unit. A small reserve can be used to react to unexpected enemy action and to conduct continued attacks on the enemy's flank.

d. The engineer effort should be centralized to support the preparation of battle positions, obstacle emplacement, and clearing routes.

e. The width of the infantry battalion sector depends on the nature of the buildings, obstacles along the street, and the time that the enemy must be delayed. Normally, narrow sectors supports delaying by alternate bounds while a wide sector better supports delaying by successive bounds.

3203. Defense of a Village. An infantry battalion assigned a defensive sector that includes a village may decide to incorporate the village as a strongpoint. This use is most common where the village stands astride a high-speed avenue of approach or lies between two difficult obstacles. To incorporate such an area into its defense, the infantry battalion must control the high ground on either side of the village to prevent the enemy from firing into the village. The battalion commander normally deploys one rifle company in the village to prepare a strongpoint defense

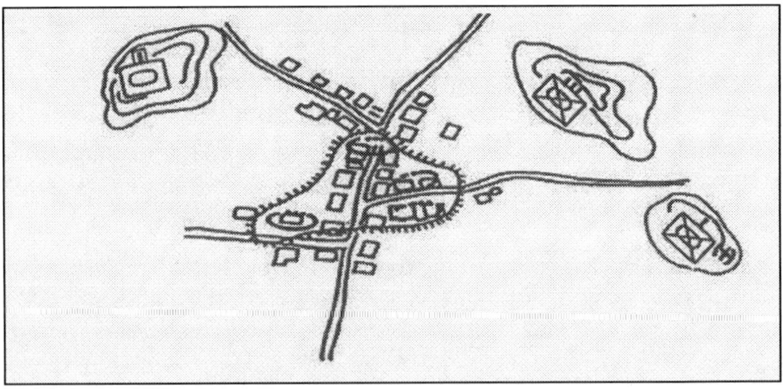

while the other rifle companies are overwatching the village, blocking bypass routes, and protecting the strongpoint's flanks. The strongpoint defense should provide a location where the enemy can be stopped and counterattacks can be launched (Figure 3-11).

Figure 3-11. Defense of a Village

a. Armored Vehicles and antiarmor weapons should be employed where the maneuver room is the greatest. This is usually on key terrain to the flanks of the village. As the security force withdraws and rifle companies become engaged, armored vehicles and weapons support by fire.

b. Although the infantry battalion's disposition should prevent large enemy forces from threatening the rear and flanks of the village, the danger of small-unit enemy infiltration requires the village be prepared for all-around defense.

c. Engineers should be assigned to the rifle company in the village strongpoint to provide continuous engineer support in preparation for the enemy's attack. The priority of engineer support should go to the strongpoint.

Section III

Rifle Company Operations

This section addresses some of the missions the infantry battalion could assign to the rifle company. For the purpose of this publication, the rifle company is a task-organized and/or reinforced unit. Assigned missions are normally part of a larger battalion operation.

3301. Defense of a Village. Once the commander has completed his reconnaissance of the village and the surrounding terrain, he develops his plan for the defense (Figure 3-12). The commander must decide whether to defend on the leading edge of the village or within the village.

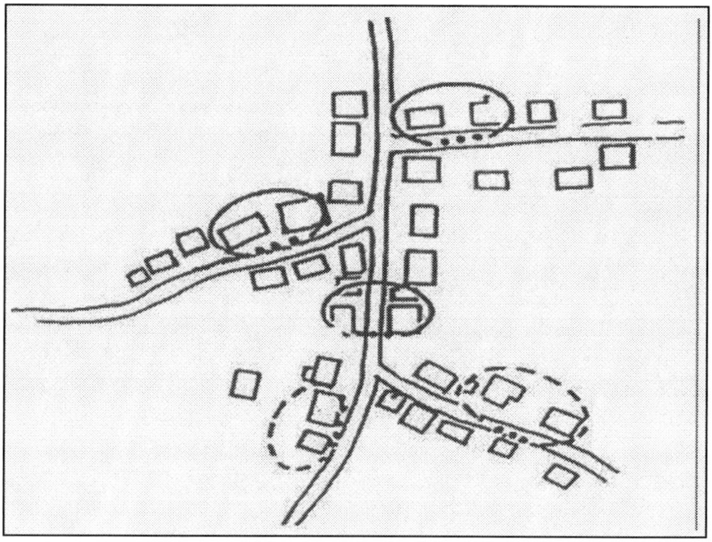

Figure 3-12. Defense of a Village

a. Several factors influence his decision. First, he must know the type of enemy that he will defend against. If the threat is mainly dismounted infantry, the greatest danger is in allowing them to gain a foothold in the town. If the threat is armor or motorized infantry, the greatest danger is that the company's defensive positions can be destroyed by direct fire. The commander must also consider the terrain forward and to the flanks of the village from which the enemy can direct fires against his positions.

b. The commander places platoons in a small group of buildings where each platoon leader establishes mutually supporting squad-sized positions. This positioning increases the area that the rifle platoon can control and hampers the enemy's ability to isolate or bypass a platoon. A rifle platoon can be responsible for the road through a village. The rest of the rifle company is positioned to provide all-around security and defense in depth.

c. Mortar positions must be protected from direct fire and allow for overhead clearance.

d. Combat vehicles should be assigned primary, alternate, and supplementary positions as well as primary and secondary sectors of fire. They should be positioned in defilade behind rubble and walls or inside buildings for cover, concealment, and movement into and out of the area.

e. If the company is reinforced with antiarmor weapons, the reinforcements should be positioned on high ground in and around the town to attain good fields of fire.

f. If the company is reinforced with tanks, the commander could place the tanks along the leading edge complement the other antiarmor weapons. The tank platoon leader selects exact firing positions and assign sectors of fire. If attacked by enemy infantry, the tanks should move to alternate positions protected by infantry. The alternate positions should allow the tanks to engage to the front and flanks with as little movement as possible. Tanks can be used as a mobile reserve after withdrawal from the leading edge of the town.

g. Obstacles are easily constructed in a built-up area. Commanders must stop enemy vehicles without interfering with his own movement in the village. Command detonating cratering charges at key street locations may be used to hamper enemy mobility. Mines should be laid on the outskirts of the town and along routes that the company will not use.

h. Engineers assist rifle platoons to prepare defensive positions. Engineers use explosives to make firing ports, mouseholes, and demolition obstacles. They also assist in the emplacement and recording of the minefields.

i. Ammunition expenditure is increased when fighting in an urban environment. Ammunition should be stockpiled at each squad position.

Other factors that the company commander may consider are:

- Resupply
- MEDEVAC
- Communications
- Firefighting

.3

- Sleep and alert plans
- Security
- Limited visibility
- Control of noncombatants.

j. The company should install a wire net and develop a plan for pyrotechnic signals to ensure adequate communications. Backup wire should be laid in case primary lines are cut by vehicles, fires, or the enemy. The commander also plans for the use of messengers .

3302. Delay in a Built-Up Area. A rifle company may be assigned the mission to delay in a built-up area.(Figure 3-13). To perform this mission, a rifle company can destroy enemy reconnaissance elements forward of the outskirts of the town, prevent their penetration into the built-up areas, and gain and maintain contact with the enemy to determine the strength and

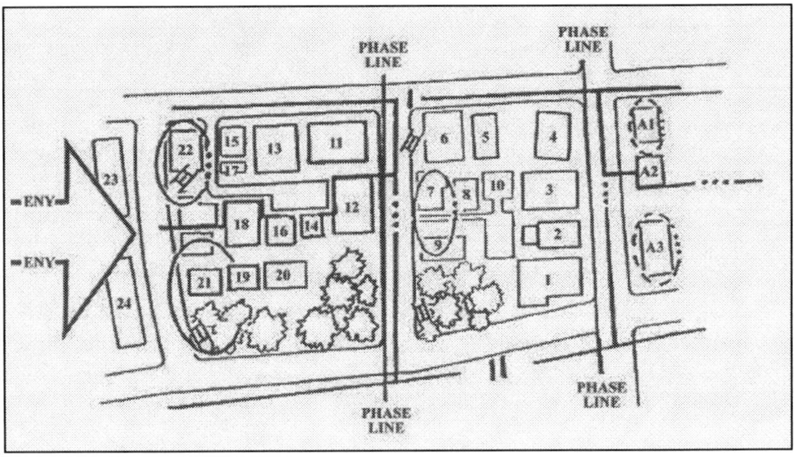

location of the main attack.

Figure 3-13. Delay in a Built-Up Area

a. Platoons delay by using patrols, OPs, and ambushes. Following each action, a platoon will disengage and withdraw.

b. The company's sector should be prepared with obstacles to enhance the delay. Engineers should prepare obstacles on anticipated enemy main routes of advance.

c. Antiarmor weapons may be positioned on the outskirts of the town to destroy the enemy at maximum range. They should be located in defilade positions or in prepared shelters. They fire at visible targets and then fall back or proceed to alternate positions.

d. Tanks in support may be used to engage enemy tanks, provide reinforcing fires, aid the disengagement of platoons, and cover obstacles by fire.

3303. Defense of a City Block. The rifle company may be assigned to defend a city block. The company actions should be coordinated with those of the security forces charged with delaying to the front of the company's position. The defense should take advantage of the protection of buildings that dominate the roads.

a. A well-organized company defense:

- Stops the attack of the enemy on the roads by using obstacles and enfilade fire
- Destroys the enemy by ambush and direct fire from prepared positions
- Ejects the enemy from footholds or fixes them for the battalion's counterattack.

b. The defensive forces should prepare ambushes on the avenues of approach, cover the obstacles by fire, and build a defense inside the buildings (Appendix D). The reserve can be tasked:

- To reinforce the fires of the defense
- To react to a danger on the flank
- To counterattack to throw the enemy from a foothold.

c. Engineers should be employed in GS to construct obstacles, prepare access routes, and assist in preparing defensive positions.

d. Tanks attached to the rifle company may be used to direct-fire support, engage enemy tanks, and support a counterattack.

3304. Defense of a Traffic Circle or Key Intersection. A rifle company may be assigned the mission to defend a key traffic circle or road junction (Figure 3-14).

a. The commander analyzes enemy avenues of approach and buildings that dominate those avenues. He plans to focus all possible firepower on the traffic circle and its approaches. He also plans for the all-around defense of the buildings that dominate the traffic circle to prevent encirclement. The commander prepares as many covered and concealed routes between these

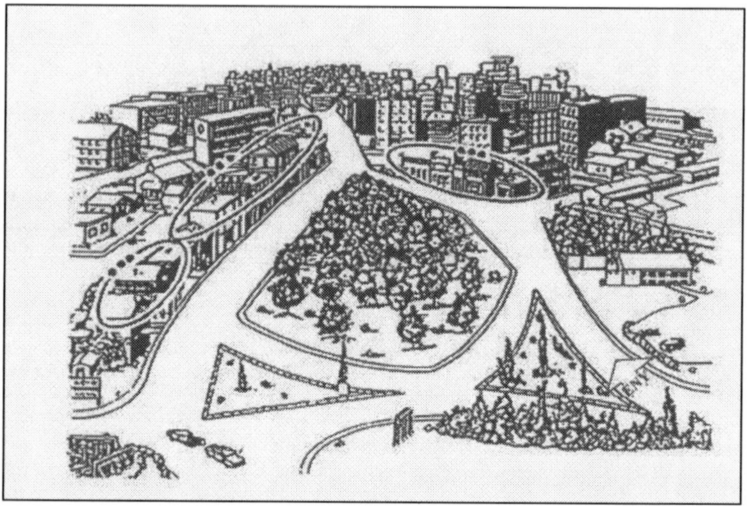

buildings as possible. These routes make it easier to mass or shift fires and to execute counterattacks.

Figure 3-14. Defense of a Traffic Circle

b. Obstacles can be used to deny the enemy the use of the traffic circle. In this case, obstacle planning should consider whether friendly forces require access to the traffic circle. Antiarmor weapons can be employed across the traffic circle if designated fields of fire exist. Tanks can be used to engage enemy tanks and provide direct-fire for counterattacks.

Section IV

Rifle Platoon Operations

This section discusses some missions the rifle company might assign to a rifle platoon.

3401. Defense of a Strongpoint. A rifle platoon may be given the mission to provide or conduct the strongpoint defense of a building, part of a building, or a group of small buildings (Figure 3-15). The rifle platoon's defense is normally integrated into the rifle company's defense. The rifle platoon must keep the enemy from gaining a foothold in its assigned buildings. It makes the best use of its weapons and supporting fires, organizes all-around defense, and counterattacks or calls for a rifle company counterattack to eject an enemy that has a foothold. The platoon leader analyzes his defensive sector to recommend to the company commander the best use of obstacles and supporting fires. (See Appendices B and D.)

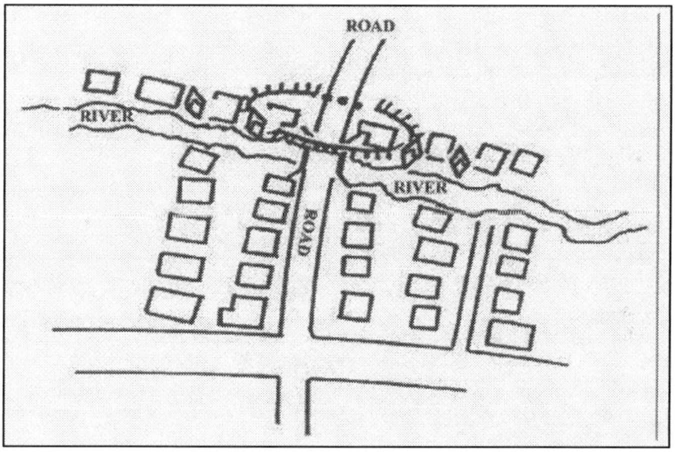

Figure 3-15. Defense of a Strongpoint

a. The rifle platoon should be organized into a series of firing positions located to cover avenues of approach, cover obstacles, and provide mutual support. Snipers may be located on the upper floors of the buildings. Unengaged units should be ready to counterattack, fight fires, or reinforce other elements of the rifle platoon.

b. Depending on the length of the mission, the rifle platoon should have available:

- Pioneer equipment (axes, shovels, hammers, picket pounders)

- Barrier material (barbed wire, sandbags)
- Munitions (especially grenades)
- Food and water
- Medical supplies
- Firefighting equipment.

3402. Defense Against Armor. Urbanized terrain is well suited to an infantry's defense against mechanized infantry and armored forces. Armored forces will usually avoid built-up areas but may be forced to pass through them.

a. Built-up areas have certain aspects that favor infantry antiarmor operations.

(1) Buildings provide cover and concealment for defending infantry.

(2) The streets restrict armor maneuver, fields of fire, and communications, thereby reducing the enemy's ability to reinforce.

(3) Rubble in the streets can be used to block enemy vehicles, conceal mines, and cover and conceal defending infantry.

(4) Rooftops, alleys, and upper floors provide good positions for flanking and overhead fire.

(5) Sewers, storm drains, and subways provide underground routes for infantry forces.

b. To prepare for antiarmor operations, a platoon should:

(1) Choose a Good Engagement Area. The best way for infantrymen to engage tanks is one at a time so that they can destroy one tank without being vulnerable to the fires of another. Good engagement areas include narrow streets, turns in the road, "T" intersections, bridges, tunnels, split level roads, and rubbled areas. Less obvious locations can also be created by using demolitions or mines to create obstacles.

(2) Select Good Weapons Positions. The best weapons positions are places where the tank is weakest (tank deadspace) and the infantry is most protected. Forcing tanks to button up greatly reduces their ability to see and fire on close-in street and overhead targets. Figure 3-16 (on page 3-38) shows close-in tank weapon vision and engagement limitations (tank buttoned up) at street and overhead levels.

.3

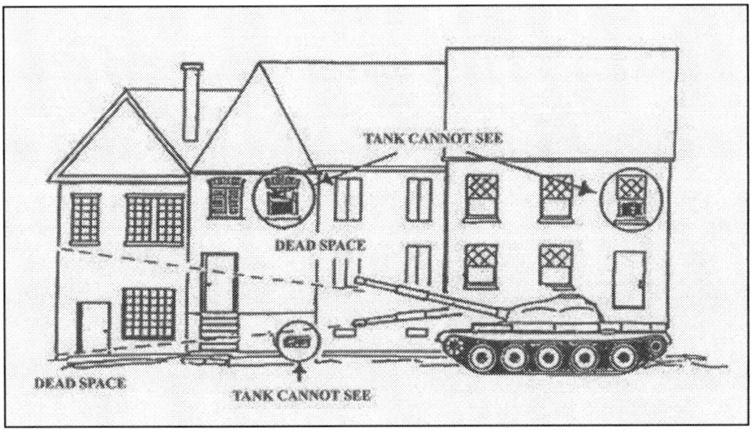

Figure 3-16. Close-in Tank Limitations at Street and Overhead Levels

(3) Assign Target Reference Points and Select Method of Engagement. After selecting the weapons positions, target reference points should be assigned to ensure adequate coverage and control fires. The target reference points should be clearly visible through the gunner's sights and should be relatively resistant to battle damage (for example, large buildings or bridge abutments, but not trees or cars). The three types of antiarmor engagement techniques are frontal, crossfire, and depth. Frontal fire is the least preferred because it exposes the gunner and it attacks the tank where the armor is the thickest. (For more information on target engagement techniques, see MCWP 3-15.5 *Antiarmor Operations*.)

Infantry forces should fire at tanks on the flanks and rear from ground level or at the top of tanks from an elevated position. A suitable antiarmor defense might be set up as shown in Figure 3-17. For a safe engagement from an elevated position, Marines should allow the tank to approach to a range three times the elevation of the weapons. However, engagement at longer ranges risks counterfire because the attacking weapon's position will not be in the tank's overhead dead space.

(4) Coordinate Target Engagement. Tanks are more vulnerable to infantry operations when they are buttoned up. The first task is to compel the tanks to button up by using direct and indirect fire. The next task is to coordinate the fires of the antitank weapons so that all targets in the engagement area are attacked simultaneously (Figure 3-18).

Figure 3-17. A Rifle Platoon's Antiarmor Defense

Figure 3-18. Coordinated Antiarmor Ambush

c. Infantry often accompany armored vehicles in built-up areas to provide them with close-in protection. Antiarmor weapons should be supported by an effective antipersonnel defense.

d. At a planned signal (for example, the detonation of a mine), all targets in the engagement area are attcked at the same time. If that is not possible, then the most dangerous target should be engaged first. Although tanks present the greatest threat, BMPs are also dangerous because their infantry can dismount and destroy friendly antiarmor positions. Rubble and mines should be used to reduce target mobility so that a target's time in the engagement area is increased.

3403. Conduct of an Armored Ambush. A rifle company can use an attached tank platoon to conduct an armored ambush (Figure 3-19). The tank platoon should be reinforced with infantry.

Figure 3-19. Armored Ambush

a. An armored ambush requires maneuver on obstacle-free roads. Obstacles outside the ambush area can be used to canalize and delay the enemy.

b. Ambushing tanks should hide approximately 1,000 meters from the expected enemy avenue of approach. A security post, located at a chokepoint, observes and reports the approach, speed, security posture, and activity of the enemy. When the enemy is reported at a trigger point or target reference point, the tank platoon leader orders the ambush.

c. The tanks move quickly from their hiding positions to firing positions, taking advantage of all available concealment. They attempt to fire flank shots on the approaching enemy at an average range of 300 to 400 meters. Firing at these ranges does not expose tanks to enemy infantry. After the engagement, the tank breaks contact and moves to a rally point with close security provided by infantry. They may then move to a new ambush site.

Chapter 4

Combat Support

Combat support units provide critical combat functions in conjunction with combat arms units to secure victory (MCRP 5-2A).

4001. Introduction. Success against a determined enemy is achieved in part by generating decisive combat power through the application of *combined arms*. This is especially true in MOUT. The MAGTF commander must effectively integrate all the capabilities of the command element (CE), GCE, ACE, and CSSE to win the battle in the shortest amount of time with the fewest casualties. The MOUT environment requires special considerations pertaining to combat support forces and capabilities.

Section I

Fire Support

4101. Indirect and Direct Fire Support

a. General. MOUT operations present unique fire support considerations. On urbanized terrain, buildings provide excellent cover and concealment to the enemy while limiting friendly observation and targeting efforts. Targets are generally exposed for brief periods of time and are often in close proximity to friendly forces. Observers will experience difficulty in finding OPs with adequate fields of observation. Terrain masking by tall buildings may restrict the delivery of indirect fires. Collateral damage and rubble effects must be considered during ammunition selection. The importance of effective communications, ROE, control measures, and procedures to prevent fratricide is magnified.

b. Offensive Support. In the offense, fire support plans should include fires to isolate the objective area, support the assault, and support the clearing action. Fires are delivered to isolate and fix the enemy and deny him the use of avenues of approach into and out of the built-up area. Most fires are normally planned and executed at the GCE or higher echelons. Fires are employed to rupture the enemy's established defenses and screen friendly maneuver, in order to maintain the momentum of the attack. Fire support is also allocated to units involved in clearing operations. Fire support plans should incorporate the employment of aerial observers and UAVs to compensate for restrictions to observation and to assist in the delivery of deep fires. Procedures for designating the forward line of own troops (FLOT), marking targets, shifting fires, and communicating in the urban environment should also be considered.

c. Defensive Support. In the defense, fire support plans address fires to disrupt and slow the enemy attack. Fires are delivered at maximum ranges along avenues of approach to separate armor and infantry forces, to canalize the enemy into killing zones, and to deny the enemy use of key facilities in the city, such as communications and transportation facilities. Most fires are planned and executed at the GCE or higher echelon. Defensive fires planned at lower echelons support fighting within the built-up area. Defensive fire support plans identify fire support coordination procedures necessary to execute the delivery of fire support.

4102. Artillery

a. Employment of Artillery. The mission of artillery does not change in MOUT. However, the method of employment is tailored to the unique requirements posed by the urban battlespace. MCWP 3-16.3, *Marine Artillery Support,* provides a detailed discussion on the employment of artillery. Considerations for MOUT include:

(1) Organization for Combat. Standard command relationships for tactical missions are used to organize for combat. During the initial phases of offensive MOUT, particularly isolation of the objective area and the initial assault to secure a foothold, control of artillery is more centralized than in other types of offensive operations. Centralized control provides better responsiveness in shifting and massing fires. As the attack progresses into the clearance phase, control of artillery becomes more decentralized with direct support (DS) artillery providing direct and indirect fires. In the defense, control of artillery should be sufficiently centralized to enable the rapid massing of fires.

(2) Movement and Positioning. Key considerations for the positioning of artillery in MOUT include:

- Mobility restrictions
- Limited availability of firing positions
- Masking of fires
- Security
- Enemy counterbattery fires

(a) Offense. Artillery is initially positioned outside or in the outskirts of the city. Artillery may displace to suitable firing positions within the city as the attack progresses. Positions should be selected that minimize masking, provide several routes for escape, and afford as much cover and concealment as possible. Potential firing positions can be found in parks, racetracks, sports fields, school yards, stadiums, parking lots, and similar areas. Special techniques for the emplacement of howitzers may be required if the ground is not suitable for normal emplacement. Some artillery positions may be maintained outside the city (see appendix B). Artillery generally makes few displacements in MOUT because it requires extensive route reconnaissance and security. If and when artillery is moved, it should be given priority of movement during displacements. Movement is conducted during night or periods of reduced visibility.

(b) Defense. Artillery is positioned on the friendly outskirts of the built-up area. Artillery displaces along predetermined routes to avoid traffic congestion.

(3) Delivery of Fires. The commander, in coordination with the fire support coordinator (FSC), considers the following when planning fire support in a built-up area:

(a) Target acquisition may be more difficult because of the increased cover and concealment afforded by the terrain. Ground observation is limited; therefore, FOs should be placed near tops of buildings.

(b) For targets in defilade, fire may be adjusted laterally so that the rounds impact on the street perpendicular to the observer's position. Airburst rounds are more suitable for this adjustment. Adjustments may have to be made by sound. When rounds impact on a perpendicular street, they are adjusted for range. When the range is correct, a lateral shift is made onto the target and the observer fires for effect.

(c) Special consideration must be given to shell and fuze combinations when effects of munitions are influenced by buildings. These include:

- Careful use of variable time is required to avoid premature arming.
- HE ammunition with impact or delay fuzes may create unwanted rubble.
- The close proximity of enemy and friendly forces requires careful coordination.
- White phosphorus (WP) may create unwanted fires and smoke.
- Fuze delay should be used to penetrate fortifications.
- Illumination rounds can be effective. However, friendly positions should remain in shadows and enemy positions should be highlighted. Tall buildings may mask the effects of illumination rounds.
- HE/variable time, HE/time fuze, and improved conventional munitions are effective for clearing enemy positions, observers, and antennas from rooftops.
- Swirling winds may degrade smoke operations.
- FASCAM may be used to impede enemy movements. FASCAM effectiveness is reduced when delivered on a hard surface.

(4) Targeting is difficult on urbanized terrain because the enemy has many covered and concealed positions and movement lanes. The enemy may be on rooftops and in buildings and may use sewer and subway systems. Targets should be planned on rooftops to clear enemy FOs, communications, and radar equipment. Targets should also be planned on major roads, at road intersections, and on known or likely enemy fortifications. Employing artillery in the direct-fire mode to destroy fortifications should be considered. Also, restrictive fire support coordination measures (such as a restrictive fire area or no fire area) may be imposed to protect civilians and critical installations.

(5) The 155-mm howitzer can be effective in neutralizing concrete targets by direct fire. A 155-mm round can penetrate 36 inches of concrete at ranges up to 2,200 meters. Restrictions may be placed on the type of ammunition used in order to reduce rubbling on friendly avenues of approach.

(6) FOs must be able to determine the location and size of dead space. Dead space is the area in which indirect fires cannot reach the street level because of buildings. This area is a safe haven for the enemy because he is protected from indirect fires. For low-angle artillery, the dead space is about five times the height of the building. For mortars and high-angle artillery, the dead space is about one-half the height of the building.

(7) Aerial observers and UAVs are effective in providing observation behind buildings immediately to the front of friendly forces. They are extremely helpful in adjusting fire. Aerial observers can also relay calls for fire when communications are degraded because of power lines or building masking.

(8) Weapons locating radars are effective in locating enemy artillery and mortar targets in an urban environment because of their high percentage of high-angle fires. However, LOS must be maintained in order to exploit this advantage. Radars should not be placed in the midst of a city due to masking. They are best located on the outskirts of the city

(9) The use of airburst fires is an effective means of clearing snipers and manpads from rooftops. HE shells with delay fuzes may be effective against enemy troops in the upper floors of buildings, however, because of the overhead cover provided by the building, such shells have little effect on the enemy in the lower floors. (The planning and employment of artillery in offensive and defensive operations is also addressed in Chapters 2 and 3.)

b. Local Security. Artillery crews and howitzers require close security when used in the direct-fire mode. Artillery is particularly vulnerable when units are employed in isolated firing positions.

c. Communications and Information Systems (CIS). CIS must be able to satisfy the unique C2 requirements for artillery in MOUT. CIS should provide the commander with the ability to collect, process, analyze, and exchange information. Requirements such as massing fires and call for fire must be carefully planned and adapted to meet the requirements of

MOUT. For example, small units conducting independent operations must have the communications capabilities and maps to allow them to call for fire.

4103. Mortars

a. Employment of Mortars. Mortars are often the primary indirect-fire weapon for forward units in the assault or defense of a built-up area. Their high rate of fire, short minimum range, and high trajectory give mortar units the ability to fire in the tight confines of the city. Mortars can provide obscuration, neutralization, suppression, or illumination fires.

(1) Organization for Combat. Normally, 81-mm mortars are employed in GS of the battalion. The 60-mm mortar is normally used at the discretion of the rifle company commander

(2) Movement and Positioning. 81-mm mortars employed in GS normally occupy a position within or near the infantry battalion reserve. Key considerations for the selection of positions include:

- The minimum range of the weapon
- Mask and overhead clearance
- Terrain suitable for setting base plates
- Dispersion, and accessibility.

Special techniques can be used to position and lay the 81-mm mortar. These techniques include placing buffers under base plates, using curbs and sandbags to anchor or brace mortars, and filling cans with dirt or using sandbags to emplace aiming posts. Fixed objects may be used as distant aiming points. When the depth of the defensive position is shallow, or when suitable firing positions are not available, mortars may have to be positioned behind the reserve. Positioning of mortars behind buildings greatly enhances survivability. Displacement is often executed by section. Mortars are usually not placed on top of buildings because the location lacks cover and concealment and recoil/concussion could collapse roofs.

(3) Delivery of Fires. Mortar fires are effective in the offense as well as in the defense. The indirect fires are extremely responsive. Mortars are well suited for combat in built-up areas because of their high rate of fire, steep angle of fall, and short minimum range. Mortar fires can be used to inhibit enemy fires and movement, allowing friendly forces to maneuver to a position of advantage. Effectively integrating mortar fires with dismounted maneuver is key to successful combat in a built-up area.

(a) HE Ammunition. HE is used more than any other type of ammunition. Although mortar fires are often targeted against roads and other open areas, the natural dispersion of indirect fires will result in many hits on buildings. Leaders must use care when planning mortar fires to minimize collateral damage. Considerations when using HE include:

- HE ammunition is particularly effective when used on lightly built structures within cities. However, it does not perform well against the reinforced concrete found in larger urban areas.

- Point-detonating fuzes should be used. The use of proximity fuzes should be avoided because the nature of built-up areas causes proximity fuzes to function prematurely. However, proximity fuzes are useful in attacking targets such as OPs on tops of buildings.

- During World War II, the Middle East conflicts, and most recently in Bosnia, mortar HE fires have been used extensively to deny the use of streets, parks, and plazas to enemy and civilian personnel.

(b) Incendiary Fires. Employment of these munitions is planned in detail to prevent the burning of buildings and surrounding terrain that may create obstacles to friendly forces.

(c) Illumination. In the offense, illumination rounds should burst above the objective to put enemy troops in the light. Buildings reduce the effectiveness of the illumination by creating shadows. If the illumination is behind the objective, the enemy troops would be in the shadows rather than in the light. In the defense, illumination is planned to burst behind friendly forces to put them in the shadows and place the enemy forces in the light. Continuous illumination requires close coordination between the FO and the fire direction center (FDC) to produce the proper effect.

(d) Other Considerations. When planning the use of mortars, commanders must consider the following:

- FOs should be positioned on tops of buildings so target acquisition and adjustments in fire can best be accomplished.

- Commanders must understand ammunition effects to correctly estimate the number of volleys needed for the specific target coverage. Also, the effects of using white or red phosphorus may create unwanted smoke screens or limit visibility, which could interfere with the tactical plan.

- Mortar sections should plan to provide their own security.

- Commanders must give consideration to where, when, and how mortars are to displace in order to maintain immediate indirect fire support. Combat in built-up areas may adversely affects the ability of mortars to displace because of rubbling.

b. Communications. An increased use of wire, messenger, and visual signals will be required. However, wire should be the primary means of communication between the FOs and the FDC because elements are close to each other. Also, frequency-modulated (FM) radio transmissions in built-up areas are likely to be erratic. Remote siting of antennas to upper floors or roofs may improve communications and enhance operator survivability. The use of radio retransmissions can also be effective. Both in the offense and defense, commanders consider using existing civilian systems to supplement the unit's capability.

c. Magnetic Interference. Anything that uses or generates electricity generates its own magnetic field. In an urban environment, all magnetic instruments are affected by surrounding structural steel, electrical cables, and automobiles. Minimum distance restrictions for the use of the M-2 aiming circle will be difficult to overcome. However, an azimuth may be obtained from a distant aiming point.

4104. Naval Surface Fire Support (NSFS) and Naval Gunfire (NGF)

a. General. When NSFS ships are available, naval surface fires and NGF can provide some fire support.

b. Employment of Naval Surface Fire Support. The urban environment severely limits NSFS, and NGF in particular, due to its flat trajectory. NSFS can be used to supplement existing fire support but should not be planned as the primary fire support asset for MOUT. NSFS is most effective when firing on targets to isolate the battlefield and engage targets located on the seaward side of a city. Only when the ROE allows extensive collateral damage can the full effect of NSFS be brought to bear on units occupying a city.

(1) Organization for Combat. NSFS is best employed under centralized control because of the limited use of NSFS within the confines of the city.

(2) Positioning. Many of the fires delivered by NSFS ships require a careful orientation of the gun target line and may limit the size of a fire support area.

(3) Delivery of Fires. NGF can deliver a variety of ammunition in MOUT. NGF can deliver HE, incendiary, and illuminating projectiles. The NGF high-capacity projectile is very effective in producing fragmenting and penetration effects. Offensive and defensive fire support plans can incorporate NSFS fires on bridges, roads, reinforcements, artillery positions, and command and control centers. NSFS can be effectively delivered on targets along the outskirts of the city to rupture established defenses. These fires are particularly effective on buildings and on targets with a linear axis such as streets and airfields. Because of its flat trajectory, NGF employment is limited on targets within the city. Firing reduced charges at high angles does enable NGF to hit more targets but will reduce its range and accuracy.

4105. Aviation

a. Employment of Aviation. Both fixed- and rotary-wing aircraft can provide fire support in MOUT. The aviation combat element (ACE) provides the MAGTF commander with a variety of options in ordnance delivery. Aerial delivered ordnance is particularly effective against hard targets. Cluster bomb units, rockets, cannons, laser-guided bombs, and electro-optically guided missiles can be used against a variety of targets. However, the presence of civilians, the requirement to preserve key facilities, and the residual effects caused by rubble and collateral damage must be taken into consideration when planning for and using aviation in an urban environment. (See MCWP 3-24 *Assault Support Helicopter Tactical Manual* (under development), MCWP 3-23.1 *Close Air Support* (under development),and NWP 55-3-AH1, *AH1 Tactical Manual,* for greater detail on MOUT considerations for aviation.)

(1) Fixed-Wing Aircraft

(a) Offensive/Defensive Operations. During offensive operations, fixed-wing aircraft can isolate, interdict, and reduce enemy strongpoints throughout the battlespace. During defensive operations, they attrite enemy forces as they approach, attack forces attempting to bypass, or canalize the enemy to take avenues of approach that are advantageous to the defender.

(b) Other Considerations

1 **Fixed-Wing Attack of a Built-Up Area.** Air strikes can reduce or destroy vital supplies and facilities supporting the enemy. The resulting shock, concussion, and psychological effects of such attacks can reduce the efficiency and fighting spirit of enemy forces.

2 **Rubble and Debris.** The rubble and debris resulting from air attacks may increase the defender's cover and create obstacles to the movement of friendly forces.

3 **Proximity of Friendly Forces.** The proximity of opposing forces to friendly forces increases the potential for fratricide. This may require the use of precision guided munitions (PGMs) or the temporary disengagement of friendly forces in contact.

4 **Locations of Indigenous Civilians or Key Facilities.** The use of aerial delivered ordnance may be restricted by the ROE because of the presence of civilians or the requirement to preserve key facilities within a city.

5 **Limited Ground Observation.** Limited ground observation may require the use of forward air controllers (airborne) (FAC(A)).

(2) Rotary-Wing Aircraft. Attack helicopters provide a distinct advantage to the commander in MOUT. They can rapidly maneuver within the built-up area to attack a variety of targets. Urban structures offer ideal cover for attack helicopters. The ability to deliver PGMs, rockets, and 20 mm cannon fire makes attack helicopters valuable assets in reducing strongpoints, breaking up enemy concentrations, and providing suppressive fires for attacking forces. Attack helicopters also provide a highly manueverable antiarmor capability.

(a) Attack Helicopter Fire Support During Offensive/Defensive Operations

1 Provide close air support in support of the commander's scheme of maneuver

2 Provide suppression and/or destruction of enemy forces within buildings and strongpoints

3 Conduct counterattacks

4 Isolate the built-up area or objectives within the built-up area

5 Conduct armed reconnaissance and patrols.

6 Conduct rapid concentration of fires.

4106. Armored Vehicles

a. Employment of Armored Vehicles. Tanks, LAVs, and AAVs are primarily mobility assets that function best in an unrestricted, mobile environment. However, these vehicles can play an important role inside the city by using their firepower in support of offensive and defensive operations. LAVs and AAVs highly are susceptible to armor-piercing rounds and manpacked antiarmor weapons. All vehicles are susceptible to top-attack munitions. Some of the fire support roles armored vehicles may provide include:

- Suppression and/or destruction of enemy forces within buildings and strongpoints (tanks/LAVs/AAVs)
- Provide machinegun suppression fires
- Reserve or counterattack suppression fires (tanks and/or AAVs)
- Creating entry/exit points in buildings
- Isolation of the built-up area or objectives within the built-up area (tanks/LAVs/AAVs)

- Artillery-towing vehicle (LAVs/AAVs)
- Antiarmor fires
- Breaching obstacles in a direct fire mode.

b. Special Consideration. The existence of waterways, canals, and rivers in and around urban areas can provide avenues of ingress into designated objective areas. If used in conjunction with supporting fires, smoke, and/or deception operations, AAVs can capitalize on their amphibious capability and make use of waterways previously viewed as obstacles.

4107. Antiarmor Weapons. Antiarmor weapons have more utility in the defense than in the offense. Their primary task is defeating enemy armor. Secondary tasks may involve blasting holes in structures and fortified positions. As with other weapons systems, antiarmor weapons are normally employed in a decentralized fashion at the small unit level. Enemy armor should be engaged in predetermined locations that afford the most advantage to the defender. Weapon positions should be selected that provide maximum range fires. TOWs and Dragons may be fired from an enclosed area if it is a sturdy ventilated structure of appropriate size. For a more detailed discussion of antiarmor weapons, see MCWP 3-15.5, *Antiarmor Operations*.

4108. Employment of Snipers

a. General. Snipers are highly trained in field skills and marksmanship and can deliver long-range precision fire at selected targets from concealed positions. The method by which snipers are employed in urban warfare will be governed by the nature of the terrain/weather, number of snipers available, and whether the enemy employs snipers. A sniper can play an important role in MOUT. Employment of snipers influences the action by:

- Creating adverse psychological affects on the enemy by negating feelings of security
- Inflicting casualties on enemy leadership
- Enabling the infantry to seem to cover a large area, regardless of whether the terrain is physically occupied or not
- Detailing enemy positions and activity through the use of advanced optical equipment and observation techniques
- Controlling fire support
- Conducting reconnaissance in conjunction with their sniper role
- Disrupting enemy movement, observation, and infiltration, and negating the possibility of surprise.

b. Offensive Employment. Snipers should be used to gain and maintain contact with the enemy. This enhances security and prevents surprise by keeping constant and unrelenting pressure on the enemy. Some of the considerations for sniper employment include:

(1) As the "eyes" of a commander, a sniper increases the commander's flexibility by gathering and transmitting information on fire support targets.

(2) Snipers attack those targets that affect the enemy's ability to wage battle and those that will cause the maximum amount of confusion on the battlefield. Potential targets for a sniper are:

- Officers and noncommissioned officers
- Enemy snipers and marksmen
- Engineer personnel
- Personnel manning crew-served weapons
- Communications personnel and equipment
- Fire support observers and/or equipment
- Commanders of armored vehicles.

c. Combat on Urbanized Terrain

(1) Snipers can be employed to operate for extended periods of time from hidden positions to dominate and establish a "no-man's land," screen flanks, protect the rear, and deny the enemy access to certain areas or avenues of approach. Snipers can operate with the covering element to deliver accurate fires in support of search elements. Optical devices enhances their ability to detect movement and engage it instantly.

(2) Snipers can also support (by fire) infantry movement across streets. They can provide immediate precision fire on enemy machine gun nests and enemy snipers. The best countersniper weapon is the sniper. Snipers are assigned the following supporting missions:

- Delaying and inflicting casualties on the advancing enemy
- Observing and reporting potential targets
- Covering (by fire) avenues of approach and obstacles

.3

- Harassing the enemy and causing him to deploy prematurely
- Directing supporting arms fire.

The urban environment provides an abundance of secure firing positions that are not highly accessible to countersniping.

d. Defensive Employment. In defensive operations, the sniper is best employed forward of the FEBA to provide early warning of enemy approach, disrupt it, and if possible, cause the enemy to deploy prematurely. Snipers are employed to:

- Prevent enemy infiltration
- Detect and destroy infiltrators
- Protect unit at the FEBA from surprise attacks
- Protect patrols from ambush
- Screen the flanks and rear of the defensive positions.

More detailed information regarding the employment of snipers can be found in Marine Corps Warfighting Publication (MCWP) 3-15.3, *Scout Sniping* (under development).

Section II

Other Combat Support

4201. Engineers

a. General. The nature of the urban battlespace requires extensive use of engineers. In both the offense and defense, the proper employment of engineers can be decisive. Engineers must also be prepared to repair and maintain the internal support facilities of the city in order to support further operations. MAGTF commanders plan to use all available engineer assets, including engineers in the GCE, ACE, and CSSE, as well as any naval construction forces (NCFs).

b. Engineer Organization. Engineers are almost always task organized. The compartmentalized, closed-in nature of the city, which limits observation, mobility, and communications, will dictate the best organization for the mission. Teams should be formed to perform specific tasks for small units and must be positioned for quick employment. In MOUT, engineer support is key to maintaining the mometum of the attack. Engineer teams will usually move with the units they are supporting. Two-man demolition teams may be assigned to support a rifle platoon or squad to breach obstacles or to neutralize booby traps and improvised explosive devices (IEDs).

Commanders must carefully analyze mission requirements, the alignment of enemy forces, the condition of the battlespace, and available assets, to ensure an organization that will provide optimum engineer support. Most of the actual engineer work is done at the small unit level.

c. Engineer Tasks

(1) Offense. Urban combat is by nature a very slow, deliberate affair. Even if the enemy has not prepared the area for a defense, it will take time to secure the city. A well-constructed urban defense will slow or stop any attack unless impediments to movement can be quickly removed or bypassed. Properly organized and positioned engineers can reduce delays by performing a variety of tasks which include:

(a) Engineer reconnaissance to determine the level of effort required to penetrate outer and inner defenses, clear obstacles, and identify bypass routes.

(b) Bridge repair and construction on critical avenues of approach. Bridge construction may also open additional avenues of approach that the enemy is not prepared to defend.

(c) Breaching obstacles both outside and inside the city. This includes breaching minefields and neutralizing booby traps and IEDs.

(d) Assault breaching into buildings.

(e) Reducing enemy strongpoints using demolitions and heavy equipment.

(f) Preparing helicopter landing zones (HLZs).

(g) Assist in the preparation of defensive positions in the event of a counterattack.

(2) Defense. An urban area can be made nearly impregnable with proper organization of the battlespace and skillful preparation by engineers. Obstacles in outlying areas can delay/disrupt the enemy and canalize him into engagement areas. The streets and alleys inside a city can also be use to canalize enemy forces into engagement areas and kill zones. In preparing the defense of an urban area, engineers perform several critical tasks to include:

(a) Engineer preparation of the urban battlespace. This should begin well outside the city with complex obstacles in support of security forces. Selective demolition of bridges and roads and other countermobility measures can force the enemy into taking avenues of approach that favor the defense.

(b) Constructing obstacles forward of and between defensive positions to block, disrupt, fix, and turn enemy attacks. A majority of the obstacles will be directed at interfering with the movement of personnel and vehicles, but some effort must be directed at preventing vertical and subterranean movement. Material designed especially for obstacles and barricades, such as barbed wire and tetrahedron(s), is necessary, but improvised barriers can be just as effective and are readily available. Vehicles and building rubble are the most common improvised barriers.

(c) Enhancing mobility by opening and/or marking preplanned routes into attack positions for counterattacks or alternate positions to protect a flank.

(d) Providing technical advice on which buildings are more suitable for use as strongpoints or which rooftops are strong enough for helicopter landings can be vital. Any reinforcement of building roofs should be supervised by the engineers.

(e) Preparing tank positions that offer them cover and concealment.

(f) Assisting infantry units in clearing clear fields of fire, especially for longer range weapons like the TOW.

(g) Playing a potential role in the maintenance of critical urban services such as power, water, and communications. This function may also be performed by naval construction units.

(h) Assisting in the preparation and maintenance of railroads, airfields, and ports. This function may also be performed by naval construction units.

4202. Assault Support Helicopters

a. Combat Assault Transportation. The mobility provided by helicopters can be a significant advantage in the urban environment. In many modern cities, office buildings may have helipads on their roofs. Other buildings, such as parking garages, are usually strong enough to support the weight of a helicopter. The delivery of Marines onto a building can also be accomplished by rappelling, fastroping, or jumping out of the helicopter as it hovers just above the roof. The rooftops should be inspected to ensure that no obstacles exist, such as electrical wires, telephone poles, antennas, or mines and wire. The use of helicopters in this role is usually predicated on the antiair threat. Helicopterborne forces are highly vulnerable to small arms fire, AAA and manpads.

b. Small-Scale Helicopterborne Assaults. Small units may be helo-lifted onto the rooftop of a key building. Depending on the construction of the roof, rappelling or fastroping Marines from the helicopter may be more advantageous than landing them on the rooftop. Fastroping provides a rapid and safer means of insertion than jumping from a low hover. With practice, Marines can accomplish a fastrope insertion with minimum exposure.

c. Large-Scale Helicopterborne Assaults. For large-scale assaults, rooftop landings are not practical; however, open spaces (parks, parking lots, sports arenas) within the built-up area may be used. Depending on METT-T, helicopterborne forces may assault into an LZ in one wave or scheduled waves.

d. Mobility of Forces and Supplies. In the urban environment, movement of forces by helicopters may become a major requirement. Units engaged in house-to-house fighting normally suffer more casualties than units fighting in open terrain. These casualties must be evacuated and replaced quickly with new forces. At the same time, roads are likely to be crowded with resupply and evacuation vehicles and may also be blocked with craters or rubble. Helicopters provide a means of overcoming these obstacles.

e. Helicopterborne Assault Techniques. Helicopter assaults into enemy-held territory are extremely difficult (Figure 4-1 on page 4-16). One technique is to fly nap-of-the-earth down a broad street or commercial ribbon while attack helicopters and assault helicopter door gunners suppress buildings on either side of the street. Artillery preparations can be incorporated into the helicopter assault plan to suppress rooftop and building antiair threat. Feints and demonstrations in the form of false insertions can deceive the enemy as to the intended objective. Helicopters often deliver forces to the last covered position, short of the fighting, then return without exposure to enemy direct fire. Similar flight techniques can be used for transportation of supplies and casualties.

Figure 4-1. Helicopterborne Assault of a Built-Up Area

f. Combat Roles. Below are some of the roles that assault support helicopters may play in urban warfare:

- Conduct insertion and extraction of forces (See MCRP 3-11.7A, *Helicopter insertions and Extractions* [under development])
- Provide command/control and aerial retransmission
- Support CSS operations
- Provide air reconnaissance.

4203. Employment of Reconnaissance Forces. Reconnaissance units traditionally make maximum use of rural camouflage and terrain. Area, zone, and route reconnaissance are conducted from well-concealed vantage points through clandestine insertion by foot, vehicle, rubber boat, or helicopter. These methods of insertion may still apply in village and strip areas for collection of information, sensor employment, conducting OP radio relays, or a combination of missions. In major urban areas, undetected insertions are difficult to accomplish because of the high density civilian population.

The use of reconnaissance assets must be used wisely. Reconnaissance assets must not be risked indiscriminately if the information can be collected by other means. Consideration should be given to flank and rearward employment of ground reconnaissance units, in addition to close (0 to about 2,000 meters), distant (approximately 2,000 - 20,000 meters), and deep (from about 20,000 meters and beyond, plus preassault) missions. Reconnaissance of bypassed pockets of resistance, avenues of approach, canals, bridges, subways, and sewer systems (possibly in conjunction with combat engineers) is feasible. Observation, sniper, and listening posts manned by reconnaissance personnel are also missions that reconnaisance personnel perform. The limited combatant diving capability of division reconnaissance may be employed to advantage in many coastal cities.

4204. Military Police (MP). MPs can play a significant role in the urban environment. They may conduct battlefield circulation control, area security, enemy prisoner of war (EPW) and law and order operations. These operations require continuous coordination with MAGTF units as well as with host nation military and civilian police.

a. MP units support area damage control operations. With the increased possibility of rubbling, MP units report on and block off affected areas and reroute movement to alternate road networks.

b. When conducting EPW operations, MPs collect prisoners as far forward as possible and transport them to the rear. They operate temporary collecting points and holding areas to briefly retain EPWs and civilian internees. EPW operations are important in built-up areas because the rate of capture can be higher than normal.

c. Commanders must realize that MP support may not always be available and that infantry units may have to assume certain MP functions. These functions include:

(1) Providing convoy escort and security for lines of communications.

(2) Controling roads, waterways, and railroad terminals that are critical chokepoints in the main supply routes.

(3) Providing security of critical sites and facilities, including communications centers, government buildings, water and electrical supply facilities, command and control nodes, airfields, storage facilities, and other essential areas.

(4) Provide refugee control in close cooperation with host nation civil authorities.

(5) Collect and escort EPWs.

4205. Communications

a. General. Manmade structures create many challenges for CIS in an urban environment. Other CIS challenges include extensive commercial and communications networks and public service radio nets. Commercial communication networks are commposed of miles of underground protected cable which connect and control telephones and computer networks. Public service radio nets, (including police, fire, civil defense, and taxi) compete with existing communication systems. To communicate on an effective, continuous basis in the urban environment, commanders must minimize communication limitations imposed by this environment and maximize the advantage of existing civil communications capabilities.

b. Radio. Manmade structures will decrease the effectiveness of VHF radio communications. The physical locations of radios are critical for effective communications. Some specific locations to avoid include powerlines, underground areas, and steel bridges.

Very high frequency (VHF) radios provide the primary communication capability at the battalion level. It is imperative that these radios be employed as effectively as possible. Communicators can use the following techniques to enhance VHF communications equipment:

- Relocation of VHF radios to overcome phase cancellation due to numerous dead spots
- Use of the higher end (50 - 75 MHz) of the available frequency spectrum
- Use of retransmission facilities to improve the reliability of communications
- Use of radio remotes.
- Use of field expedient antennas.

In setting up retransmission stations, antennas should be camouflaged to blend in with urban surroundings and to prevent them from being targeted by enemy fires. Antennas can be concealed by such objects as water towers and steeples.

In larger cities with many tall buildings, it may be necessary to construct a terrain profile of the area of operations. A communications plateau can be deduced by calculating the average height of the structures in a zone. Operation of antennas above the median line of this plateau will result in better wave propagation and aid in retransmission.

c. Wire Communications. Wire may be used extensively in urban environments. It offers two primary advantages over radio communications:

- It is not susceptible to interference and absorption suffered by radio systems.
- Security of the overall system is greater because wire systems are essentially point-to-point.

Maximum use should be made of existing facilities for concealment. Sewers, subway tunnels, cable stairwells, elevator shafts, intact buildings, and pole lines offer ideal concealment for military wire systems. Precautions must be taken to lessen the possibility of vehicles tearing lines and disrupting communications. Existing telephone poles and tunnels can be used to keep wire above or below the street level.

Maximum use of existing commercial facilities should be made to supplement or replace military CIS. Telephone exchanges are linked together by underground cable systems and constitute a key resource in the urban communications system. These facilities are normally windowless buildings constructed of steel-reinforced concrete and provide ideal CPs and communications centers. The physical location of these exchanges, along with circuit diagrams and telephone directories, should be determined as soon as possible. Action should be initiated to secure these facilities early. Other facilities that contain extensive internal/external telephone systems may also be used as command centers. These include most municipal buildings, department stores, local police stations, and large business offices.

d. Messengers. Messengers provide a secure and flexible means of communication, particularly in an electronic warfare (EW) environment. However, careful selection of messenger routes is necessary to avoid pockets of enemy resistance. Disorientation in an urban environment occurs more often and more easily than in a rural environment. Marking facilities, transportation networks, and city streets on overlays often facilitates messenger service. Selected primary and alternate routes can often interface with designated logistics/resupply routes. Early restoration of street signs and recognizable landmarks are is essential.

e. Telephone Networks. Captured civil or military communications facilities can be utilized to augment organic communications capabilities. A civilian phone system can provide a reliable means of communication. If a telephone encryption device is not available, codes and authentication tables should be used.

Telephone technology differs from place to place, but there are universal concepts and practices in their operation. The most important concept to the military planner is network hierarchy.

Network hierarchy is a concept of organization by which a telephone company can rapidly and expeditiously allocate calls through a particular group of offices. As an ancillary function, it represents a control feature by which traffic can be rerouted to avoid congested areas or damaged facilities. This network hierarchy could play an important role in the planning and execution of the scheme of maneuver and fire support.

4206. Employment of Sensors

a. Sensor systems (seismic, infrared, acoustic, or magnetic) are normally employed in a rural context but also have traditional applications in village and strip area terrain. Use of sensors in urban areas is largely undocumented. In major urban areas, the capability for emplacement by air over select areas is reduced. Emplacement in, under, or around urban structures may also

.3

reduce microphone efficiency. Tactical employment will be degraded by such obstacles as concrete and steel structures, high-density population, and traffic.

b. Training in camouflage techniques for microphones and sensor terminals in an urban environment should be conducted. Operators should receive sufficient training to distinguish background-generated clutter from enemy foot, vehicular, and mechanized traffic. Use of and reliance on sensors may be seriously reduced during offensive operations in major urban areas. However, greater use of sensors will occur in subterranean defensive operations. Sensors should continue to be employed as far forward as possible and be centrally controlled.

c. Sensor employment should be coordinated with counterintelligence (CI) efforts for placement, readouts, and recovery. Sensors can also play a role in rear areas to augment rear security operations. Sensors, in themselves, are only component elements of an integrated intelligence system. Their employment is limited only by the imagination.

4207. Employment of Air Defense. Basic air defense doctrine does not change when units operate on urbanized terrain. The employment principles of mix, mass, mobility, and integration all apply to the employment of air defense assets. The commander must consider the following when developing his air defense plan:

a. Enemy air targets such as principal lines of communication and road and rail networks and bridges are often found in and around built-up areas.

b. Good firing positions may be difficult to find and occupy for long-range air defense missile systems in the built-up areas.

c. Movement between positions is normally restricted in built-up areas.

d. Long-range systems can provide air defense cover from positions on or outside of the edge of the city.

e. Radar masking and degraded communications reduce air defense warning time for all units. Air defense control measures must be adjusted to permit responsive air defense within this reduced warning environment.

f. Positioning of air defense weapons in built-up areas may be limited to more open areas without masking, such as parks, fields, and rail yards.

g. Stinger teams provide protection for infantry battalions the same as in any operation. When employed within the built-up area, rooftops normally offer the best firing positions.

h. Heavy machine guns emplaced on rooftops can also provide additional air defense.

Chapter 5

Combat Service Support and Legal Aspects of Combat

During combat in built-up areas, the terrain and the nature of operations create unique demands on the CSS system. Meeting these demands will require innovative techniques and in-depth planning.

Section I

Combat Service Support

5101. Introduction. CSS is "the essential capabilities, functions, activities, and tasks necessary to sustain all elements of operating forces in theater at all levels of war" (Joint Pub 1-02). The CSS organization throughout the MAGTF, down to battalion level, must provide a responsive CSS system that can perform all functions and tasks associated with meeting identified CSS requirements. This responsibility does not change during urban operations. Forward support for combat forces continues to be the basic concept governing CSS operations. No significant changes in doctrine or organization are required. However, the characteristics and nature of urban combat do affect how CSS is provided.

5102. Combat Service Support Resupply, Maintenance/Repair, and Replacement

a. Resupply of Ammunition. Combat in built-up areas is characterized by extremely high ammunition expenditure rates. Not only do individual Marines expend more ammunition, but they also use greater quantities of munitions such as smoke, concussion, and fragmentation grenades; AT4s; claymore mines; demolitions; and so on. The ammunition consumption rate for the first day of combat in a built-up area can be up to four times the normal rate. Even though this rate decreases during succeeding days, consumption remains high. Leaders should plan to meet these high consumption rates. The plan must include how ammunition and demolitions are to be moved forward to the companies. AAVs should be designated for the movement of ammunition if rubble or glass prevents wheeled-vehicle traffic. Marines may need to organize into carrying elements if streets are blocked by rubble.

b. Resupply of Fuel. The amount of bulk fuel needed by forces during combat in built-up areas is greatly reduced. Combat vehicles normally use less fuel in built-up areas because they travel shorter distances and perform less cross-country traveling. Engineer equipment and power generation equipment may use more fuel, but requirements are small. A unit may not use much fuel daily, but when it does need fuel, a problem exists in delivering bulk fuel to the vehicle. In open terrain, a vehicle that has run out of fuel can be recovered later. In built-up areas, the same vehicle is probably going to be lost quickly. Leaders should plan for and provide the means of moving limited amounts of bulk fuel forward to combat units.

c. Maintenance and Repair. Maintenance contact teams must operate well forward to support units fighting in built-up areas. Although some maintenance operations may be consolidated in civilian facilities, many vehicles will have to be repaired near fighting positions. Battle damage assessment (BDA) and repair procedures allow mechanics to analyze, repair, and return damaged vehicles to a serviceable condition. Other considerations:

(1) Combat in built-up areas generates a high demand for tires.

(2) The dust and rough handling that are characteristic of combat in built-up areas also place great strains on communications and night vision devices.

(3) The unit armorers and their small-arms repair kits provide only limited maintenance. S-4s should plan for increased weapons maintenance demands.

d. Replacements. Units conducting combat in built-up areas must expect high casualty rates. Casualty reports must be prepared scrupulously and forwarded via the S-1 to the battalion personnel officer located at the consolidated administration center (CONAD). Other considerations:

(1) The S-4s must plan to expedite the evacuation of wounded from the built-up area. Location of battalion aid stations and evacuation routes must be planned and disseminated to the lowest level. Higher casualty rates should be expected and may require the stockpiling of medical supplies and augmentation of additional HSS personnel.

(2) The personnel officer quickly processes replacements and coordinates their movement forward via the S-1 officer.

(3) Proper accountability of personnel at all levels is required. Timely and accurate personnel accountability and strength reporting support decisionmaking and initiate the replacement cycle. Leaders maintain accountability through the use of battle rosters and by establishing procedures for periodic reporting of numbers to higher commands. During combat, strength reports are provided on request or as significant changes occur.

5103. Critical Classes of Supply. Requirements for supply and resupply will vary from those encountered during operations in other tactical environments. The critical classes of supply listed below will assume greater importance during combat inside a built-up area.

a. Class I (Rations). The process of ordering and moving rations to forward units is complicated by the dispersed nature of combat in built-up areas and its increased caloric demands on Marines. Hot meals should be provided when practical.

(1) Combat in built-up areas not only causes great stress on Marines, but also requires great physical exertion. This combination of stress and exertion quickly causes dehydration. Unless potable water is continuously provided, Marines will seek local sources, which are usually contaminated by petroleum, oils, and lubricants (POL) runoff, sewage, bacteria, or unburied corpses. Marines who are not provided sufficient quantities of potable water become casualties as a result of drinking from contaminated sources or from dehydration. Waterborne contaminants can quickly render entire units combat ineffective.

(2) Water and other liquid supplements such as coffee, tea, or soup that must be forwarded to exposed positions may need to be backpacked at night.

b. Class II (General Supplies). Combat in built-up areas increases wear and tear on combat uniforms and footgear. Supply officers should increase on-hand stocks of uniforms, boots, and individual combat equipment such as protective masks and armored vests. Nuclear, biological, and chemical (NBC) protective suits become damaged quickly when worn in the urban environment. Extra stocks of these and protective mask filters should be kept on hand. (For further information on NBC considerations, see Appendix G.) Limited amounts of other Class II and IV (barrier materials) items may be available locally. These should be gathered and used if authorized and practical. Local shops may provide such items as hand tools, nails, bolts, chains, and light construction equipment, which are useful in preparing a defense or reducing enemy-held positions. The unit's organic wire communications net may be augmented with locally obtained telephone wire and electrical wire.

c. Class III (POL). Bulk fuel may have to be brought forward from fuel tankers by using 5-gallon cans. One man can carry a fuel can long distances, even over rubble, if it is lashed to a pack frame. Supplies of bulk Class III items and some prepackaged POL may be available at local gas stations and garages. These may be contaminated or of poor quality. The S-4s should coordinate to have fuel tests performed.

d. Class IV (Barrier Materials). If a unit is defending a built-up area, the requirements for Class IV materials are less than in other areas. This class of supply is probably the most available locally. After coordinating the effort with higher headquarters, S-4s and supporting engineer officers gather materials for use in strengthening the defense. Cargo trucks, wreckers, or recovery vehicles from maintenance platoons or engineer units can be used to load and move barrier materials. The defense of a built-up area may require concertina wire and/or barbed wire to restrict the enemy infantry's movements. Barriers can be built of abandoned cars and buses, which are dragged into position, turned on their sides, and chained together through the axles.

e. Class V (Munitions). Combat in built-up areas causes ammunition to be expended at extremely high rates. Commanders should plan for early resupply of explosives, grenades, and ammunition for small arms, direct fire, and indirect fire.

(1) In the defense, the S-4 should prestage as much ammunition as practical in dispersed storage areas. These storage areas should be protected, and they should be easily accessible to the forward defensive positions. In the offense, attacking forces should not be overburdened with excessive ammunition. Mobile distribution points may be set up as low as at company level.

(2) Leaders should plan to continuously deliver ammunition to the leading units as they advance. Ammunition may be carried by armored vehicles close behind the advancing forces or by designated carrying elements. Modern ammunition, particularly missiles, is characterized by extensive amounts of packing material. S-4s must remove the ammunition depot overpack before the ammunition is transported forward. Resupply by helicopter (prepackaged slingloads) may be feasible.

(3) Removing the overpack from large amounts of ammunition can be a time-consuming process. It may require being augmented by available Marines. If carrying elements are used to move ammunition forward, a Marine can carry about 75 to 90 pounds by using a pack frame. Bulky and heavier loads can be carried by lashing them to litters and using teams of two to four men. Loads of up to 400 pounds can be carried moderate distances using four-man teams.

Note: DO NOT use medical corpsmen to carry ammunition forward as described above—it is a violation of the Geneva Accords.

f. Class VIII (Medical Supplies). Because of the decentralized nature of combat in built-up areas, medical supplies should be dispersed throughout the unit, not just consolidated with the aid station and the individual corpsman. Marines should carry additional bandages, and units should have additional splints and stretchers.

5104. Health Service Support. The regimental and battalion surgeons are responsible for planning and executing HSS within their respective units. The most critical functions during combat in built-up areas include preventive medicine, trauma treatment, and evacuation. In addition, there should be a plan for the treatment and evacuation of NBC-related casualties that could occur in combat in built-up areas.

a. Combat in built-up areas exposes Marines not only to combat wounds, but also to the diseases endemic to the AO. Commanders must enforce preventive measures against the spread of infectious diseases. The unit surgeon advises the commander on how best to implement the use of prophylactics.

b. Corpsmen should be placed at strongpoints, at battle positions, and in units likely to be fighting in somewhat isolated positions.

c. Battalion aid stations must be placed farther forward than in operations in more open terrain. Protection offered by urban structures will permit this forward location, but the limited range of observation will require that personnel involved in evacuation be aware of their surroundings and the threat.

d. Corpsmen attached to rifle platoons are trained in the treatment of traumatic injuries; however, they can quickly become overwhelmed by the number of casualties needing care. If time and training are available before the conduct of urban operations, the commander should increase first aid training. Immediate first aid support by a buddy to an injured Marine may be the difference between life and death and will greatly assist the corpsmen. The aid station should plan to care for the increased casualties that are inherent to combat in built-up areas, as the incidence of crushing injuries, eye injuries, burns, shrapnel wounds, and fractures increases.

e. The difficulties encountered when evacuating casualties from urbanized terrain are many and require innovative techniques and procedures. The planning for evacuation on urbanized terrain must include special equipment requirements, use of litter teams, use of air MEDEVAC, use of an ambulance shuttle system, communications requirements, and techniques for locating casualties.

(1) Special equipment requirements include ropes, pulleys, skid litters, axes, crowbars, and other tools used to break through barriers.

(2) Although litter teams are labor intensive, they are required for evacuation from buildings, where casualties can occur on any level. Also, rubble in the streets, barricades, and demolition of roads will require a heavy reliance on litter teams.

(3) Communications presents one of the biggest obstacles to casualty evacuation. Urbanized terrain renders LOS radios ineffective. Also, individual Marines normally do not have access to radios. Therefore, a wounded Marine within a building may be difficult to find and evacuate. The unit SOP should contain alternate forms of communication such as colored panels or other forms of markers that can be displayed to hasten rescue when the battle is over. Also, a systematic search of the area after the battle may be required to recover casualties.

f. The use of local medical facilities, hospitals, professional medical help, and medical supplies may be available during combat in large, built-up areas. The commander must adhere to the guidelines established within the theater as to when and how these facilities can be used. If civilians are wounded in the unit's area, the commander is responsible for providing them with aid and protection without disrupting military operations. A commander cannot confiscate civilian medical supplies unless he makes provisions to provide adequate replacements if civilians are wounded.

g. The commander is responsible for the evacuation of deceased personnel to the nearest mortuary affairs collection point, whether they are U.S., allied, or enemy personnel, or civilians. (See Joint Pub 4-06, *JTTP for Mortuary Affairs in Joint Operations*.) Some general considerations for the handling of deceased personnel include:

(1) The theater commander is the approval authority for hasty burial.

(2) The deceased person's personal effects must remain with the body to assist in the identification of the body and to facilitate shipment of personal effects to the next of kin.

Retention of personal items is considered looting and is, therefore, punishable by the Uniform Code of Military Justice (UCMJ).

(3) When operating under NBC conditions, the bodies of deceased personnel should be decontaminated before removal from contaminated areas to prevent further contamination and casualties.

(4) Care must be exercised when handling deceased personnel. Improper handling of deceased personnel can result in a significant decrease in unit and civilian morale.

5105. Personnel Services. Timely and accurate personnel services are just as important during combat in built-up areas as in any other operation.

a. The S-1 plans for all personnel services that support and sustain the morale and fighting spirit of the battalion. Among the most important of these services are:

- Replacement operations
- Strength accounting
- Casualty reporting.

These functions allow the commander to maintain accountability of his forces and to redistribute personnel or change taskings for subordinate units to maintain combat power and tempo.

Chapter 6

Noncombatant Considerations in Urban Operations

Noncombatants can have a significant impact on the conduct of military operations.

Section I

6101. Introduction. Commanders must be well educated in the legal aspects of military operations involving noncombatants. The urban environment contains numerous legal pitfalls in dealing with the civilian population. Adherence to the ROE and the standards of the Geneva Accords will keep the commander on firm legal ground.

6102. Noncombatant Impact. Large concentrations of civilians can greatly impede tactical operations. Noncombatants may have the following effects on military operations:

a. Mobility. Noncombatants civilians, attempting to escape the battlespace, can block military movement. Commanders plan routes to be used by civilians and seek the assistance of the civil police in refugee control.

b. Firepower. The presence of noncombatants can restrict the use of firepower. Areas may be designated no-fire areas to prevent noncombatant casualties. Other areas may be limited to small-arms fire and grenades. The control of fire missions may be complicated by the requirement for positive target identification. Detailed guidance on the use of fires in the presence of noncombatants is promulgated by the MAGTF commander. In the absence of guidance, the general rules of the law of land warfare apply.

c. Security. The presence of noncombatants increases security requirements in an urban environment to preclude:

- Noncombatants entering defensive areas
- Pilferage of equipment
- Sabotage
- Terrorism

d. Obstacle Employment. The presence and movement of Noncombatants will influence the MAGTFs commander's obstacle plan. Minefields may not be allowed on designated refugee routes or, if allowed, must be guarded until the passage of refugees is completed. The use of boobytraps may be curtailed until noncombatants have been evacuated. Commanders need to review and understand current national and international treaties concerning the employment of mines and boobytraps.

6103. Command Authority. The limits of command authority over government officials and noncombatants must be established. The host government's responsibility for its populace and territory can affect the commander's authority in civil-military matters. In less secure areas, where the host government may be only partly effective, the commander may have to assume greater responsibility for noncombatants.

6104. Source Utilization. Operations in highly populated areas may require the diversion of men, time, equipment, and supplies to accomplish humanitarian tasks. If host government agencies collapse, the impact on military resources could be substantial.

6105. Health and Welfare. The disruption of noncombatant health and sanitary services may sharply increases the risk of disease among both noncombatant and military personnel.

6106. Law and Order. Marines may have to augment civilian forces to protect life and property and to restore order. Marines may also have to secure vital government facilities for the host nation. (For more information on how to control civilians violating civil law, see FM 9-15, *Civil Disturbances*.)

6107. Public Affairs Officer and Media Relations. Free media access to units in the field is desirable; however, operational security, existing guidelines, and ROE considerations have priority. All members of the media should have an escort officer. Escort officers may be detailed from line units when public affairs personnel are unavailable. Enforcing established guidelines helps prevent negative publicity which could jeopardize the operation or national and strategic objectives.

6108. Civil Affairs Activities and Psychological Operations (PSYOP).

a. General. CA and PSYOP are mutually supporting and may have essential roles in MOUT. They can be critical force multipliers. CA and PSYOP may offer the possibility of victory without the destruction, suffering, and horror of protracted urban conflict.

(1) CA activities are activities performed by commanders, staffs, department of defense elements and units, and foreign military forces that (1) embrace the relationship between military forces and civil authorities and populations in areas where military forces are present; and (2) involve application of civil affairs functional specialty skills, in areas normally the responsibility of civilian government, which enhance conduct of civil-military operations (Joint Pub 1-02). CA activities are inherently civil-military in nature. They assist in coordinating military efforts to strengthen host government legitimacy, and help prevent or reduce violence by bridging critical gaps between civil and military sectors. MAGTF responsibilities for CA activities are determined during planning. U.S. Army CA units can be requested to support the MAGTF. The Marine Corps Civil Affairs Group (CAG), located in the Marine Corps Forces Reserve, may be activated and assigned to the MAGTF.

(a) CA activities are characterized by applications of functional specialties in areas normally the responsibility of indigenous government or civil authority. CA activities may extend to assumption of government functions required in an occupied territory during or immediately subsequent to hostilities. Refer to Joint Pub 3-57, *Doctrine for Joint Civil Affairs*.

(b) CA support to a MAGTF would normally include procuring local goods and services for military use, reducing civil interference in military operations, and assisting a MAGTF commander in fulfilling moral and legal obligations to the local government and noncombatant populace. See MCWP 3-33.1, *MAGTF Civil Affairs*, for more detailed information.

(2) Psychological operations are planned operations to convey selected information and indicators to foreign audiences to influence their emotions, motives, objective reasoning, and ultimately, their behavior. The purpose of psychological operations is to induce or reinforce foreign attitudes and behavior favorable to the originator's objectives (Joint Pub 1-02).

PSYOPs are planned and conducted by the MAGTF. PSYOPs are integrated and coordinated down to the lowest tactical level. PSYOPs can weaken enemy resistance and reduce its combat effectiveness. PSYOPs can also help prevent civilian interference with military operations.

PSYOPs may include the use of television, radio, posters, leaflets, and loud speakers, to disseminate propaganda and information. Television, including video tapes, is one of the most effective mediums for persuasion. In areas where television is not common, televisions may be distributed to public facilities and selected individuals. See MCWP 3-36.2/FM 33-1, *Psychological Operations*, for more detail.

6109. Refugee Control. The control of refugee movements in combat operations is primarily a host nation function.

The most important principle in refugee control is reliance on host government forces (military/civilian) whenever possible. U.S. forces will have limited refugee control capabilities. If a host government is incapable of performing the function, MAGTFs and MPs may be required to conduct any or all of the following refugee control measures:

- Enforce standfast orders or prohibit other unauthorized movements
- Screen and take custody of enemy agents or other hostile personnel discovered among the refugees
- Screen refugees for suspicious activities (including detection of infiltration by guerrillas, enemy agents, and escaping members of hostile forces)

.3

- Perform traffic control on designated refugee routes
- Escort refugee columns
- Direct or divert refugee columns to secondary roads and through areas not used by military elements
- Perform police duties such as traffic control or security at refugee assembly areas, collecting points, and centers or camps
- Enforce prescribed restrictive measures such as curfews, identification and pass systems, restricted area restraining lines, no-passage lines, checkpoints and roadblocks, travel permits or manifests, prohibition on the use of vehicles, priorities of movement, and limits on personal belongings.

Section II

6201. Commander's Legal Authority and Responsibilities

a. General. Leaders at all levels of command are responsible for protecting noncombatants and minimizing collateral damage to their property. Looting, vandalism, and brutal treatment of civilians are strictly prohibited by the Geneva Accords. Individuals who commit such acts may be severely punished. Civilians, and their religions and customs, must be treated with respect. Women are especially susceptible to many forms of abuse, and special care must be taken to protect them. Urban warfare offers many opportunities for abuses of the civilian population. This section focuses on those civilian-military confrontations that are most common in the urban environment.

b. Control Measures. Commanders may enforce control measures to conduct operations, maintain security, or ensure the safety and well-being of the civilians.

(1) Curfew. A commander with the mission of defending a town could establish a curfew to maintain security or to aid in control of military traffic. However, a curfew would not be legal if imposed strictly as punishment.

(2) Evacuation. A commander can require civilians to evacuate towns or buildings if the purpose of the evacuation is to use the town or building for essential military purposes, to enhance security, or to safeguard those civilians. The commander must specify and provide security for those being evacuated. Food, clothing, and sanitary facilities should be provided at the destination.

(3) Forced Labor. The Geneva Accords prohibit the use of civilians in combat operations. However, they may be required to perform some forms of forced labor prior to and after the conduct of combat operations. Guidelines for use of civilian labor should be published. The commander may force civilians over 18 years of age to work if the work does not oblige them to take part in military operations. An example of forced labor that is permitted would be maintenance of public utilities as long as those utilities are not used in the general conduct of the war. Jobs can also include services to the local population such as care of the wounded and burial. Civilians can also be forced to help evacuate and care for military wounded, as long as doing so does not involve any physical danger. Prohibited forced labor include digging entrenchments, constructing fortifications, transporting supplies or ammunition, or acting as guards. Only civilians who are volunteers can be employed in such work.

c. Civilian Resistance Groups. Another situation that commanders might encounter is combat with a civilian resistance group. The combatant status of these groups varies with their organization, methods, and adherance to the Geneva Accords. For example:

(1) Civilians accompanying their armed forces (with an identity card authorizing them to do so) are treated as EPWs when captured. Examples would include civilian members of

military aircraft crews, war correspondents, supply contractors, and members of labor units or of service organizations responsible for the welfare of the armed forces.

(2) Captured civilians of a nonoccupying territory who have taken up arms against an invading enemy without time to form regular armed forces are considered EPWs if they wear a distinctive insignia that can be seen at a distance, carry their weapons openly, and operate according to the rules and customs of warfare. Other civilians who provide assistance to such groups may not be entitled to status as combatants, depending on whether they are actually members of the resistance group. They are normally best treated as combatants until a higher authority determines their status.

(3) Armed civilian groups that do not meet the criteria of a legal resistance group or individuals caught in the act of sabotage, terrorism, or espionage are not considered legal combatants. If captured, they will be considered criminals under the provisions of the law of land warfare. They should be detained in a facility separate from EPWs and should be quickly transferred to the MPs. Reprisals, mass punishments, hostage taking, corporal punishment, pillaging, or indiscriminate destruction of property are prohibited.

(4) The law of land warfare lets the commander control the civil population under the conditions already described using his own organic resources. However, language and cultural differences between U.S. and foreign personnel make it good practice to use local authorities such as the police for such purposes. Use of the police does not relieve a commander of his responsibility to safeguard civilians in his area.

d. Protection of Property. Like civilian personnel, civilian buildings and towns normally have a protected status. As such, they are not legal targets. They lose their protected status if authorities determine that the enemy is using them for military purposes. All doubt as to whether a town or building is a military target must be resolved before it can be targeted. For example:

(1) If the enemy is using a building or a portion of the town for military purposes—such as a supply point or a strongpoint—that building or that portion of the town is a legal target. However, engagement of the target should be limited to the buildings or portion of the town being occupied by the enemy.

(2) Normally, religious, historical, and cultural objects and buildings are not legal targets. They are sometimes marked with symbols to signify their status. Medical facilities are identified under the internationally recognized Red Cross, Red Crescent, Red Lion, or Red Star of David symbols. The fact that such symbols are absent does not relieve a commander of his responsibility to protect objects he recognizes as having religious, cultural, medical, or historical value.

(3) The misuse of such objects by the enemy is grounds to disregard their protected status. A demand should be made for the enemy to stop his misuse of the protected object before any action is taken. However, if an enemy FO or sniper uses a church for an OP, a commander would be justified in destroying it immediately because of the immediate danger posed to his troops. Another example is a religious shrine that is used as a telephone switchboard; in this case a warning would be appropriate because it would take some time to dismantle the system. Attacks on these types of targets should be limited to the least amount necessary to neutralize the enemy installations.

(4) The destruction, demolition, or military use of other buildings is permitted under the law of land warfare if required by clear military necessity. Destroying a house to obtain a better field of fire would be a legal act. But destroying it as a reprisal would not. However, firing on any houses that are occupied or defended by an enemy force is usually legal.

Chapter 7

The Urban Environment and Restrictions to Operations

"The responsibility of officers . . . and the training necessary are of a very different order from their responsibilities and training in ordinary military duties. Instead of striving to generate the maximum power with the forces available, the goal is to gain decisive results with the least application of force and the consequent minimum loss of life."

USMC, 1940 (FMFRP 12-15, *Small Wars Manual*)

Section I

Military Operations Other Than War

7101. Introduction. One of the most likely missions that U.S. Marines will undertake abroad will be military operations other than war (MOOTW). These missions typically will take place in the Third World and are normally associated with urban areas. MOOTW focuses on deterring war, resolving conflict, promoting peace, and supporting civil authorities in response to domestic crises. It is important to remember that political considerations permeate at all levels. Although some types of MOOTW, such as peace enforcement, may have the same characteristics of war, they normally have more restrictive rules of engagement (ROE).

Urban environments are inherently complex, and sometimes unpredictable. Local governments may have limited functions. Services may be overburdened, leading to allocation on the basis of personal connections and corruption. Services may be provided by a mixture of formal, high-technology and informal, labor-intensive systems. Large portions of the population may be poor, young, and underemployed. Nevertheless, there may be a high degree of social organization and interaction. Word of mouth and informal organizations may be as important as mass media and formal organizations to the communication structure of the area. Indigenous populations may have experienced centuries of foreign interventions and become quite adept at innovative ways of communicating. These factors and others often combine to make MOOTW a challenging and frustrating endeavor.

7102. Types of MOOTW. MOOTW encompasses a wide variety of military operations, that may take place in an urban environment. Forward deployed MAGTFs are ideal forces to respond to contingency operations involving MOOTW. These operations include:

- Combatting Terrorism
- DOD support to counterdrug operations
- Enforcement of Sanctions/MIO
- Enforcing Exclusion Zones

- Humanitarian Assistance
- Military Support to Civil Authorities
- Nation Assistance/Support to Counterinsurgency
- NEO
- Peace Operations
- Recovery Operations
- Show of Force Operations
- Strikes and Raids.

7103. Recent Experience and Lessons Learned. Recent Marine Corps urban experiences— Lebanon (1982 - 1984); Grenada (1983); Panama (1989); and Somalia (1992 - 1994)— are examples of MOOTW. Although each operation was unique, the lessons learned are similar. For detailed information, commanders may obtain after action reports through the Marine Corps Lessons Learned System (MCLLS). Some of the lessons learned include:

(1) Demonstrate a powerful military presence. Where and when possible, use tanks, armored vehicles, heavy automatic weapons, fixed-wing aircraft, and attack helicopters.

(2) PSYOP, CI, HUMINT teams, and linguists, are essential to communications and intelligence gathering.

(3) Topographic detachment is useful to provide necessary sketches and maps.

(4) UAVs can be used for security, reconnaissance, and surveillance.

(5) Whenever entering an AO, isolate the area, patrol, and use intelligence to evaluate specific buildings or groups of buildings for search-and-clear operations. PSYOP, CI, HUMINT teams, and linguists can be used, as appropriate, to support patrols.

(6) Take advantage of the night by using night vision devices.

(7) The most effective means of dealing with snipers is with counter snipering operations.

(8) Snipers are also a tremendous asset to overwatch patrol movements, provide intelligence, and direct fire support. (See MCWP 3-15.3.)

(9) Random patrolling, traffic control points, random checkpoints/roadblocks, and ambushes disrupt enemy movements, reduce looting, seize curfew violators, and convince the population that order has been restored.

(10) Use armored vehicles to move quickly to new checkpoints and set up roadblocks or ambushes. They also increase protection for infantry. Their night vision sights also provide increased security and recognition during periods of limited visibility.

(11) Extensive patrolling provides current combat information used to maintain situational awareness. It helps to familiarize units with the AO and presents a strong military presence. (See Appendix A, Section VII)

(12) Tanks, armored vehicles, or attack helicopters increase military presence and can provide immediate, accurate gunfire.

(13) Before search and clear of a building, use loudspeakers to explain your intentions to noncombatants and enlist their cooperation.

(14) Use a minimum amount of force when dealing with noncombatants and their property. This will assist you in obtaining them to cooperate and may prevent turning them into active combatants. Assign engineers to search-and-clear forces, as necessary.

(15) MPs can instruct small unit leaders on the proper method of conducting vehicle, individual, and room search techniques. In some cases, it may be appropriate to attach MPs to search-and-clear forces. (See MCWP 3-34.1, *Military Police in Combat.*)

(16) Maintain a strong, mobile reserve.

(17) Avoid becoming predictable.

7104. Urban Terrorism. Terrorism is "the calculated use of violence or threat of violence to inculcate fear; intended to coerce or to intimidate governments or societies in the pursuit of goals that are generally political, religious, or ideological" (Joint Pub 1-02). Terrorism is a criminal act that is symbolic in nature. Its intent is to influence an audience other than the victim.

The likelihood of Marine involvement with terrorism increases in the urban environment. Terrorism in an urban environment can occur across the spectrum of conflict. Terrorists are an integral element in an insurgency and also play a major role in unconventional warfare. Terrorists can disrupt economic functions, expose a government's incompetence, eliminate opposition leaders, and elevate social anxiety. (See MCRP 3-02D, *Combating Terrorism*; and MCRP 3-02E, *The Individual's Guide for Understanding and Surviving Terrorism.*)

In countering this threat, Marines should determine whether it is internally or externally directed terrorism. Terrorism rooted externally must be severed from its roots. Against internal terrorism,

Marines should attempt to penetrate the infrastructure and destroy the leadership of the terrorist groups.

The keys to defeating terrorists are awareness, education, and intelligence in order to deny, deter, delay, and detect terrorist acts. Rapid coordination among military units, military services, local police, and other government agencies is essential in denying the terrorists targets and refuge.

a. Nature of Urban Terrorism. Terrorism is generally an urban phenomenon. Incidents of terrorism are increasing in the international community, and in the United States, and pose a significant threat for U.S. military forces. Military personnel in noncombatant roles are considered terroist targets. For example, the attack on the Marine compound in Lebanon can be considered a terrorist attack instead of a military action because the Marines were officially in a noncombat activity (a multinational peacekeeping activity).

b. Terrorists in the Urban Area. Cities are target rich environments for terrorists. They offer ample cover and concealment. They contain large amounts of resources, transportation, communications, and potential recruits or sympathizers to their cause. Cities also provide the media attention essential for terrorists to accomplish their mission.

c. Combating Terrorism

(1) Counterterrorism. Counterterrorism includes the full range of offensive measures to prevent, deter, and respond to terrorism (Joint Pub 1-02). Population protection and control to combat the terrorist threat may be a major effort within the urban environment. The ability to protect the general population from attack and intimidation is essential.

(2) Antiterrorism. Antiterrorism includes defensive measures used to reduce the vulnerability of individuals and property to terrorist acts, to include limited response and containment by local military forces (Joint Pub 1-02). The extent of these measures varies based on assessments of the local threat. These include personnel awareness and knowledge of personal protection techniques. They also include crime prevention and physical security programs to "harden" the target, making installations and personnel less appealing as terrorist targets.

Section II

Restrictive Conditions

7201. Introduction. Commanders can expect to encounter restrictions on the use of force in MOOTW, especially in the urban environment. While we tend to focus on a traditional image of urban combat, MOUT extends beyond the house-to-house, close-combat effort to seize a city. Recent history provides many examples of the need for precise application of force to minimize collateral damage. MOOTW generally requires the application of force in varying degrees. Future missions in urban areas will likely follow this same pattern. While Marine forces must be prepared for more intensive urban combat, they must also train to apply only the level of force necessary to accomplish the mission. *Our tactics may have to rely more on innovation than firepower.*

Basic tenets of manuever warfare remain the same in MOOTW, but the tactics, techniques, and procedures may have to be modified to stay within established ROE and to avoid unnecessary collateral damage.

7202. Rules of Engagement. ROE are "directives issued by competent military authority which delineate the circumstances and limitations under which United States forces will initiate and/or continue combat engagement with other forces encountered" (Joint Pub 1-02). ROE for Marine forces are usually set by the JTF commander. They are based on National Command Authorities (NCA) guidance, the mission, the threat, the laws of war, and host nation constraints on force deployment.

MOOTW focuses on deterring war, resolving conflict, promoting peace, and supporting civil authorities in response to domestic crises. It is important to remember that political considerations permeate at all levels. Although some types of MOOTW, such as peace enforcement, may have the same characteristics of war, they normally have more restrictive rules of engagement (ROE). Restraint requires the careful balancing of the need for security, the conduct of operations, and the political objective. Excessive force antagonizes those parties involved, thereby damaging the legitimacy of the organization that uses it while possibly enhancing the legitimacy of the opposing party. Ensure all personnel understand the ROE and incorporate the practice into training exercises at all levels.

ROE must be fully integrated in the MAGTFs plans. As an example, table 7-1, presents the ROE used in Operation United Shield:

ROE Used for Operation United Shield
Nothing in these Rules of Engagement limits your right to take appropriate action to defend yourself and your unit.
a. You have the right to use deadly force in response to a hostile act or when there is a clear indication of hostile intent.
b. Hostile fire may be returned effectively and promptly to stop a hostile act.
c. When US forces are attacked by unarmed hostile elements, mobs and/or rioters, US forces should use the minimum force necessary under the circumstances and proportional to the threat.
d. Inside designated security zones, once a hostile act or hostile intent is demonstrated, you have the right to use minimum force to prevent armed individuals/crew-served weapons from endangering US/UNOSOM II forces. This includes deadly force.
e. Detention of civilians is authorized for security reasons or in self-defense.
Remember: 1. The United States is not at war. 2. Treat all persons with dignity and respect. 3. Use minimum force to carry out mission. 4. Always be prepared to act in self-defense.

Table 7-1. ROE Used in Operation United Shield

7203. Planning. Commanders and planners thoroughly study each situation to identify the restrictive measures necessary to prevent needless civilian casualties and damage while also protecting the lives of the Marines who will execute the mission.

a. Mission/Command Guidance. The mission statement itself may determine much about the required level of force for a given operation. A humanitarian relief operation may require only individual weapons for security and personal protection. A raid on an insurgent communication center may need more. The requirements for an NEO conducted in a permissive environment would be different from those in a nonpermissive one. The key is to conduct a thorough analysis of the mission and determine the necessary combat power and how it must be applied to accomplish the mission. Many of these missions will be accompanied by a great deal of additional guidance concerning operational restrictions. These restrictions may come from the NCA, State Department, Department of Defense (DoD), or the joint task force (JTF) commander, and other sources. Such guidance must be carefully studied and incorporated into the plan. The commander can impose restrictions on his own forces that are more stringent then those dictated by higher authority. A careful analysis of the ROE will provide key insights into other restraints that the commander may wish to impose on his own forces.

b. Intelligence/Threat. Obtaining detailed and specific intelligence on the target and the area is essential to minimize collateral damage. Information on the physical layout of an area is often as important as information about the enemy. This is especially true for an NEO where avoiding conflict is a primary concern.

c. Location. The geographic area into which operations are to be conducted will normally dictate some restrictions on operations. In an urban environment, where the city is located in a desert, a common restriction is placed on the destruction of water producing facilities. If the operation is conducted where there is a large Hindu population, special protection may be given to cattle. If the operation is on an island nation, restrictions may be placed on destroying port or airfield facilities. In an urban environment with a large amount of shanty towns, restrictions may be placed on the use of incendiary devices.

d. Restrictions on Force Employment. The *right of self-defense is inherent in all military operations*. Marines should not be placed in a situation where they are not provided the means and authority to protect themselves or others. Beyond that, the restrictions placed on the forces conducting a particular operation depend on a variety of factors. These include:

- NCA guidance
- Type of mission
- Nature of the threat
- Political environment
- Host nation attitude.

e. Lethal Fires. Most restrictions in MOOTW are normally concerned with the application of fires. For instance, a NEO in a permissive environment may not require the landing of fire support assets However, a raid may require supporting arms to be landed with detailed instructions on their use. Aviation in MOOTW is commonly restricted in the amount and type of ordnance it can deliver because of the potential for collateral damage. However, in both a permissive and nonpermissive environment, supporting arms still should be planned. The most common types of fire support assets that will be restricted in the urban environment include artillery mortars, tanks, armored vehicles, and fixed-wing attack aircraft.

f. Search and Clear Buildings. Marine forces organize to search and clear a building under restrictive conditions in the same ways as discussed in Appendix C. However, because noncombatants may be occupying rooms within the building, clearing procedures must be modified as follows:

(1) Squad organization remains the same and is reinforced as appropriate (see Appendix A), except assault elements become search elements. Search elements enter and search rooms in the same manner as assault elements enter and clear rooms, as discussed in Appendix A.

(2) ROE must be identified and known by all personnel before entering or clearing a room or building.

(3) PSYOP or CA teams can help remove noncombatants before a search or battle starts.

(4) The building or area of search should be isolated with security forces and the intent to search the building should be communicated via loudspeakers.

(5) Assault forces move to the building and begin systematically to search room by room and floor by floor. Marines should carry ammunition and other equipment as in normal urban operations. However, Marines will be restricted in the use of their weapons and other equipment as dictated by the ROE. If permitted, a riot control agent (such as CS gas) can be used to assist in the clearing of a building. Additionally, shotguns and stun grenades are more appropriate than rifles and machine guns or fragmentation grenades for this type of restricted, close-combat operation.

g. Communications and Information Systems (CIS).

(1) CIS must be able to satisfy the C2 requirements of MOOTW. CIS systems must provide the commander with the ability to rapidly collect, process, analyze, and exchange information in order to keep the potential for violence to a minimum. These systems must make available the information needed, when it is needed, whenever it is needed. When supporting arms are on standby, fire support coordination and air traffic control nets are particularly critical.

(2) CIS officers (G/S-6) must carefully analyze the mission, situation, threat, and specific characteristics of the urban environment to design reliable, flexible, and responsive CIS. These systems should include single and multichannel radios, tactical telephones, and tactical satellite communications. A detailed backup CIS plan should also be prepared. The success of CIS in MOOTW depends on designing, planning, and employing CIS that satisfies the commander's C2 requirements.

Appendix A

Organization for Urban Combat and Fundamental Combat Skills

Successful combat operations in built-up areas depend on training in fundamental skills and proper employment techniques and procedures.

Before conducting urban operations, Marines must be trained in fundamental combat skills to conduct urban operations. Training for this unique environment will reduce casualties because properly prepared Marines will be able to fight more effectively. This appendix supersedes all other fundamental combat skills taught for fighting in urbanized terrain. Urban combat will require Marines to use their best initiative and judgment in applying the techniques and procedures described in this appendix.

Section I

Organization

1. Structure. Basic infantry organization does not change when the battlefield is moved into an urban area. However, infantry units may be task organized or reinforced down to squad level in order for small units to fight and win in this violent, three-dimensional battlespace. Several new terms that address organization and employment have been added to clearly define mission assignments. Figures A-1 through A-3 show platoon through fire team organization breakdown.

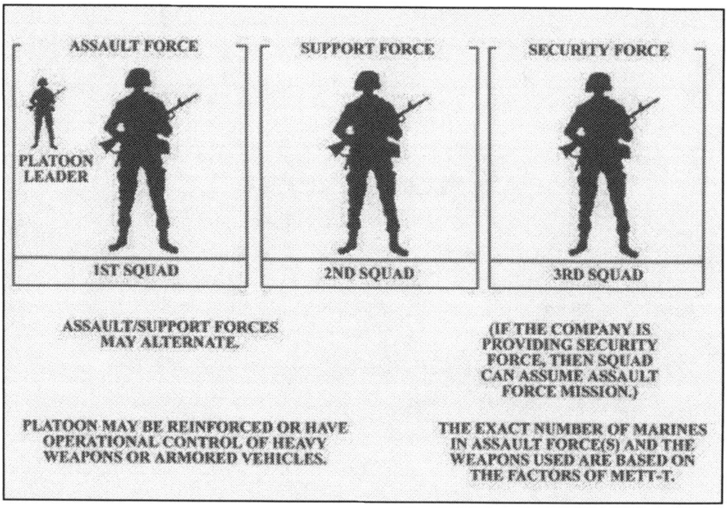

Figure A-1. Task-Organized Marine Platoon

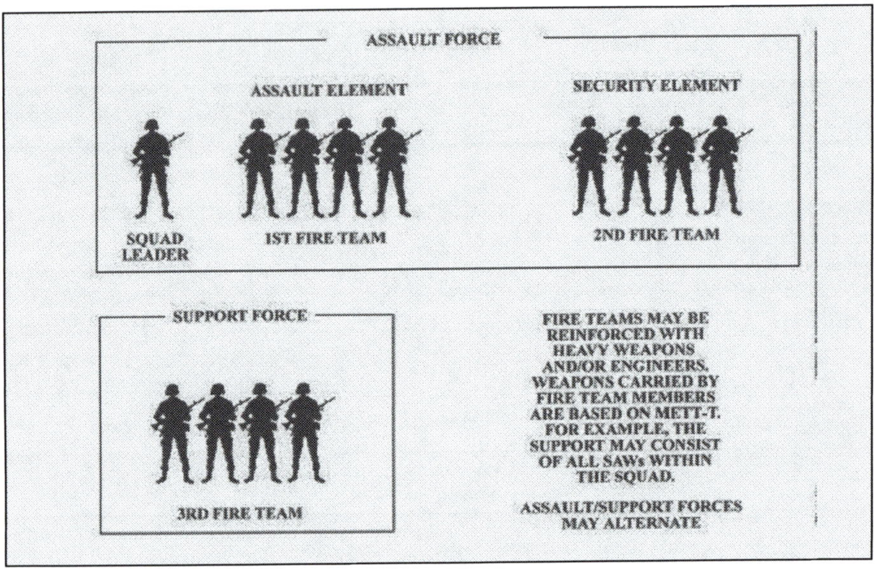

Figure A-2. Task-Organized Marine Squad

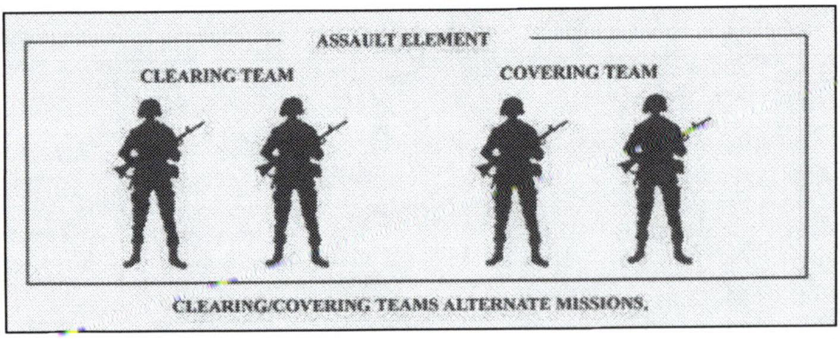

Figure A-3. Task-Organized Marine Fire Team

Section II

Weapons Handling and Firing Techniques

2. Weapons Carries. During MOUT, a Marine often finds himself very unexpectedly engaging targets at close ranges. Weapons carries provide a safe and effective method for handling the service rifle. For additional in-depth discussion of weapons handling and firing techniques, refer to MCWP 3-01X, *Field Firing of the M16A2 Rifle.*

a. Tactical Carry. The tactical carry is used when no immediate threat is present. It permits control of the rifle while moving and still allows quick engagement of the enemy. The buttstock of the rifle is placed alongside the body at approximately hip level, and the barrel is angled upward approximately 45 degrees in the general direction of the enemy (Figure A-4).

Figure A-4. Tactical Carry

b. Alert Carry. The alert carry is used when enemy contact is likely. Engagement of the enemy is faster from the alert carry than from the tactical carry. The buttstock of the rifle is placed in the shoulder with the muzzle angled down approximately 45 degrees and pointed in the likely direction of the enemy (Figure A-5).

In the alert carry, the Marine keeps both eyes open and scans for any threat with the weapon always in line with the Marine's LOS (this is called "guns 'n' eyeballs"). When a threat is spotted, the Marine quickly snaps his weapon into a firing position, looks over the rear sight aperture, and aligns the tip of the front sight post center mass on the target. Once the threat has been eliminated, the Marine returns to the alert carry and continues to scan for targets. Accuracy is the primary goal.

Figure A-5. Alert Carry

c. Ready Carry. The ready carry is employed when contact with the enemy is imminent. It allows for immediate target engagement. The buttstock of the rifle is in the shoulder with the muzzle of the rifle pointed in the direction of the enemy (Figure A-6). As in the alert carry, sight alignment and sight picture are achieved as the shot is fired. Marines should always strive for a clear tip of the front sight post and center mass hold.

Figure A-6. Ready Carry

d. Short Stocking. Rifles may be "short stocked" to increase their maneuverability in an enclosed area and to reduce possible "target indicators" for the enemy (muzzles sticking around corners, for example). The weapon is held in such a manner as to reduce the length of the weapon without sacrificing too much accuracy (Figure A-7). The stock is positioned so that the pistol grip is behind the Marine's head. The Marine may use his index finger or thumb to manipulate the trigger. The handguard of the weapon is placed against the Marine's cheek, and a firm stockweld is attained. The Marine should use the tip of the front sight post. Short stocking may be conducted from either side. However, care should be taken not to mask the ejection port cover. The short stocking technique may be used in the tactical, alert, or ready carry.

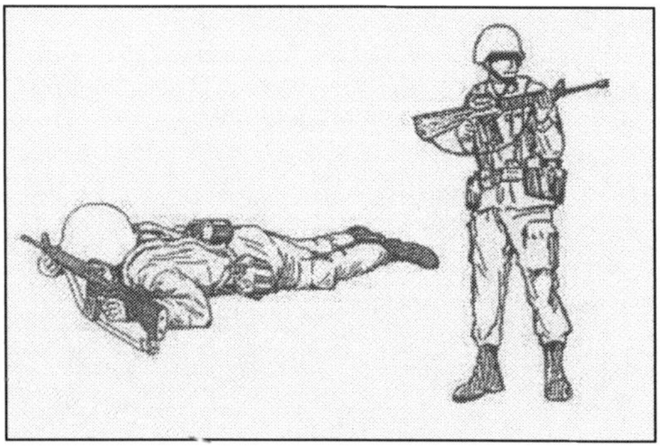

Figure A-7. Short Stocking

e. Weak-Hand Carries. Marines may carry their weapon on the weak-hand side to obtain maximum cover and fire from the weak-hand position to avoid exposing themselves. However, firing from the weak-hand position may reduce a Marine's ability to fire accurately. This may be acceptable when Marines are providing supporting fires, but in close kill-or-be-killed engagements, rapid, accurate firing from the strong-hand position is preferred.

3. Firing Techniques. The speed and uncertainty of combat require Marines to act without hesitation while achieving accurate target engagement. Several firing techniques are listed below. For an in-depth discussion of firing techniques that can be applied to MOUT operations, see MCWP 3-01X.

a. Pieing. Pieing is an effective technique for clearing dead space inside rooms and buildings to gain security of hallways, stairwells, mouseholes, and so on. It is conducted by using the ready carry position or short stocking technique. The weapon is aimed at a sector of a window, doorway, corner, or hallway and slowly moved at different angles, sectoring off the window, doorway, corner, or hallway until each sector is cleared of any threat. As soon as a hostile threat is seen, immediate, accurate fire can be placed on the threat. Figures A-11, A-34, A-35, A-38, A-47, and A-49 show examples of the pieing technique.

b. Aimed Quick Fire. The Marine's initial focus is on the target. As the rifle is brought up, the firing eye looks through or just over the rear sight aperture, and the front sight post is used to aim at the target. The Marine fires two quick shots. Focus remains on the front sight post throughout the aiming process.

.3

Immediately after target engagement, the Marine scans for additional targets. To scan, perform the following steps:

(1) Lower the weapon to look over the sights.

(2) Place the trigger finger straight along the receiver.

(3) Scan the area for targets and assess the situation. Wherever the head moves, the muzzle moves (eyes, muzzle, target). Keep both eyes open to increase the field of view.

c. Pointing Quick Fire. The pointing system is based on the phenomenon that when a person looks at an object and simultaneously points a finger at it, the finger aligns itself on the point of focus of the eyes with no conscious effort on the part of the individual. When a Marine looks at an object and simultaneously brings his rifle to his shoulder, the rifle in effect becomes an extension of the pointed finger. Consequently, it aligns itself naturally with the object on which the shooter is focusing.

When a target appears, the Marine will keep both eyes open, concentrating intensely on a small, specific focal point near the base of the target mass. The rifle is brought simultaneously to the hollow of the shoulder. The head is held high, stock welded to the jaw. The eyes are 2 - 3 inches over the top of the sights, staring intently at the target. As soon as the rifle is brought to the shoulder, two quick shots are fired. Focus on the target is not broken during the interval between initially seeing the target and discharging the rifle.

d. Instinctive Shooting. There may be situations in which a Marine is surprised and may need to react immediately. If possible, the Marine should engage the threat by using the tip of the front sight post. However, *speed* may be more important. The Marine's weapon and body are quickly "pointed," and the target is engaged. It is important that the body be turned with the weapon in order to achieve a natural point of aim. Simply pointing the weapon will usually result in a miss. Once the first two shots have been fired and the Marine regains the initiative, the weapon should be quickly moved to the Marine's shoulder and the tip of the front post used for sighting subsequent shots.

Section III

Movement

Movement techniques within built-up areas are the first fundamental skills that Marines should master. These movement techniques should be practiced until they become habitual. To reduce exposure to enemy fire, Marines should avoid silhouetting themselves, avoid open areas, and select their next covered position before moving.

4. Crossing a Wall. Each Marine should learn the correct method of crossing a wall (Figure A-8). After reconnoitering the other side, the Marine quickly rolls over the wall, keeping a low silhouette. The speed of his movement and a low silhouette deny the enemy a good target.

Figure A-8. Crossing a Wall

5. Observation Around Corners. The area around a corner should be observed before the Marine moves beyond it. The most common mistake the Marine makes at a corner is allowing his weapon to extend beyond the corner before observing, thereby exposing his position. Using the short stocking technique (Paragraph 2.d.) reduces exposure. Additionally, Marines can use a technique called "popping the corner" to reduce exposure time (Figure A-9). The Marine will get into a prone position near the corner of a building or obstacle around which he needs to observe. The weapon is short stocked, and the muzzle is pointed in the direction the Marine is looking. This allows the Marine to engage a target, if necessary, when he observes around a corner. The Marine will crawl to the corner but not expose himself. He will raise his upper body onto his elbows. The Marine will then push his body forward with his feet and legs without moving his elbows. His upper body, with the weapon ready, will move forward. The final position will expose the weapon, the Marine's helmet, and a minimal amount of the Marine's face. The Marine's forearms will come to rest on the deck giving him a low profile, the ability to observe around the corner, and the immediate capability to engage targets with his weapon.

Figure A-9. Correct Technique for Observing Around a Corner

6. Movement Past Windows. Windows present another hazard to Marines. The most common mistake made when passing a window is exposing the head to enemy observation from inside the room. If a Marine shows his head, an enemy gunner inside the building could engage him through the window without exposing himself to return fires.

a. The correct technique for passing a window is for the Marine to stay below the window level. He makes sure he does not silhouette himself in the window; he "hugs" the side of the building. An enemy gunner inside the building would have to expose himself to covering fires to engage the Marine (Figure A-10A).

Figure A-10A. Moving Past Windows

b. The most common mistake Marines make when passing a basement window is not being aware of it. A Marine should not walk or run past a basement window because he presents a good target to an enemy gunner inside the building. When using the correct procedure for negotiating a basement window, the Marine stays close to the wall of the building and steps or jumps past the window without exposing his legs (Figure A10-B).

c. An alternate method is for the pointman to pie off the window (or other opening) as he approaches it. As the pointman moves toward the window, the second Marine (behind the pointman) steps out to provide security to the direct front. The pointman continues to cover the opening by pieing, all the while using the near edge for cover. The pointman covers and sectors off that part of the interior that can be viewed from the outside. The second Marine moves with the pointman and maintains security to the front, eventually moving past the pointman as the pointman clears the inside corners of the opening. Once the opening is cleared, the second man may assume the duties of point, or when subsequent Marines move up to secure the opening, the original pointman may reassume those duties. Figure A-11 illustrates this alternate method.

Figure A-10B. Marine Passing Basement Windows

7. Use of Doorways. Doorways should not be used as entrances or exits because they are normally covered by enemy fire. If a Marine must use a doorway as an exit, he should move quickly through it to his next position, staying as low as possible to avoid silhouetting himself (Figure A-12). Preselection of positions, speed, a low silhouette, and the use of covering fires should be emphasized when exiting doorways.

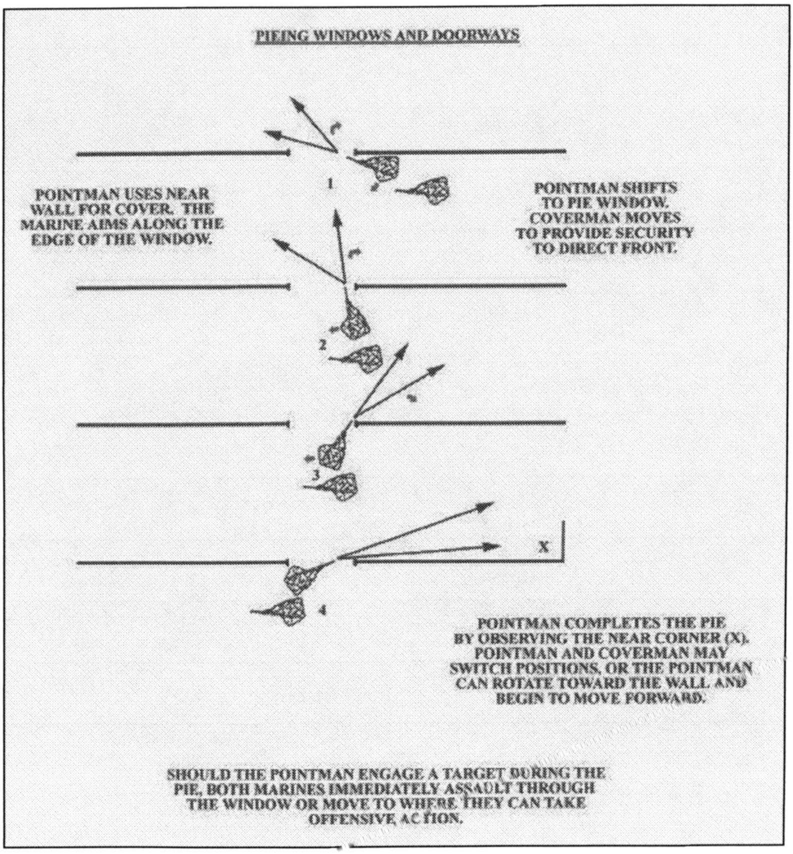

Figure A-11. Alternate Technique for Moving Past Windows

8. Movement Parallel to Buildings. Marines and small units may not always be able to use the inside of buildings as a route of advance. In such cases, they should move on the outside of the buildings (Figure A-13). Smoke, covering fires, and cover and concealment should be used to hide movement. When correctly moving along the outside of a building, the Marine hugs the side of the building, stays in the shadow, presents a low silhouette, and moves rapidly to his next position (Figure A-14). If an enemy gunner inside the building fires on a Marine, he exposes himself to fire from other squad/fire team members who are covering him. Furthermore, an enemy gunner farther down the street may have difficulty detecting and engaging the Marine.

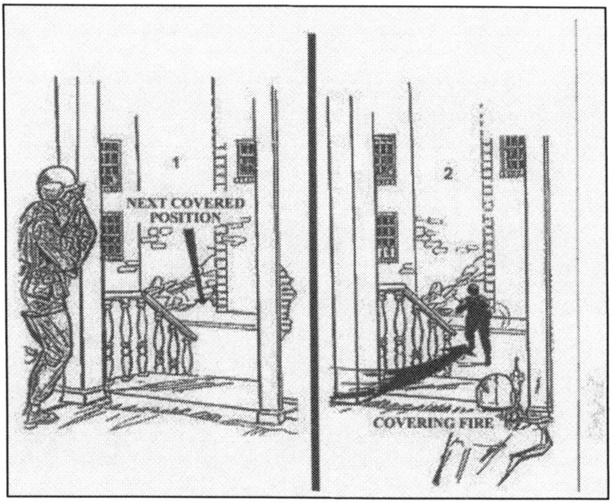

Figure A-12. Exiting a Doorway

Figure A-13. Moving Outside a Building

Figure A-14. Selection of the Next Position

9. Crossing Open Areas. Open areas such as streets, alleys, and parks should be avoided. They are natural kill zones for enemy crew-served weapons. They can be crossed safely if certain fundamentals are applied by Marines and small-unit leaders.

a. When using the correct procedure for crossing an open area, the Marine develops a plan for his own movement. (Smoke from hand grenades or smoke pots should be used to conceal the movement of all Marines.) The Marine runs the shortest distance between buildings and moves along the far building to the next position. By doing so, he reduces the amount of time during which he is exposed to enemy fire.

b. Before moving to another position, the Marine should make a visual reconnaissance and select the position that offers the best cover and concealment. At the same time, he should select the route that he will take to get to that position.

c. When moving from position to position, each Marine should be careful not to mask his supporting fires. When he reaches his next position, he should be prepared to cover the movement of other members of his assault force or element.

10. Assault Element Employment. Moving as an assault element from building to building or between buildings presents a problem because an assault element presents a large target to enemy fire. When moving from the corner of one building to another, the assault element should move across the open area in groups of two, maintaining their integrity within their respective clearing or covering team (Figure A-15). The assault element leader (fire team leader) is responsible for the secure movement of his unit. Smoke should be used to screen the assault element's movement. The covering team will move first to establish security on the far side. Their movement is protected by the fire of the clearing team. Once the covering team is in position, they provide cover as the clearing team moves across and prepares, then enters, the structure. Moving from the side of one building to the side of another presents a similar problem, and the technique of movement employed is the same (Figure A-16).

There may be situations in which it is best to move in larger groups. An example is that Marines crossing in pairs may key enemy snipers in to the movement, setting them up for good shots at the next pair to cross. Leaders should remain aware of the situation and threat and plan movements accordingly.

Figure A-15. Assault Element Movement

11. Movement Inside a Building. When moving within a building that is under attack (Figure A-17), Marines should avoid silhouetting themselves in doors and windows. If forced to use a hallway (Figure A-18), Marines should move by hugging the wall to avoid presenting targets to the enemy.

Figure A-16. Movement to an Adjacent Building

Figure A-17. Movement Within a Building Under Attack

Figure A-18. Moving Down a Hallway

Section IV

Entry Techniques

Marines should enter a building with minimum exposure. They should select an entry point before moving toward the building; avoid windows and doors; use smoke to conceal their advance to the building; use demolitions, tank rounds, and so on to make new entrances; precede the entry of the clearing team with a grenade; enter immediately after the grenade explodes; and be covered by the covering team.

12. Upper Building Levels. Clearing a building from the top down is the preferred method. Clearing or defending a building is easier from an upper story. Gravity and the building's floor plan become assets when throwing hand grenades and moving from floor to floor.

a. An enemy who is forced to the top of a building may be cornered and fight desperately or escape over the roof, but an enemy who is forced down to the ground level may withdraw from the building, thus exposing himself to friendly fires from the outside.

b. Various means such as ladders, drainpipes, vines, helicopters, or the roofs and windows of adjoining buildings may be used to reach the top floor or roof of a building. In some cases, one Marine can climb onto the shoulders of another and reach high enough to pull himself up. Another method is to attach a grappling hook to the end of a scaling rope so that a Marine can scale a wall, spring from one building to another, or gain entrance through an upstairs window.

13. Use of Ladders. Ladders offer the quickest method to gain access to the upper levels of a building (Figure A-19). Units can get ladders from local civilians or stores or obtain material to build ladders through supply channels. If required, ladders can be built with resources that are available throughout the urban area; for example, lumber can be taken from inside the walls of buildings (Figure A-20). Although ladders may not permit access to the top of some buildings, they may offer security and safety through speed.

14. Use of Grappling Hook. A suitable grappling hook and rope are selected to meet the need. The grappling hook should be sturdy, portable, easily thrown, and equipped with hooks that can hold inside a window. The scaling rope should be 5/8 of an inch to 1 inch in diameter and long enough to reach the objective window. Knots are tied in the rope at 1-foot intervals to make climbing and holding easier. Marines should follow the procedures outlined below.

a. When throwing the grappling hook, the Marine should stand as close to the building as possible (Figure A-21). If the Marine stands close to the building, he is less exposed to enemy fires and he does not have to throw the hook as far.

Figure A-19. Using Ladders To Get to Upper Levels

Figure A-20. Getting Lumber From Inside the Walls

Figure A-21. Grappling Hook Thrown at Close Range

b. Making sure there is enough rope to reach the target, the Marine should hold the hook and a few coils of rope in the throwing hand. The remainder of the rope, in loose coils, should be in the other hand. The rope should be allowed to play out freely. The throw should be a gentle, even, upward lob of the hook, with the other hand releasing the rope as it plays out.

c. Once the grappling hook is inside the window (or on the roof), the Marine should pull on the rope to obtain a good hold before beginning to climb. When using a window, the hook should be pulled to one corner to ensure a good "bite" and to reduce the Marine's exposure to lower windows during the climb.

d. The use of grappling hooks is the least preferred method for gaining entry to upper levels of buildings. Grappling hooks should be used only as a last resort and away from potential enemy positions. This method may potentially be used on adjacent buildings that offer concealed locations and a roof that connects to enemy positions.

15. Scaling Walls. When forced to scale a wall during exposure to enemy fire, all available means of concealment should be used. Smoke and diversionary measures improve the chances of a successful exposed movement. When using smoke for concealment, Marines should plan for wind direction. They should use fire, shouts, and fake movement to distract the enemy.

a. A Marine scaling an outside wall is vulnerable to enemy sniper fire. Marines moving from building to building and climbing buildings should be covered with friendly fire. Areas between buildings offer good fields of fire to the enemy. Properly positioned friendly weapons can suppress and eliminate enemy fire. The M203 grenade launcher is effective in clearing the enemy from rooms inside buildings (Figure A-22).

Figure A-22. Employment of the M203 Grenade Launcher

b. A Marine scaling a wall with a rope should avoid silhouetting himself in windows of uncleared rooms or exposing himself to enemy fires from lower windows. He should climb with his weapon slung over the firing shoulder so that he may quickly bring it to a firing position. He should clear the lower room with a hand grenade before going outside the window. Before doing this, the Marine first loosens the safety pin so that he needs only one hand to throw the grenade. The objective upper story window should not be entered before a hand grenade has been thrown in.

c. The Marine enters the objective window with a low silhouette (Figure A-23). Entry can be head first; however, a preferred method is to hook a leg over the windowsill and enter sideways, straddling the ledge.

16. Rappelling. Rappelling (Figure A-24) is an entry technique that Marines can use to descend from the rooftop of a tall building into a window. (See Army manual TC 21-24 for more information on rappelling.)

17. Entry at Lower Levels. Buildings should be cleared from the top down. However, it may be impossible to enter a building at the top; therefore, entry at the bottom or lower level may be the only course of action. When entering a building at the lower level, Marines avoid entry through windows and doors because both can be easily boobytrapped and are usually covered by enemy fire.

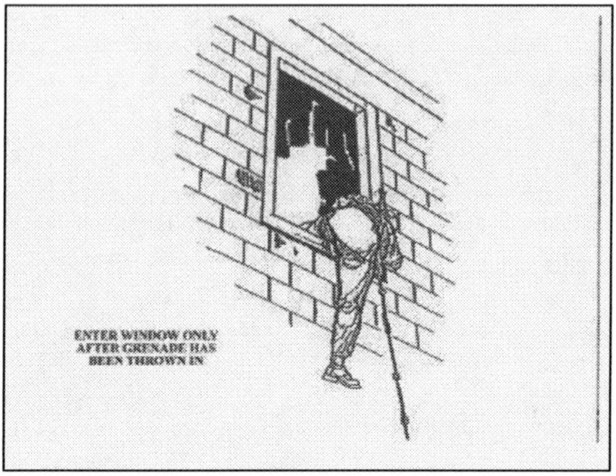

Figure A-23. Entering the Objective Window

Figure A-24. Rappelling

a. Ideally, when entering at lower levels, demolitions, artillery, tank fire, antiarmor weapons fire, or similar means are used to create a new entrance to avoid boobytraps. Quick entry is required to follow through on the effects of the blast and concussion.

b. When the only entry to a building is through a window or door, supporting fire should be directed at that location. (Specific lower level entry techniques are shown in Figure A-25.) If no supporting fire is available, AT4s can be employed instead.

c. Before entering, Marines throw a hand grenade into the new entrance to reinforce the effects of the original blast. When making a new entrance in a building, they consider the effects of the blast on the building and adjacent buildings. If there is the possibility of a fire in an adjacent building, Marines coordinate with adjacent units and obtain permission before starting the operation. In wooden-framed buildings, the blast may cause the building to collapse. In stone, brick, or cement buildings, supporting fires are aimed at the corner of the building or at weak points in the building construction.

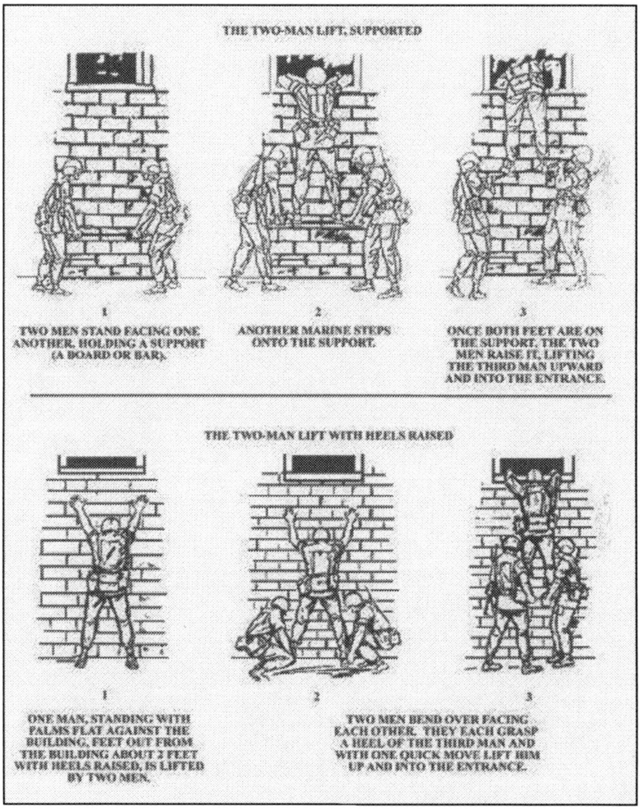

Figure A-25. Lower Level Entry Techniques

Figure A-25. Lower Level Entry Techniques (Continued)

Figure A-25. Lower Level Entry Techniques (Continued)

18. Hand Grenades. Combat in built-up areas (mainly during the attack) requires extensive use of hand grenades. The Marine should throw a grenade before entering rooms or negotiating staircases, mouseholes, and so on. This usually requires the use of both hands and employs both the overhand and underhand methods of throwing.

a. Two techniques may be used in the throwing of a grenade. The *preferred technique* involves throwing the grenade hard enough that it bounces or skips around, making it difficult to pick up. The hard-throw, skip/bounce technique may be used by Marines in training and combat. The *least preferred technique*, which involves cooking-off the grenade for two seconds then throwing it to prevent the enemy from grabbing the grenade and tossing it back, may be used by Marines, as appropriate, during actual combat but not in training.

b. Nonverbal or verbal alerts are used, as appropriate, before throwing a grenade. A visual showing (preferred) of the grenade to be thrown is made to assault element members, and a visual acknowledgment from them is received. A nonverbal alert may ensure that the enemy is surprised when the grenade is thrown. If the situation demands, a voice alert can be used, but the element of surprise may be lost. When or if a voice alert is used, the voice alert is "**FRAG OUT**;" when an enemy grenade has been identified, friendly forces shout, "**GRENADE**."

c. The construction material used in the building being cleared influences the use of grenades. In some situations, concussion grenades may be preferred over fragmentary grenades during offensive operations or when defending from hasty defensive positions. If the walls of a building are made of thin material, such as sheetrock or thin plyboard, the Marine should either lie flat on the floor with his helmet pointing toward the area of detonation or move away from any wall that might be penetrated by grenade fragments.

d. Marines should throw grenades into an opening before entering a building to eliminate any enemy that might be near the entrance (Figure A-26). Using the M203 grenade launcher is the best method for putting a grenade in an upper story window.

e. When a hand grenade must be used, the Marine throwing the grenade should stand close to the building, using it for cover. At the same time, the individual and the rest of the element should have a planned area to which they can move for safety if the grenade does not go through the window, but falls back to the ground.

f. The Marine throwing the grenade should step out far enough to lob the grenade into the upper story opening. The Marine's weapon should be kept in the nonthrowing hand so that it can be used if needed. The weapon should never be laid down outside or inside the building. Once the grenade has been thrown into the opening, assaulting forces should move swiftly to enter the building. This technique should be employed only when the window has already been broken. Otherwise, the chances are high that the thrown grenade will bounce off of the window and fall back onto the ground without going into the room.

Figure A-26. Hand Grenade Thrown Through Window

WARNING

After throwing the grenade, the Marine should immediately announce, "FRAG OUT," to indicate that a grenade has been thrown. He then takes cover because the grenade may bounce back or be thrown back, or the enemy may fire at him.

g. If Marines must enter the building by using the stairs, they first look for boobytraps. Then they throw a grenade through the stairwell door, let it detonate, and move quickly inside. They can use the staircase for cover.

h. The best way to enter a building is to breach the exterior wall. Again, a grenade should be thrown through the hole while using all available cover, such as the lower corner of the building (Figure A-27).

i. Another way to enter a room is to blast mouseholes with demolitions. In moving from room to room through mouseholes, Marines should use grenades to clear the room as they do in moving through open doorways.

j. Although buildings are best cleared from the top down, this is not always possible. While clearing the bottom floor of a building, Marines may encounter stairs that must also be cleared. Once again, grenades play an important role. Before climbing the stairs from the bottom floor, Marines should first inspect for boobytraps then toss a grenade to the head of the stairs, if appropriate (Figure A-28). The use of a grenade in these situations is based on stair/building construction and building layout. *Throwing grenades up staircases is extremely* *dangerous, and extreme caution and common sense should be used.* In most situations, throwing a grenade up a staircase is not recommended. (Again, the grenade is being used in the initial building entry or clearing of the bottom floor of a building having a staircase.) If a grenade is thrown too hard or the staircase is too steep, the grenade can roll back down the staircase. For this reason, if a grenade is thrown, it should be thrown underhand to reduce the risk of it bouncing back and rolling down the stairs. A nonverbal or, as appropriate, voice alert is used when throwing the grenade. Once the first grenade has detonated, another grenade should be thrown over and behind the staircase banister and into the hallway, destroying any enemy hiding to the rear.

Figure A-27. Entering Through a Breached Hole

Figure A-28. Tossing a Grenade Up a Stairway

k. After the stairs have been cleared, assaulting forces move to the top floor and clear it. Upon clearing the top floor, forces move downstairs to clear the center and bottom floors and to continue with the mission.

Note: Because large quantities of hand grenades are used when clearing buildings, a continuous supply must be available to forces having this mission within a built-up area.

c. Before entering, Marines throw a hand grenade into the new entrance to reinforce the effects of the original blast. When making a new entrance in a building, they consider the effects of the blast on the building and adjacent buildings. If there is the possibility of a fire in an adjacent building, Marines coordinate with adjacent units and obtain permission before starting the operation. In wooden-framed buildings, the blast may cause the building to collapse. In stone, brick, or cement buildings, supporting fires are aimed at the corner of the building or at weak points in the building construction.

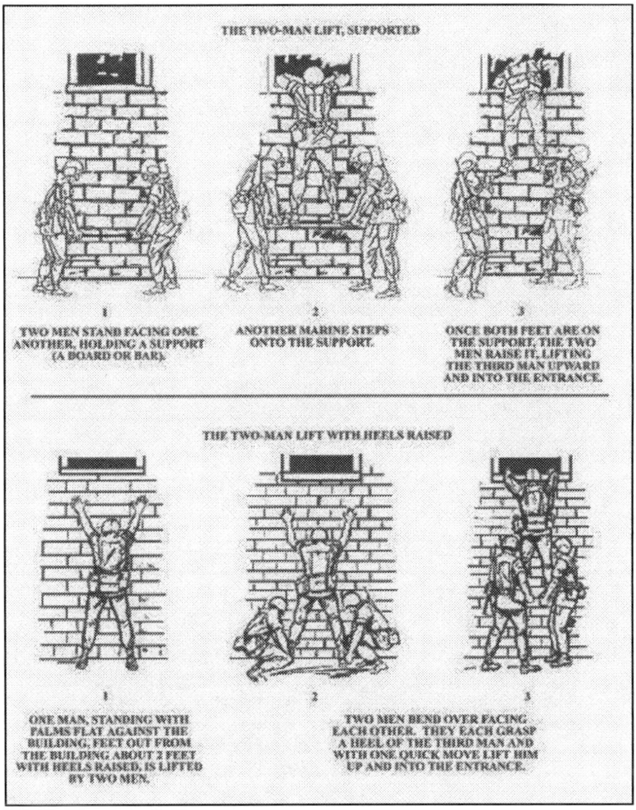

Figure A-25. Lower Level Entry Techniques

Figure A-25. Lower Level Entry Techniques (Continued)

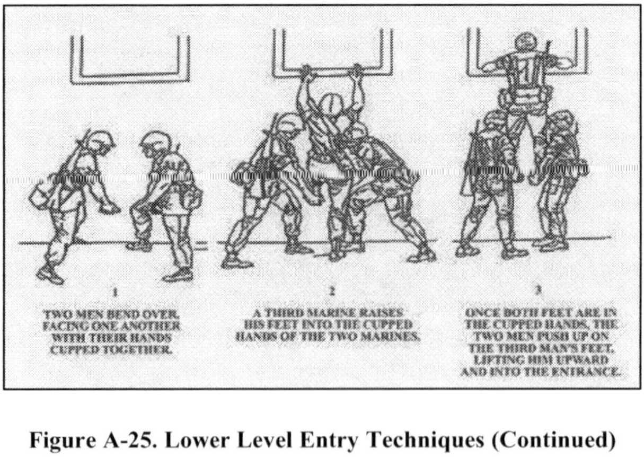

Figure A-25. Lower Level Entry Techniques (Continued)

18. Hand Grenades. Combat in built-up areas (mainly during the attack) requires extensive use of hand grenades. The Marine should throw a grenade before entering rooms or negotiating staircases, mouseholes, and so on. This usually requires the use of both hands and employs both the overhand and underhand methods of throwing.

a. Two techniques may be used in the throwing of a grenade. The *preferred technique* involves throwing the grenade hard enough that it bounces or skips around, making it difficult to pick up. The hard-throw, skip/bounce technique may be used by Marines in training and combat. The *least preferred technique*, which involves cooking-off the grenade for two seconds then throwing it to prevent the enemy from grabbing the grenade and tossing it back, may be used by Marines, as appropriate, during actual combat but not in training.

b. Nonverbal or verbal alerts are used, as appropriate, before throwing a grenade. A visual showing (preferred) of the grenade to be thrown is made to assault element members, and a visual acknowledgment from them is received. A nonverbal alert may ensure that the enemy is surprised when the grenade is thrown. If the situation demands, a voice alert can be used, but the element of surprise may be lost. When or if a voice alert is used, the voice alert is "**FRAG OUT**;" when an enemy grenade has been identified, friendly forces shout, "**GRENADE**."

c. The construction material used in the building being cleared influences the use of grenades. In some situations, concussion grenades may be preferred over fragmentary grenades during offensive operations or when defending from hasty defensive positions. If the walls of a building are made of thin material, such as sheetrock or thin plyboard, the Marine should either lie flat on the floor with his helmet pointing toward the area of detonation or move away from any wall that might be penetrated by grenade fragments.

d. Marines should throw grenades into an opening before entering a building to eliminate any enemy that might be near the entrance (Figure A-26). Using the M203 grenade launcher is the best method for putting a grenade in an upper story window.

e. When a hand grenade must be used, the Marine throwing the grenade should stand close to the building, using it for cover. At the same time, the individual and the rest of the element should have a planned area to which they can move for safety if the grenade does not go through the window, but falls back to the ground.

f. The Marine throwing the grenade should step out far enough to lob the grenade into the upper story opening. The Marine's weapon should be kept in the nonthrowing hand so that it can be used if needed. The weapon should never be laid down outside or inside the building. Once the grenade has been thrown into the opening, assaulting forces should move swiftly to enter the building. This technique should be employed only when the window has already been broken. Otherwise, the chances are high that the thrown grenade will bounce off of the window and fall back onto the ground without going into the room.

Figure A-26. Hand Grenade Thrown Through Window

WARNING

After throwing the grenade, the Marine should immediately announce, "FRAG OUT," to indicate that a grenade has been thrown. He then takes cover because the grenade may bounce back or be thrown back, or the enemy may fire at him.

g. If Marines must enter the building by using the stairs, they first look for boobytraps. Then they throw a grenade through the stairwell door, let it detonate, and move quickly inside. They can use the staircase for cover.

h. The best way to enter a building is to breach the exterior wall. Again, a grenade should be thrown through the hole while using all available cover, such as the lower corner of the building (Figure A-27).

i. Another way to enter a room is to blast mouseholes with demolitions. In moving from room to room through mouseholes, Marines should use grenades to clear the room as they do in moving through open doorways.

j. Although buildings are best cleared from the top down, this is not always possible. While clearing the bottom floor of a building, Marines may encounter stairs that must also be cleared. Once again, grenades play an important role. Before climbing the stairs from the bottom floor, Marines should first inspect for boobytraps then toss a grenade to the head of the stairs, if appropriate (Figure A-28). The use of a grenade in these situations is based on stair/building construction and building layout. *Throwing grenades up staircases is extremely* *dangerous, and extreme caution and common sense should be used.* In most situations, throwing a grenade up a staircase is not recommended. (Again, the grenade is being used in the initial building entry or clearing of the bottom floor of a building having a staircase.) If a grenade is thrown too hard or the staircase is too steep, the grenade can roll back down the staircase. For this reason, if a grenade is thrown, it should be thrown underhand to reduce the risk of it bouncing back and rolling down the stairs. A nonverbal or, as appropriate, voice alert is used when throwing the grenade. Once the first grenade has detonated, another grenade should be thrown over and behind the staircase banister and into the hallway, destroying any enemy hiding to the rear.

Figure A-27. Entering Through a Breached Hole

Figure A-28. Tossing a Grenade Up a Stairway

k. After the stairs have been cleared, assaulting forces move to the top floor and clear it. Upon clearing the top floor, forces move downstairs to clear the center and bottom floors and to continue with the mission.

Note: Because large quantities of hand grenades are used when clearing buildings, a continuous supply must be available to forces having this mission within a built-up area.

Section V

Clearing Techniques

This section will discuss basic clearing techniques; however, the fundamentals may have to be modified depending on the various situations and building/room layouts that Marines may encounter. Caution should be taken when using a stacked type of position as personnel may be concentrating in a danger area where they could be exposed to fires penetrating walls or explosive effects from boobytraps or grenades thrown by the enemy. *Marines should always remain aware of the protective, or lack of protective, capabilities of the materials comprising the walls, floors, ceilings, and doors within the areas they are clearing.*

19. Clearing a Room, Door Closed or Open, From Stacked Positions. The following discussion and series of figures will focus on the two-man clearing team in the conduct of clearing a room. The covering team is positioned by the assault element leader to provide security as the clearing team begins its procedures to enter and clear a room.

a. Shooter Number Two is behind Shooter Number One (stacked position). This positioning is the same if the door is already open or nonexistent (Figure A-29).

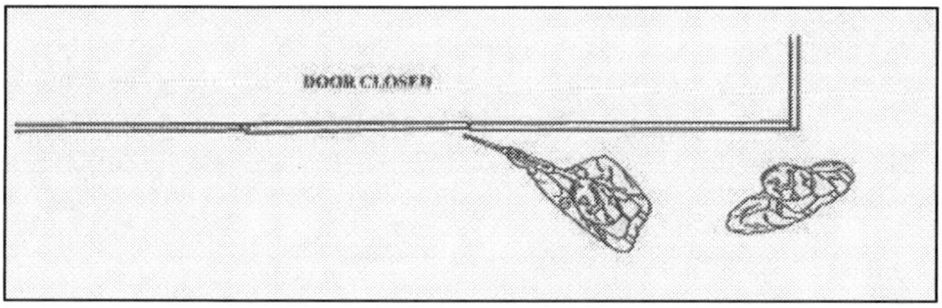

Figure A-29. Clearing a Room, Door Closed

b. Shooter Number One determines the direction in which the door opens then quickly informs Shooter Number Two and forces the door open.

When a door is the only means of entering a room, Marines should beware of fire from enemy soldiers within the room and of boobytraps. When opening a door, Marines should not expose themselves to fires through the door. The two-man clearing team forces the door open by directing a short burst of automatic fire through the door around the latch (Figure A-30) and then kicking it open. Another method of opening the door is to use a shotgun with 12-gauge slugs to shoot the hinges and/or the latch. When the doorknob is locked and/or deadbolt locks are present, shooting the hinges of the door is recommended. The door may then be opened by kicking or using a battering ram.

Figure A-30. Shooting the Door Open

Marines can also use an ax or demolitions if they are available. As a last resort, Marines can kick the door open with a front kick (Figure A-31) or mule kick. This is the least preferred technique because it is difficult and tiring to the Marine. It also rarely works the first time, thereby giving any enemy units within the room ample warning (and giving the enemy time to shoot through the door). In situations where restrictions on firing and limiting structural damage have been imposed and/or boobytraps are not expected, using the door handle to open the door may be appropriate (apply METT-T).

Figure A-31. Opening a Door With a Front Kick

c. Once the door is open, Shooter Number Two immediately throws a hand grenade into the room. Shooter Number One maintains cover for Shooter Number Two as he prepares to throw the grenade into the room. Shooter Number Two then returns to the original stacked position behind Shooter Number One (Figure A-32). If Marines have no grenades, the procedures remain the same except for throwing the grenade. *Note: When the door is already open or nonexistent, the stacked position is always used.*

WARNING

Because fragments from M67 fragmentation grenades may injure Marines outside the room, they should not be used. Marines should use MK3A2 offensive hand grenades instead. Also, the technique of cooking-off hand grenades can be deadly unless properly performed.

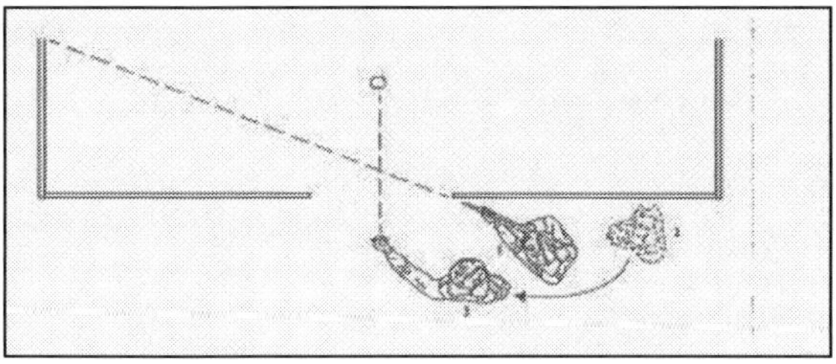

Figure A-32. Throwing a Grenade Into a Room

(1) Throwing Grenades. Two techniques are available for using grenades in room clearing. The *preferred technique* for Marines is to throw a concussion hand grenade into the room so hard that it skips and bounces, making it difficult for the enemy to pick up and throw back. The skip/bounce technique should be used by Marines during training and combat. The *least preferred technique* is to cook-off a concussion hand grenade by removing the grenade's safety pin, releasing the safety lever, counting off two seconds by thousands (one thousand and one, one thousand and two), and then throwing the grenade into the room. Cooking-off the grenade will be used only as appropriate during combat.

(2) Nonverbal and Verbal Alerts. To alert all that a grenade will be thrown, a visual showing of the grenade is made to assault element members, and a visual acknowledgment from them is received. A nonverbal alert may ensure that the enemy is surprised when the grenade is thrown. If the situation demands, a voice alert can be used, but the element of surprise may be lost. When or if a voice alert is used, the voice alert is "**FRAG OUT**;" when an enemy grenade has been identified, friendly forces shout, "**GRENADE**." This allows Marines to distinguish between warnings for outgoing and incoming grenades.

d. After the grenade explodes, Shooter Number One steps across the threshold and clears his immediate area (Figure A-33). He engages targets from the ready carry position. Shooter Number Two follows immediately behind Shooter Number One, buttonhooks, and clears his area.

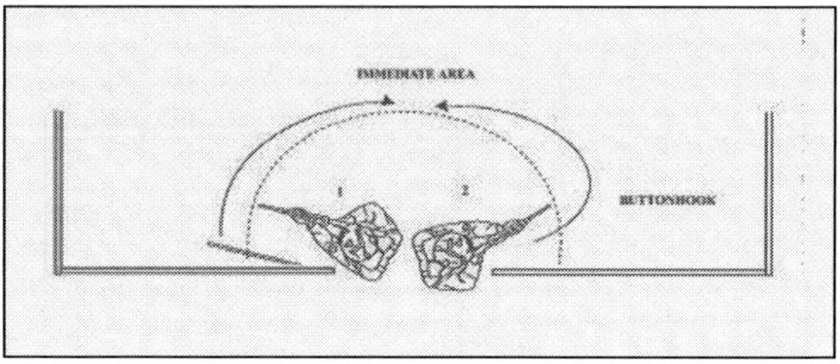

Figure A-33. Shooters Enter the Room

e. Both shooters clear the immediate area and along their respective walls, starting from the nearest respective corner and continuing to the farthest respective corner (Figure A-34). Shooters use the pieing technique to systematically clear the room by sector. (See Paragraph 3.a. of this appendix.)

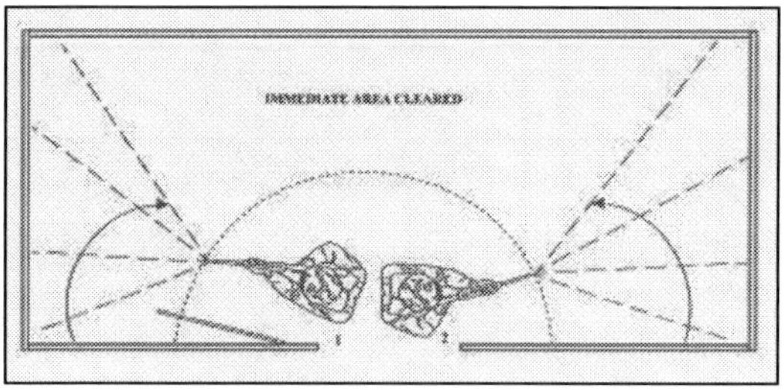

Figure A-34. Clearing a Room

f. Both shooters then establish a dominant position in the room one step away from the wall and two steps into the room and clear the room by sector, pieing to the opposite side of the room (Figure A-35). Meanwhile, the covering team (two-man team) in position outside the room being cleared provides security.

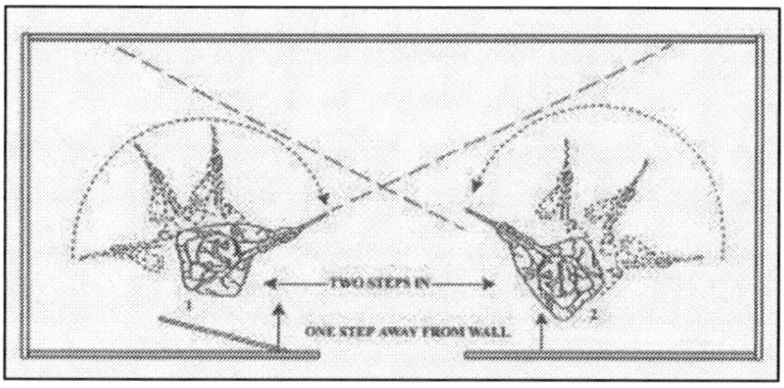

Figure A-35. Clearing a Room (Continued)

g. Another option for entering a room (situation dependent): If Shooter Number One of the clearing team enters a room without Shooter Number Two immediately following him (this could be due to room size, entering the room through a narrow opening, or a number of factors caused by the situation), then Shooter Number One will position himself inside the room to the left or right of the door and quickly scan the entire room. He then gives the command to Shooter Number Two, "**NEXT MAN IN, LEFT (RIGHT)**." Shooter Number Two shouts, "**COMING IN, LEFT (RIGHT)**," enters the room, positions himself up against the wall to the left (right) of the entrance as designated, and scans the room. Once in position, the clearing team leader can call in additional members with the "**NEXT MAN IN**" command, as the situation dictates. It is critical that all assault element members tell each other where they are to avoid fratricide.

h. When the clearing team has cleared the room and is ready to exit the room, it uses the following voice alerts. Once a room has been cleared, the clearing team yells, "**CLEAR**," to inform the covering team. Before leaving the room and rejoining the covering team, the clearing team yells, "**COMING OUT**" and waits to hear the covering team yell, "**ALL CLEAR**." The clearing team then executes its movement out of the room. The assault element then marks the room according to unit SOP.

When moving up or down a staircase, the appropriate team yells, "**COMING UP**" or "**COMING DOWN**." As before, it waits for the response "**ALL CLEAR**" before executing movement and rejoining its element. When exiting a building, the clearing team or assault element yells, "**COMING OUT**" and waits for the support force or covering team to respond with "**ALL CLEAR**."

20. Clearing a Room, Door Closed, Split Positions

a. Before opening a door, Shooter Number Two of the clearing team positions himself opposite Shooter Number One on the other side of the door and away from the wall in a safe position that allows Shooter Number One to shoot the door-opening mechanism (Figure A-36). This positioning allows Shooter Number Two to be in position to immediately move to a wall position opposite Shooter Number One. No matter what method is used to open the door, Shooter Number Two should get in a position on the side of the door opposite Shooter Number One. If the door is already open, the stacked positions in Paragraph 19.a. of this section are used.

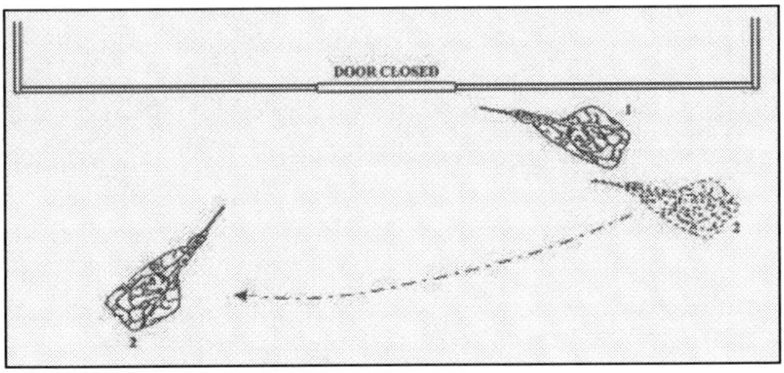

Figure A-36. Positioning To Open a Door

b. After Shooter Number One shoots the door open, Shooter Number Two immediately moves to a kneeling position against the wall (Figure A-37). Shooter Number One kicks the door open from a standing position. If the Marines have grenades, then Shooter Number Two immediately throws a grenade into the room as discussed in Paragraph 19.c. of this section. If the Marines do not have any grenades, the stand/kneel positioning of shooters places the shooters' weapons at different levels, thus allowing each shooter to engage targets inside the room while staying out of the other shooter's line of fire. The shooters scan the room for targets.

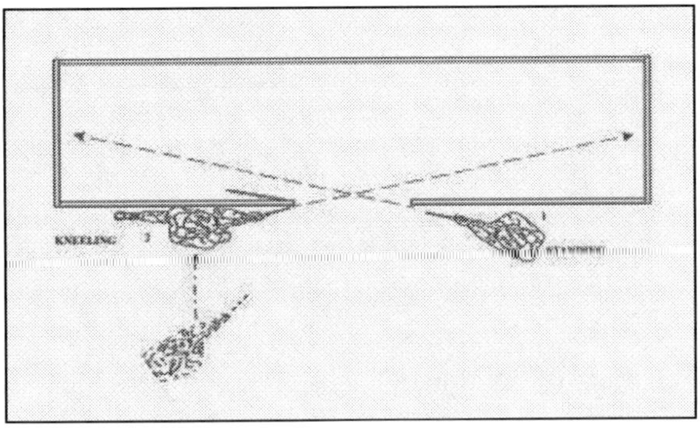

Figure A-37. Positioning To Enter a Room

c. If no grenade was thrown: Once the initial scan of the room is complete, Shooter Number One yells, "**MOVE**," and both shooters pivot from their respective positions toward each other, pieing (see Paragraph 3.a. of this appendix) while maintaining eye-muzzle-target contact. Shooter Number Two is kneeling, and Shooter Number One is still standing (Figure A-38). Both shooters clear their respective sector of fire.

If a grenade was thrown: After the explosion, Shooter Number One yells, "**MOVE**," and both shooters pivot from their respective positions toward each other, immediately enter the room, and buttonhook (Figures A-38, A-39A, and A-39B). Shooter Number Two goes from kneeling to standing as he pivots. Both shooters clear their respective sector of fire.

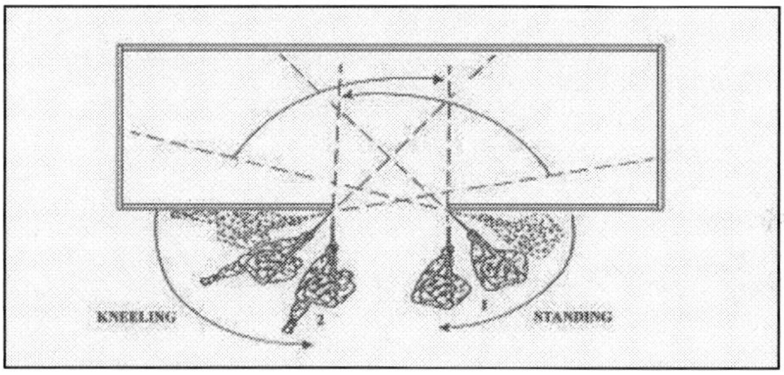

Figure A-38. Shooters Clear Sectors of Fire From Doorway

d. If no grenade was thrown: After scanning is completed, Shooter Number Two stands up and yells, "**READY**," and Shooter Number One Yells, "**MOVE**." Both shooters enter the room together (Figure A-39A).

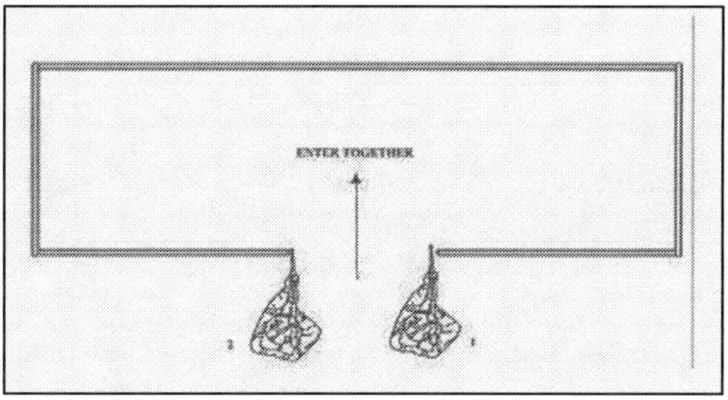

Figure A-39A. Shooters Enter the Room Together

e. Both shooters pass through the doorway together and immediately face respective corners, using the adjacent wall for cover. Both shooters buttonhook their respective areas (Figure A-39B). At this point, clearing the room is the same as discussed in Paragraphs 19.e. - 19.h.

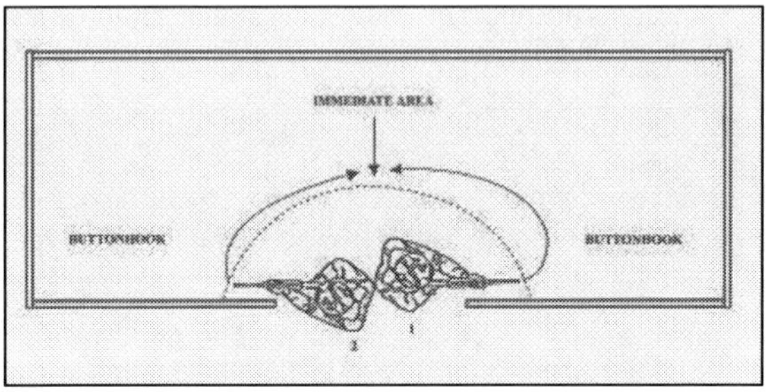

Figure A-39B. Shooters Clear Immediate Area

f. Cross Method. When employing the cross method, two Marines position themselves on either side of the entryway. Each Marine faces into the room covering the corner of the room opposite his position. On a prearranged signal, each Marine alternately enters the room. Each Marine crosses quickly to the opposite corner while covering the half of the room toward which he is moving. Once in the near corner, he assumes an outboard kneeling position to reduce his silhouette and continues to maintain coverage of his half of the room. He may change the position of his weapon to best cover the room. (Figure A-40)

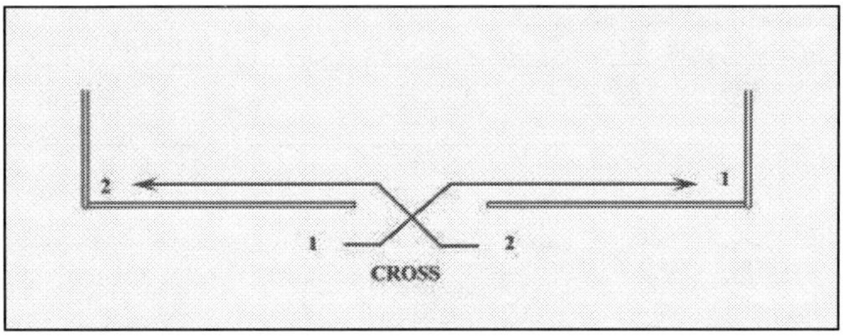

Figure A-40. Cross Method of Entering and Clearing Rooms

21. Covering Team Members Used To Support Clearing of a Room

a. Three Marines Clearing a Room. Clearing a room with three men involves the same procedures used by the aforementioned two-man clearing team, with a third Marine (from the covering team) following immediately (stacked) behind Shooters Number One and Two. Shooter Number Three moves to one side of the door and establishes a center sector of fire coverage (Figure A-41). The remaining covering team Marine provides outside security. If the split position is used, Shooters Number Two and Three are placed on the opposite side of the door.

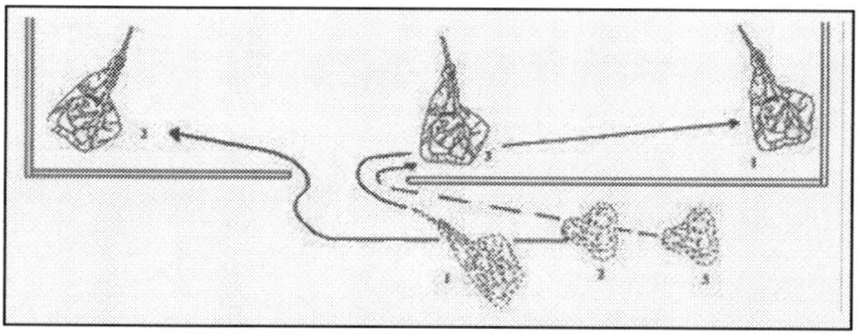

Figure A-41. Three-Marine Clearing Team (Stacked)

b. Four Marines Clearing a Room. When four Marines (entire assault element) are used to clear a room, outside security should be established by another assault element. (If outside security cannot be established, then the two- or three-Marine clearing team should be used.) The two-Marine clearing team procedures are used for initial entry into the room. Shooters Number Three and Four (split position example) stack up opposite Shooters Number One and Two. Shooters Number Three and Four follow Shooter Number Two into the room. Shooter Number Three crosses the threshold and establishes a center sector of fire coverage. Shooter Number Four follows Shooter Number Three and establishes a center sector of fire coverage opposite Shooter Number Three (Figure A-42). If the stacked position is used to enter a room, all shooters stack up on the same side of the door and follow each other in with the same positioning.

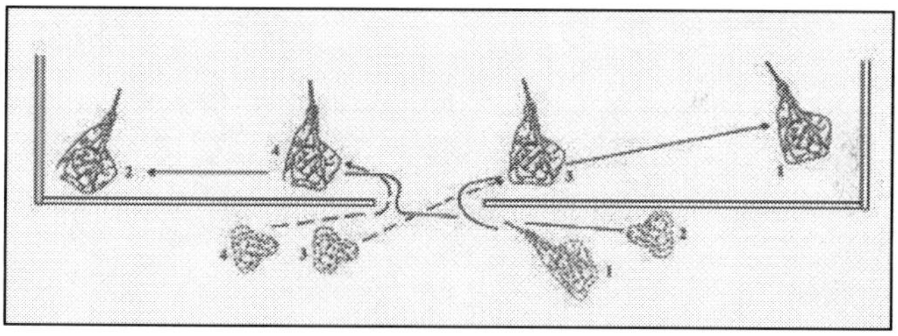

Figure A-42. Four-Marine Clearing Team (Split)

c. Single Marine Clearing a Room. This is the *least preferred* method of clearing a room. However, there may be times when this is the only method available, for example, when gaining access to an upper story window. When this is the case, the Marine gains entry to the room by shooting, kicking, ramming, or breaking a window. He then throws a grenade into the room if one is available and no restrictions apply and takes cover. Immediately after the grenade explodes, he crosses the threshold using the buttonhook technique depicted in earlier figures, scans the room for targets, and engages targets with two shots as they appear.

22. Clearing a Room, Entering Through Mousehole. Mouseholes measure about 2 feet wide and are blown or cut through a wall so that Marines can enter a room (Figure A-43). Mouseholes are safer entrances than doors, which can be easily boobytrapped and should be avoided. Procedures discussed in Paragraph 19 are used. The assault element leader positions the covering team to provide security.

a. In this situation, Shooter Number One would enter the room through a mousehole after throwing a grenade and waiting for the detonation.

b. Shooter Number One will position himself inside the room to the left or right of the mousehole and quickly scan the entire room. He then gives the command to Shooter Number Two, "**NEXT MAN IN, LEFT (RIGHT)**." Shooter Number Two shouts, "**COMING IN, LEFT (RIGHT)**," enters the room, positions himself against the wall to the left (right) of the entrance, and scans the room.

c. The clearing team continues clearing the room as discussed in Paragraph 19.

Figure A-43. Entering Through a Mousehole

23. Clearing an L-Shaped Hallway. After entering a room or building, the clearing team may encounter an L-shaped hallway that should be cleared. The following procedures are used:

a. The clearing team finds an L-shaped hallway.

b. Each shooter takes a dominant position.

c. Shooter Number Two aims his weapon to cover the dead space (Figure A-44).

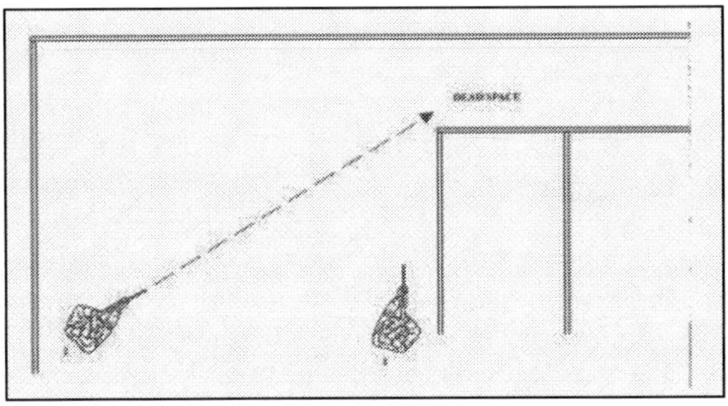

Figure A-44. Positioning To Clear an L-Shaped Hallway

d. Shooter Number Two, with his weapon covering the dead space, pies as far as possible and moves forward to increase his angle of fire further into the dead space. Shooter Number One moves accordingly with Shooter Number Two (Figure A-45).

e. Shooter Number Two stops pieing just before he reaches the position where he cannot cover the dead space. Shooter Number One stops short of the hallway and goes to a kneeling position.

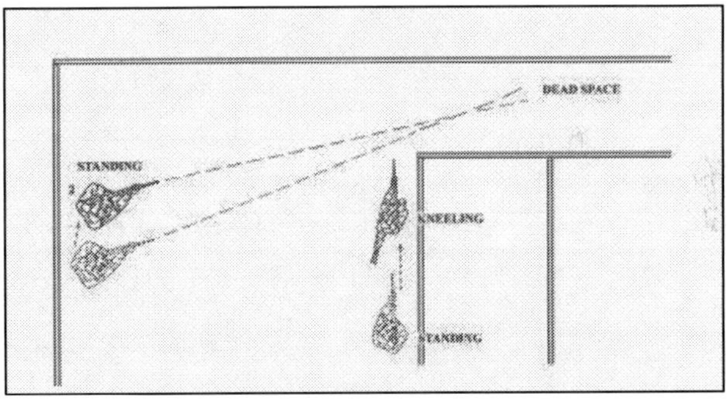

Figure A-45. Clearing an L-Shaped Hallway

f. Shooter Number One says, "**READY!**"

g. Shooter Number Two then says, "**MOVE!**" and both shooters move simultaneously to clear the hallway (Figure A-46). Shooter Number Two remains standing, and Shooter Number One remains kneeling as he pivots around the corner of the wall into the hallway.

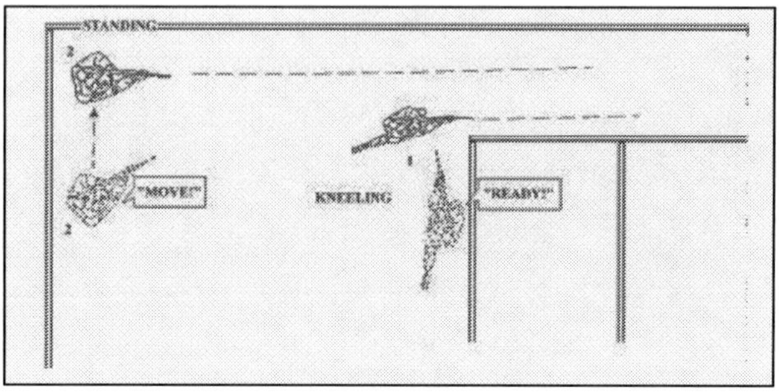

Figure A-46. Clearing an L-Shaped Hallway (Continued)

24. Clearing a T-Shaped Hallway. After entering a room or building, the clearing team may find a T-shaped hallway that should be cleared. The following procedures are used:

a. Both shooters approach the T-shaped intersection together, pieing their individual sectors while maintaining eye-muzzle-target contact (Figure A-47).

b. Both shooters move to dominant positions without entering the intersection.

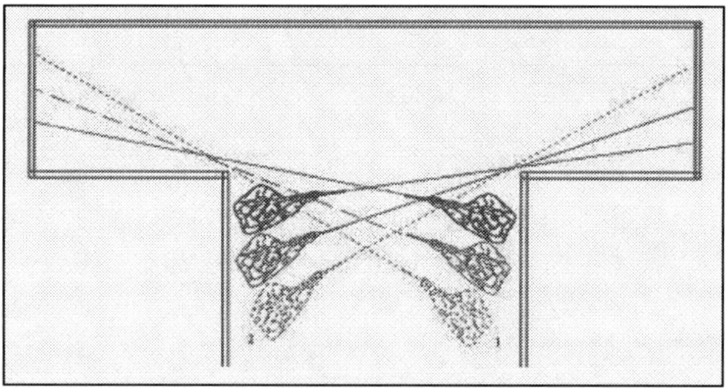

Figure A-47. Positioning To Clear a T-Shaped Hallway

c. Shooter Number One clears his sector of fire and then says, "**READY!**" (Figure A-48).

d. Shooter Number Two clears his sector of fire and then says, "**MOVE!**"

e. Both shooters buttonhook into their next positions.

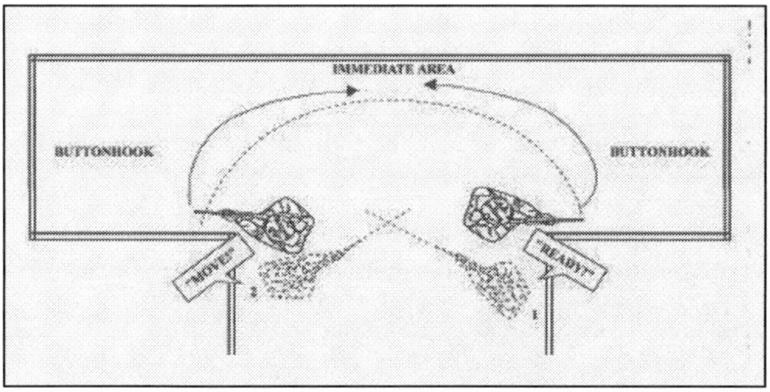

Figure A-48. Clearing a T-Shaped Hallway

f. Both shooters clear their immediate areas along their respective walls, starting from the nearest respective corner and continuing to the farthest respective corner (Figure A-49).

g. Both shooters establish dominant positions that give them control of the hallway and the doorways leading into hallways. The covering team can then be called forward to provide security while the clearing team clears one end of the hallway and associated rooms. The clearing team then clears the other end of the hallway.

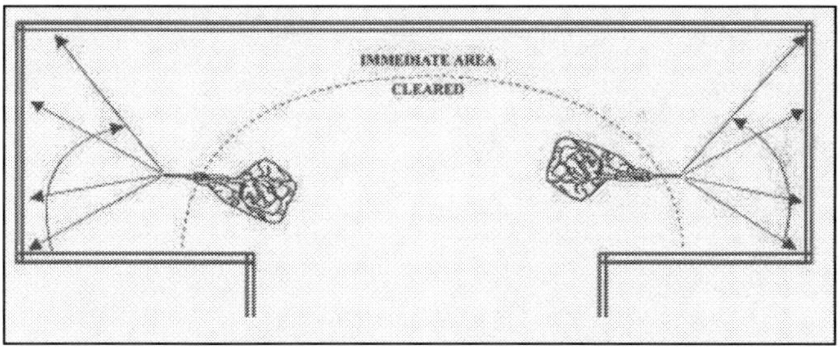

Figure A-49. Clearing a T-Shaped Hallway (Continued)

25. Clearing a Stairwell. After entering a building, the assault element may encounter stairwells. The following procedures are used:

a. Shooter Number One leads upstairs, one step ahead of Shooter Number Two (Figure A-50). Upon reaching the point just before he can be engaged from above, Shooter Number One turns around and covers overhead. From this point, Shooter Number One ascends the stairs, moving backward while covering behind and above.

b. Shooter Number Two follows Shooter Number One upstairs, one step behind and to the side of Shooter Number One. When Shooter Number One turns to cover overhead, Shooter Number Two remains oriented to the front, covering directly up the stairwell.

The clearing team's speed of movement is determined by Shooter Number One. Marines pie as much of an area as possible before ascending each step.

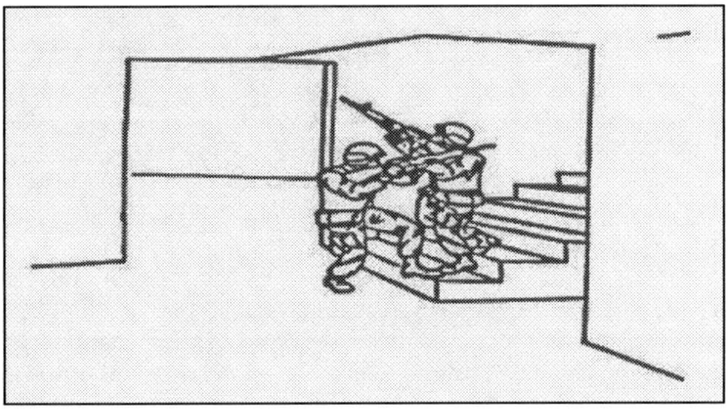

Figure A-50. Clearing a Stairwell

Section VI

Firing Positions

Whether a unit is attacking, defending, or conducting retrograde operations, its success or failure depends on the ability of the individual Marine to place accurate fire on the enemy while providing the least exposure to return fire. Consequently, Marines should immediately seek and properly use firing positions.

26. Hasty Firing Position. A hasty firing position is one that is normally occupied in the attack or the early stages of the defense. It is a position from which a Marine can place fire on the enemy while using available cover for protection from return fire. The Marine may occupy this position voluntarily, or he may be forced to occupy it because of enemy fire. In either case, the position lacks preparation before occupation. Some of the more common hasty firing positions in a built-up area involve firing around corners of buildings, firing from behind walls, firing from windows, firing from unprepared loopholes, and firing from the peak of a roof.

a. Corners of Buildings. The corner of a building provides cover for a hasty firing position if used properly.

(1) The shooter should be capable of both right-handed and left-handed firing of his weapon (using the "short stocking" technique) to be effective around corners. Short stocking the weapon will prevent the muzzle from protruding and keep the weapon ready to fire the instant visual contact with the enemy is made. Furthermore, it reduces the Marine's exposure as a target. (Figure A-51)

(2) A common mistake made when firing around corners is firing from the standing position. The shooter exposes himself at the height the enemy would expect a target to appear and risks exposing the entire length of his body as a target for the enemy.

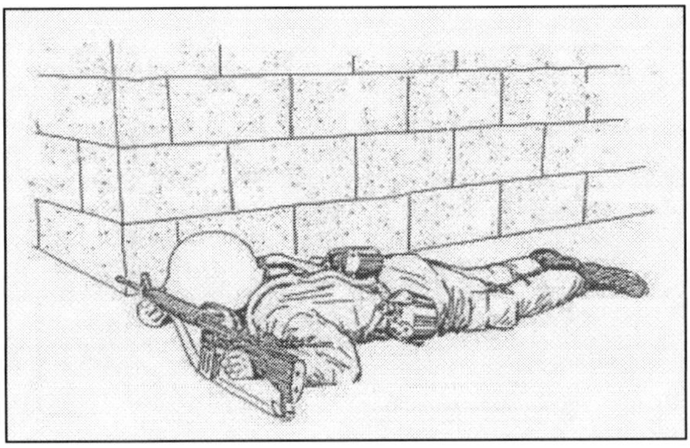

Figure A-51. Firing, Using Either Hand, Around the Corner of a Building

b. Walls. When firing from behind walls, a Marine should fire around cover—not over it (Figure A-52). Marines in this situation should be able to fire from the right or left shoulder. When firing from behind cover, a Marine should remain far enough back from that cover so that the weapon does not extend beyond it. This aids in weapon retention by ensuring that no enemy on the other side of the wall can get control of the barrel.

Figure A-52. Firing Around Cover

c. Windows. In a built-up area, windows provide convenient firing ports. A Marine should avoid firing from the standing position because it exposes most of his body to return fire from the enemy and could silhouette him against a light-colored interior beyond the window. This is an obvious indicator of the firer's position, especially at night when the muzzle flash can easily be observed. When using the proper method of firing from a window (Figure A-53), the Marine is well back into the room to prevent the muzzle flash from being seen, and he is kneeling to limit exposure and avoid silhouetting himself.

Figure A-53. Firing From a Window

d. Loopholes. A Marine may fire through a hole torn in the wall and avoid windows (Figure A-54). He stays well back from the loophole so the muzzle of the weapon does not protrude beyond the wall and the muzzle flash is concealed.

Figure A-54. Firing From a Loophole

e. Roof. The peak of a roof provides a vantage point for snipers that increases their field of vision and the ranges at which they can engage targets (Figure A-55). A chimney, a smokestack, or any other object protruding from the roof of a building can reduce the size of the exposed target and should be used.

Figure A-55. Firing From the Peak of a Roof

f. No Position Available. When a Marine is subjected to enemy fire and none of the positions mentioned are available, he should try to expose as little of himself as possible. When a Marine is in an open area between buildings (a street or alley) with no cover available and he is fired upon by an enemy from one of the buildings to his front, he should lie prone as close as possible to a building on the same side of the open area as the enemy. To engage the Marine, the enemy must then lean out of the window and expose himself to return fire.

g. No Cover Available. When no cover is available, target exposure can be reduced by firing from the prone position, by firing from shadows, and by presenting no silhouette against buildings.

27. Prepared Firing Position. A prepared firing position is one built or improved to allow the shooter to engage a particular area, avenue of approach, or enemy position, while reducing his exposure to return fire. Examples of prepared positions include: barricaded windows, fortified loopholes, sniper positions, antiarmor positions, and machine gun positions.

a. The natural firing port provided by windows can be improved by barricading the window and leaving a small hole for the firer's use (Figure A-56). The barricading may be accomplished by using materials torn from the interior walls of the building or any other available material. When barricading windows, a Marine should avoid:

(1) Barricading only the windows that will be used as firing ports. The enemy will soon determine that the barricaded windows are firing positions.

(2) Neat square or rectangular holes that are easily identified by the enemy. A barricaded window should not have a neat, regular firing port. The window should keep its original shape so that the position of the firer is hard to detect. Firing from the bottom of the window gives the firer the cover advantage of the wall because the firing port is less obvious to the enemy. Sandbags are used to reinforce the wall below the window and to increase protection for the firer. All glass should be removed from the window to prevent injury to the firer. Lace curtains permit the firer to see out and prevent the enemy from seeing in. Wet blankets should be placed under weapons to reduce dust that may create a signature. Wire mesh placed over the window keeps the enemy from throwing in hand grenades.

b. Although windows are usually good firing positions, they do not always allow the firer to engage targets throughout his sector.

(1) To avoid establishing a pattern of always firing from windows, an alternate position is required such as the prepared loophole (Figure A-57). This involves cutting or blowing a small hole in the wall to allow the shooter to observe and engage targets in his sector.

(2) Sandbags are used to reinforce the walls below, around, and above the loophole. If the position is on the second floor or higher, two layers of sandbags are placed on the floor under the firer to protect him from an explosion on a lower floor. A wall of sandbags, rubble, furniture, and so on should be constructed to the rear of the position to protect the firer from explosions in the room.

Figure A-56. Window Firing Position

(3) A table, a bedstead, or other available material provides overhead cover for the position. This prevents injury to the shooter from falling debris or explosions above his position.

(4) The position should be camouflaged by knocking other holes in the wall. This makes it difficult for the enemy to determine which hole the fire is coming from. Siding material should be removed from the building in several places to make loopholes less noticeable.

Figure A-57. Prepared Loopholes

c. A chimney or other protruding structure provides a base from which a sniper position can be prepared. Part of the roofing material is removed to allow the sniper to fire around the chimney. He should stand inside the building on the beams or on a platform with only his head and shoulders above the roof (behind the chimney). Sandbags placed on the sides of the position protect the sniper's flanks.

d. When the roof has no protruding structure to provide protection, the sniper position should be prepared from underneath on the enemy side of the roof (Figure A-58). The position is reinforced with sandbags, and a small piece of roofing material should be removed to allow the sniper to engage targets in his sector. The missing piece of roofing material should be the only sign that a position exists. Other pieces of roofing should be removed to deceive the enemy as to the true sniper position. The sniper should be invisible from outside the building, and the muzzle flash should be hidden from view.

Figure A-58. Sniper Position

e. Some rules and considerations for selecting and occupying individual firing positions are:

(1) Make maximum use of available cover and concealment.

(2) Avoid firing over cover; when possible, fire around it.

(3) Avoid silhouetting against light-colored buildings, the skyline, and so on.

(4) Carefully select a new firing position before leaving an old one.

(5) Avoid setting a pattern; fire from both barricaded and unbarricaded windows.

(6) Keep exposure time to a minimum.

(7) Begin improving a hasty position immediately after occupation.

(8) Use construction material that is readily available in the built-up area for prepared positions.

(9) Remember that positions that provide cover at ground level may not provide cover on higher floors.

f. In attacking a built-up area, recoilless weapon and ATGM crews may be hampered in choosing firing positions by the backblast of their weapons. They may not have enough time to knock out walls in buildings and clear backblast areas. They should select positions that allow the backblast to escape, such as corner windows where the round fired goes out one window and the backblast escapes from another. The corner of a building can be improved with sandbags to create a firing position (Figure A-59).

Figure A-59. Corner Firing Position

g. The rifle squad is often reinforced with attached antitank weapons during an attack on and in defense of a built-up area. Therefore, the rifle squad leader should be able to choose good firing positions for the antitank weapons under his control.

h. Various principles for employing antitank weapons have universal applications, such as making maximum use of available cover, trying to achieve mutual support, and allowing for the backblast when positioning recoilless weapons, TOWs, Dragons, and AT4s.

i. Operating in a built-up area presents new considerations. Marines should select numerous alternate positions, particularly when the structure does not provide cover from small-arms fire. They should position their weapons in the shadows and within the building.

j. Recoilless weapons and ATGMs firing from the top of a building can use the chimney for cover (Figure A-60). The rear of this position should be reinforced with sandbags.

Figure A-60. Antitank Weapon Firing From a Rooftop

k. When selecting firing positions for recoilless weapons and ATGMs, Marines should make maximum use of rubble, corners of buildings, and destroyed vehicles to provide cover for the crew. Recoilless weapons and ATGMs can also be moved along rooftops to obtain a better firing angle from which to engage enemy armor. When buildings are elevated off the ground, positions can be prepared using a building for overhead cover (Figure A-61). The backblast under the building must not damage or collapse the building or injure the crew.

Note: When firing from a slope, ensure that the angle of the launcher relative to the ground or firing platform is not greater than 20 degrees. When firing within a building, ensure that the enclosure is at least 10 feet by 15 feet, is clear of debris and loose objects, and has windows, doors, or holes in the walls to allow the backblast to escape.

l. The machine gun can be emplaced almost anywhere. In the attack, windows and doors offer ready-made firing ports (Figure A-62). For this reason, the enemy normally has windows and doors under observation and fire; they should therefore be avoided. Any openings in walls that were created during the fighting may be used. When other holes are not present, small explosive charges can create loopholes (Figure A-63). Regardless of what openings are used, machine guns should be sited within the building and in the shadows.

Figure A-61. Prepared Positions Using a Building for Overhead Cover

Figure A-62. Emplacement of a Machine Gun in a Doorway

m. Upon occupying a building, Marines board up all windows and doors. By leaving small gaps between the slots, Marines can use windows and doors as good alternate firing positions.

n. Loopholes should be used extensively in the defense. They should not be constructed in any logical pattern, nor should they all be at floor or tabletop level. Varying their height and location makes them hard to pinpoint and identify. Dummy loopholes, knocked-off shingles, or holes that are cut but not intended to be used as firing positions aid in deception. Loopholes located behind shrubbery, under doorjambs, and under the eaves of a building are hard to detect.

Figure A-63. Use of a Loophole With a Machine Gun

o. Increased fields of fire can be obtained by locating the machine gun in the corner of the building or sandbagging it under a building (Figure A-64). Available materials, such as desks, overstuffed chairs, couches, and other items of furniture, should be integrated into the construction of bunkers to add both cover and concealment (Figure A-65).

Figure A-64. Sandbagged Machine Gun Emplacement Under a Building

p. Although grazing fire is desirable when employing the machine gun, it may not always be practical or possible. Where destroyed vehicles, rubble, and other obstructions restrict the fields of grazing fire, the gun can be elevated to a position from which it can deliver plunging fire over obstacles. Firing from loopholes on the second or third story may be necessary. A firing platform can be built under the roof (Figure A-66) and a loophole constructed. Again, the exact location of the position should be concealed by knocking off shingles in isolated patches over the entire roof.

Figure A-65. Corner Machine Gun Bunker

Figure A-66. Firing Platform Built Under a Roof

28. Target Acquisition. Built-up areas provide unique challenges to acquiring targets. Buildings mask movement and the effects of direct and indirect fires. Also, the rubble from destroyed buildings provides concealment and protection for attackers and defenders alike, making target acquisition difficult.

a. The techniques of patrolling and using OPs apply in the city as well as in wooded terrain. These techniques enable units to locate the enemy, to develop targets for direct and indirect fires in the defense, and to find uncovered avenues of approach in the offense.

b. Most weapons and vehicles produce recognizable signatures. These come from design features or from the environment in which the equipment is used. For example, firing a tank main gun in dry, dusty, and debris-covered streets raises a dust cloud; a tank being driven in built-up areas produces more noise than one moving through an open field; people moving through rubble on a street or in the halls of a damaged building create more noise than in a wooded area. Marines should learn to recognize signatures so they can locate and identify targets. Seeing and hearing assist in detecting and identifying signatures that lead to target location, identification, and rapid engagement. Marines should look for targets in areas where they are most likely to be employed.

c. Target acquisition should be continuous, whether halted or moving. Built-up areas provide both the attacker and defender with good cover and concealment but usually favor the defender. This makes rapid and accurate target acquisition extremely important.

d. When a unit is moving and enemy contact is likely, the unit should employ an overwatching element. This principle applies in built-up areas as it does in other kinds of terrain except that in built-up areas the overwatching element should observe both the upper floors of buildings and the street level.

e. Stealth should be used when moving in built-up areas because little distance separates attackers from defenders. Only hand and arm signals should be used until contact is made. The unit should stop periodically to listen and observe to ensure that it is not being followed or that the enemy is not moving parallel to the unit's flank for an ambush. Routes should be carefully chosen so that buildings and piles of rubble can be used to mask the unit's movement.

f. Observation duties should be clearly issued to squad members to ensure all-around security as they move. This security continues at the halt. All of the senses should be used to acquire targets, especially hearing and smelling. Marines soon recognize the sounds of vehicles and people moving through streets that are littered with rubble. The smell of fuel, cologne, or food cooking can disclose enemy positions.

g. OPs are "positions from which military observations are made, or fire directed and adjusted, and which possess appropriate communications" (Joint Pub 1-02). They are positions from which Marines can watch and listen to enemy activity in a specific sector. They warn the unit of enemy approach and should be positioned in the upper floors of buildings to give Marines a better vantage point than they would have at street level.

h. In the defense, a platoon leader positions OPs for local security as ordered by the company commander. The platoon leader selects the general location, but the squad leader establishes the OP (Figure A-67). Normally, there is at least one OP per platoon. An OP consists of two to four men and is within small-arms supporting range of the platoon. Leaders look for positions that allow good observation of the target sector. Ideally, an OP has a field of observation that overlaps those of adjacent OPs. The position selected for the OP should provide cover and concealment for units moving to and from the OP. The upper floors of houses or other buildings should be used. The squad leader should not routinely select obvious positions such as water towers or church steeples because those positions naturally attract the enemy's attention.

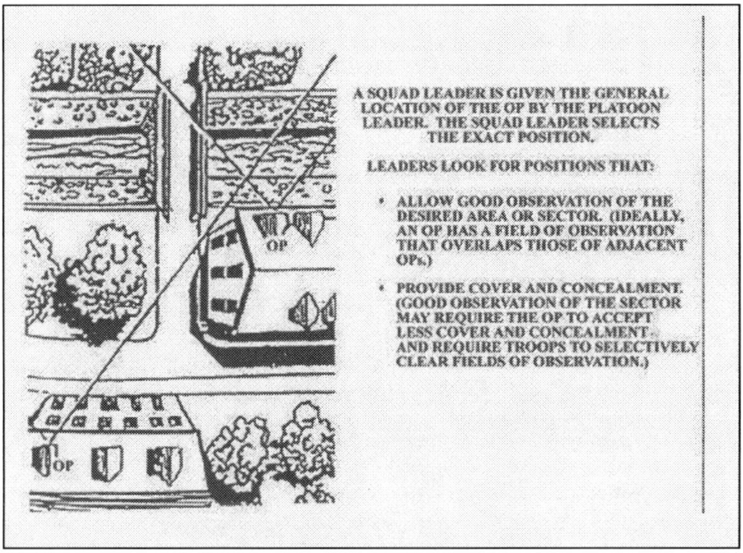

Figure A-67. Selection of an OP Location

i. The Marine should be taught how to scan a target area from OPs or from fighting positions. Use of proper scanning techniques enables observers to quickly locate and identify targets. The Marine searches quickly without optics for obvious targets, using all of his senses to detect target signatures. If no targets are found and time permits, he makes a more detailed search (using binoculars, if available) of the terrain in the assigned sector using the 50-meter method. First, he searches a strip 50 meters deep from right to left. He then searches a strip from left to right that is farther out, overlapping the first strip. This process is continued until the entire sector has been searched. In the city core or core periphery where the observer is faced with multistory buildings, the overlapping sectors may extend up rather than out.

j. Marines who man OPs and other positions should employ target acquisition devices. These devices include binoculars, image intensification devices, and thermal sights. All of these devices can enhance the unit's ability to detect and engage targets, especially at night or during periods of reduced visibility.

k. Target acquisition techniques used at night are similar to those used during the day. At night, whether using daylight optics or the unaided eye, the observer should not look directly at an object but a few degrees off to the side. The side of the eye is more sensitive to dim light. When scanning with off-center vision, the Marine moves his eyes in short, abrupt, irregular movements. At each likely target area, he pauses for a few seconds to detect any motion.

l. At night, sounds and smells can aid in acquiring targets. They transmit better in the cooler, damper night air. Running engines, vehicles, and people moving through rubble- covered streets can be heard for great distances. Odors from diesel fuel, gasoline, cooking food, burning tobacco and aftershave lotion can reveal enemy and friendly locations.

29. Firefighting Planning and Operations. Incendiary ammunition, special weapons, and the ease with which incendiary devices can be constructed from gasoline and other flammables make fire a true threat in operations in built-up areas. During defensive operations, firefighting should be a primary concern. Proper steps should be taken to reduce the risk of a fire that could make a chosen position indefensible.

a. Marines choose or create positions that do not have large openings. These positions provide as much built-in cover as possible to prevent penetration by incendiary ammunition. All unnecessary flammable materials are removed, including ammunition boxes, furniture, rugs, newspapers, curtains, and so on. The electricity and gas coming into the building should be shut off.

b. A building of concrete block construction with concrete floors and a tin roof is an ideal place for a fighting position. However, most buildings have wooden floors or subfloors, wooden rafters, and wooden inner walls that may require improvement. Inner walls are removed and replaced with blankets to resemble walls from the outside. Sand is spread 2 inches deep on floors and in attics to retard fire.

c. All available firefighting gear is prepositioned so that it can be used during combat. For the individual Marine, such gear includes entrenching tools, helmets, sand, and blankets. These items are supplemented with fire extinguishers.

d. Fire is so destructive that it can easily overwhelm personnel regardless of extraordinary precautions. Marines should plan routes of withdrawal so that evacuation can be accomplished from their fighting positions. This allows Marines to exit through areas that are free of combustible material but that still provide cover from enemy direct fire.

e. The confined space and large amounts of combustible material found in built-up areas may influence the enemy to use incendiary devices. First-aid problems that are expected include burns and smoke/flame inhalation. These can easily occur in buildings, thereby rendering the victims combat ineffective. Although there is little defense against flame inhalation and lack of oxygen, smoke inhalation may be greatly reduced by wearing the individual protective mask. Regardless of the fire hazard, defensive planning for combat in built-up areas should include location of corpsmen. Corpsmen must reach victims and their equipment and should stock extra supplies for the treatment of burns and inhalation injuries.

f. Offensive operations also require plans for firefighting because the success of the mission can easily be threatened by fire. Poorly planned use of incendiary munitions can make fires so extensive that they become obstacles to offensive operations. The enemy may use fire to cover his withdrawal and to create obstacles and barriers to the attacker.

g. Every Marine participating in an attack should be ready to deal with the problems of fire. The normal firefighting equipment available includes the entrenching tool, helmet (for carrying sand or water), and blankets (for snuffing out small fires). Fire extinguishers are normally available on vehicles.

30. Employment of Snipers. The value of the sniper to a unit operating in a built-up area depends on several factors. These factors include the type of operation, the level of conflict, and the ROE. Where ROE allow destruction, snipers may not be needed because other weapons systems available to the infantry commander have a greater destructive effect. However, snipers can contribute to the fight. Where the ROE prohibit collateral damage, snipers may be the most valuable tool the commander has. (See MCWP 3-15.3 (under development).)

a. Sniper effectiveness depends in part on the terrain. Control is decentralized by the characteristics of an urban area. To provide timely and effective support, the sniper should have a clear picture of the commander's mission and intent.

b. Where possible, snipers should be positioned in masonry buildings. These buildings should also offer long-range fields of fire and all-around observation. The sniper has an advantage because he does not have to move with or be positioned with lead elements. He may occupy a higher position to the rear or flanks of, and some distance away from, the element that he is supporting. By operating far from the other elements, a sniper avoids decisive engagement but remains close enough to kill distant targets that threaten the unit. Snipers should not be placed in obvious positions such as church steeples and rooftops because the enemy often watches these and targets them for destruction. Also, snipers should not be positioned where there is heavy traffic; these areas also invite enemy observation.

c. Snipers should operate throughout the AO, moving with and supporting the companies as necessary. Some teams may be designated to operate independently of other forces. They may be tasked to search for targets of opportunity, especially for enemy snipers. The sniper team may occupy multiple positions. A single position may not afford adequate observation for the entire team without increasing the risk of detection by the enemy. Separate positions should maintain mutual support. Alternate and supplementary positions should also be established in urban areas.

d. Snipers may be assigned the following tasks:

- Killing enemy snipers (countersniper fire)

- Killing targets of opportunity (These targets should be prioritized by the commander. Types of targets might include enemy snipers, leaders, vehicle commanders, radio men, sappers, and machine gun crews.)

- Denying enemy access to certain areas or avenues of approach (controlling key terrain)

- Providing fire support for covering barricades and other obstacles

- Maintaining surveillance of flank and rear avenues of approach (screening)

- Supporting local counterattacks with precision fire.

Section VII

Advancing/Patrolling Along City Streets

31. Advancing/Patrolling Along City Streets. During MOUT operations, skills for advancing/ patrolling along city streets will be required. The platoon, squad, or fire team mission may be such that building-to-building clearing is neither required nor desired until contact with the enemy has been made. This section will discuss advancing/patrolling along urban streets. The standard tactics, techniques, and procedures for the conduct of a patrol, as outlined in MCWP 3-11.6 *Scouting and Patrolling* (draft), apply except as otherwise discussed within this section.

a. Organization. The platoon's three-squad organization or the squad's three-fire team organization is already structured to conduct an advance/patrol down an urban street. The typical patrol is organized into assault, support, security, and command and control elements. Security is divided into front, flank, and rear security responsibilities. The platoon or squad can move down the street in a double or single column.

b. Formation and Movement

(1) Double Column. The double column is the *preferred method* for moving along an urban street (Figures A-68 and A-69). The double column provides 360 degrees of security and allows for mutual support and interlocking sectors of fire. The double column is flexible but can make control more difficult. There should be at least two squads or fire teams to conduct a double column. The two squads or fire teams move down the sides of the street using the walls of the buildings for cover and concealment. The pointmen for the two columns should always stay abreast of each other and not move in front of or behind the opposite column. The Marines following in trace of the pointmen stagger themselves accordingly. The platoon or squad leaders move where they can best control their units. If a squad is conducting the advance/patrol along the street, the trail fire team is never split.

(a) Sectors of Responsibility. Every Marine moving in a column has a particular sector for whose security he is responsible. The pointmen are responsible for security to the direct front and for pieing off windows and doorways. They are also responsible for staying in line with one another.

The Marines immediately behind the pointmen are designated as the covermen. They are responsible for providing guidance to their respective pointmen from the leader of the unit and for covering the 45-degree oblique angle across from one another. The second coverman is also responsible for covering the direct front when the pointman is pieing and checking for the enemy. The third and fourth covermen in the column are responsible for the far-side flank security; one covers the upper level and one covers the lower level building windows, doors, and stairwells. Marines will alternate providing security on upper level and lower level windows, doors, and stairwells. The fourth Marine also provides rear security if he is the last man in the column or interlocks his sector of fire with that of the lead Marine in the next fire team in the column. Should the second coverman have to take the point while the pointman pies a window or door, the third coverman would automatically shift his sector of fire to the 45-degree oblique angle. If the second coverman has to assist the pointman in pieing a large opening, the third coverman would provide security to the direct front while the fourth coverman would take the 45-degree oblique angle. Marines following in the column would adjust their sectors of fire accordingly.

The two columns provide mutual support for each other and the greatest all-around security. Teamwork and coordination between the two columns are vital.

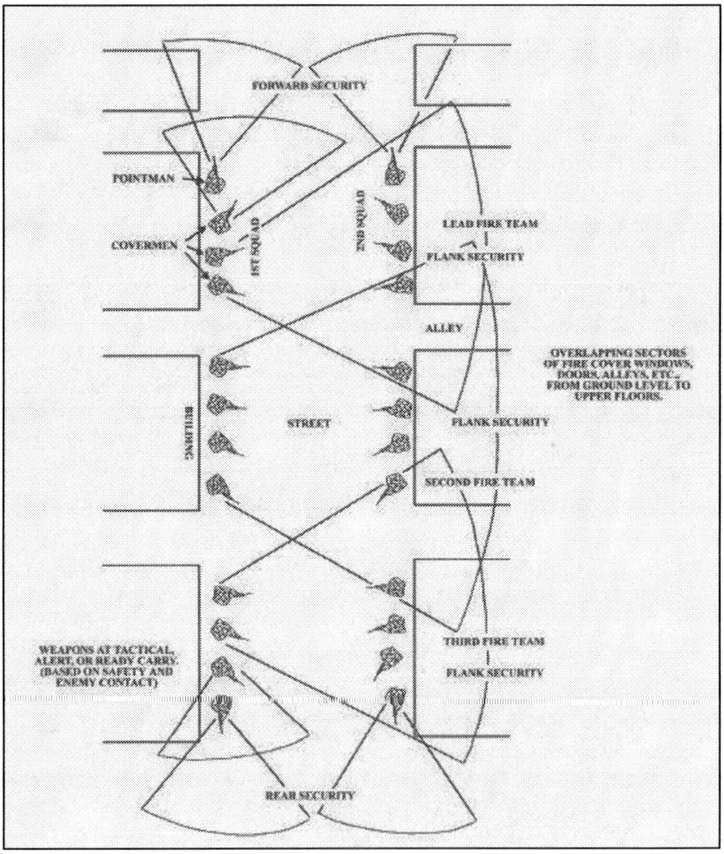

Figure A-68. Advancing/Patrolling Along an Urban Street (Squad, Double Column)

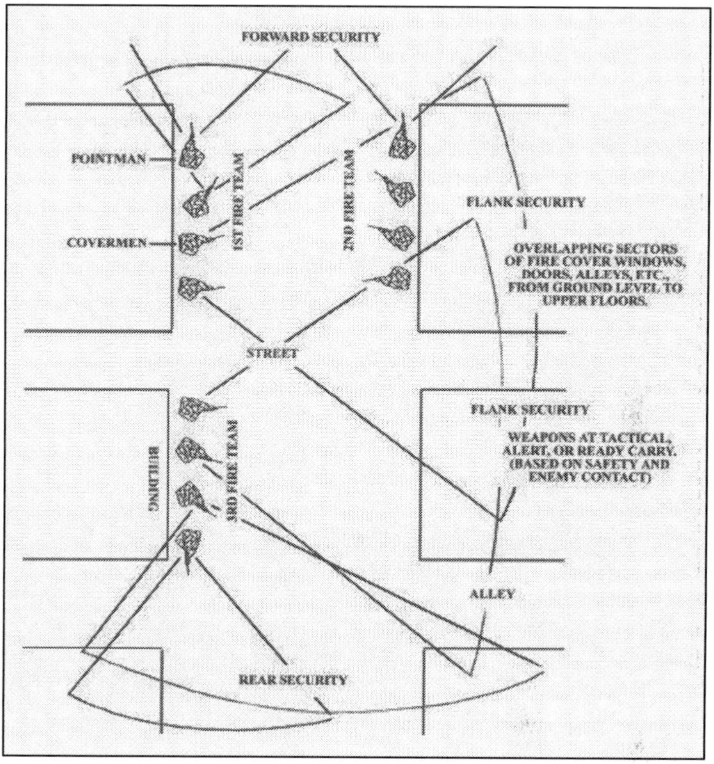

Figure A-69. Advancing/Patrolling Along an Urban Street (Fire Team, Double Column)

(b) Rear Security. The last two Marines in the squad of each column are responsible for the rear security of each of their formations. They will move in bounds in order to allow a Marine to always provide active security to the squad's rear (Figures A-70 and A-71). Two methods of bounding may be used: *alternate* or *successive bounding*.

<u>1</u> **Alternate Bounding.** In this method, the two rear-security Marines alternate being the last Marine in the squad. The last Marine in the formation will have his weapon at the ready and will be in the kneeling position along the wall of the building. The second to last Marine will move in trace of the unit until he has moved approximately 10 to 15 meters. At this point, he will step out one pace from the wall. He will then assume a good kneeling firing position and announce, "**SET**." The last Marine will rise from his position, turn around, and move next to the wall to catch up with the column. It is important that the Marine moving back

does not step out and mask the other Marine's field of fire. At this point, the Marine covering the rearward movement of the last Marine can assume the rear guard position and move back to the wall as the Marine passes by his position (Figure A-70). The process is repeated continuously. The rear fire team leaders coordinate the movements of the rear guard to keep them abreast of each other and ensure that the column does not move too fast and leave the rear guard behind.

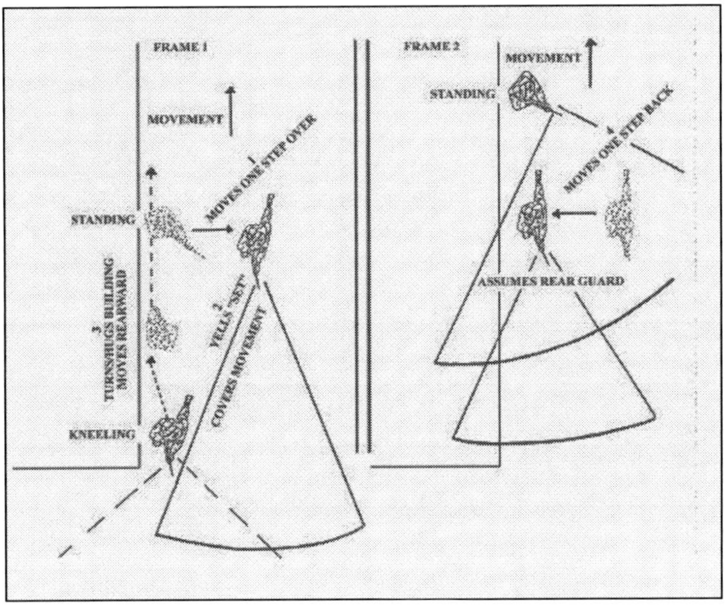

Figure A-70. Rear Security (Alternate Bounds)

2 **Successive Bounding.** In this method, the last Marine in the squad continues to be the rear guard after bounding. The same procedures are employed as in alternate bounding; however, the rear-guard Marine bounds abreast of the Marine providing cover, turns around, drops to a good kneeling position, provides rear-guard security, and announces, "**SET**." The covering Marine moves behind the rear-guard Marine and continues to move 10 to 15 meters before repeating the process (Figure A-71). The rear fire team leaders coordinate the movements of the rear guard as in alternate bounding.

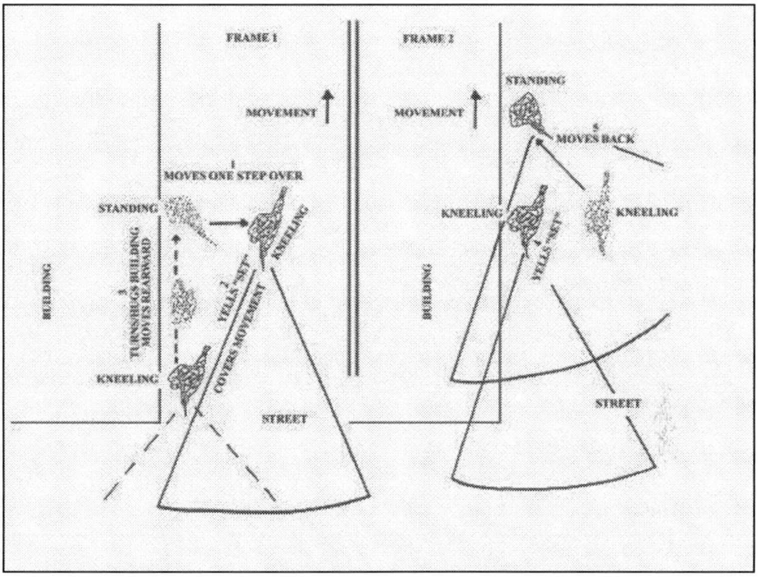

Figure A-71. Rear Security (Successive Bounds)

(2) Single Column. The single column is used to move rapidly along a street or when streets are restricted to a single column. The single column is the least preferred advance/patrol method of moving along an urban street. The same principles apply as in the double column, except the single column does not provide 360-degree security. The single column is normally used when speed is desired and contact with the enemy is considered to be unlikely.

32. Crossing a Street Intersection. Four-way intersections are very dangerous areas that should be crossed quickly. Security at an intersection is extremely important for the protection of crossing units. When moving, the pointmen notify the unit leaders, through the covermen, of an approaching intersection. Crossing the intersection will force the three elements of a unit moving along a city street to rotate positions and assume appropriate responsibilities.

a. First Fire Team To Cross. The lead (1st) fire teams in the two columns will simultaneously set security around the corners of the building. They will provide mutually supporting security to the direct front, security to the building on the far side of the intersection, and intersection security (Figure A-72). The middle (2nd) fire teams provide mutually supporting flank and rear security. On order, the rear (3rd) fire teams move forward, cross the intersection, and assume the lead (1st) fire teams' forward security mission.

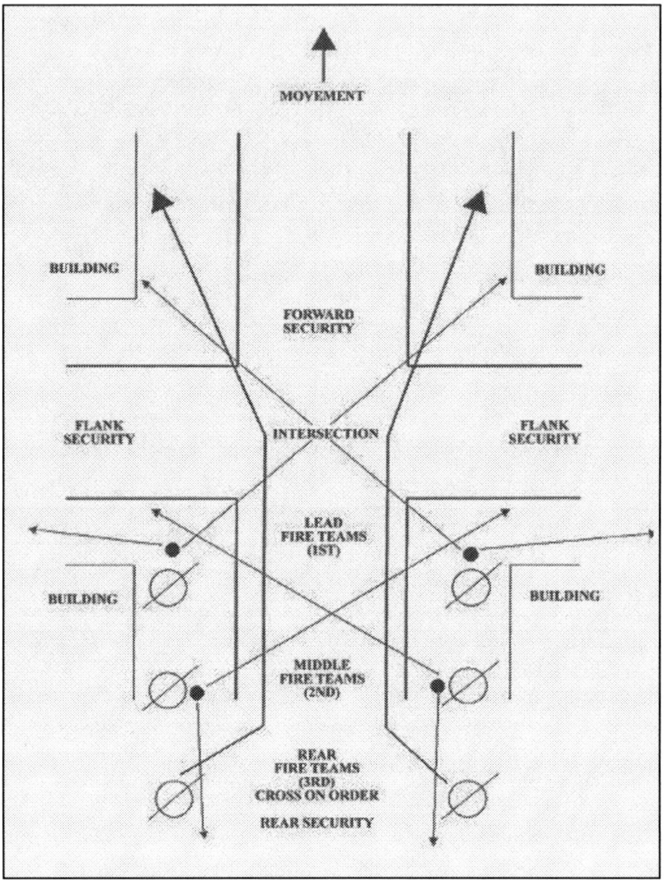

Figure A-72. Crossing an Intersection (Frame 1)

b. Second Fire Team to Cross. The 1st fire teams (now the middle fire teams) continue to provide intersection security and assume rear area security. The 2nd fire teams (now rear fire teams) cross the intersection on order. The second fire teams position themselves to provide rear and intersection security to support the covering of the last fire teams (Figure A-73).

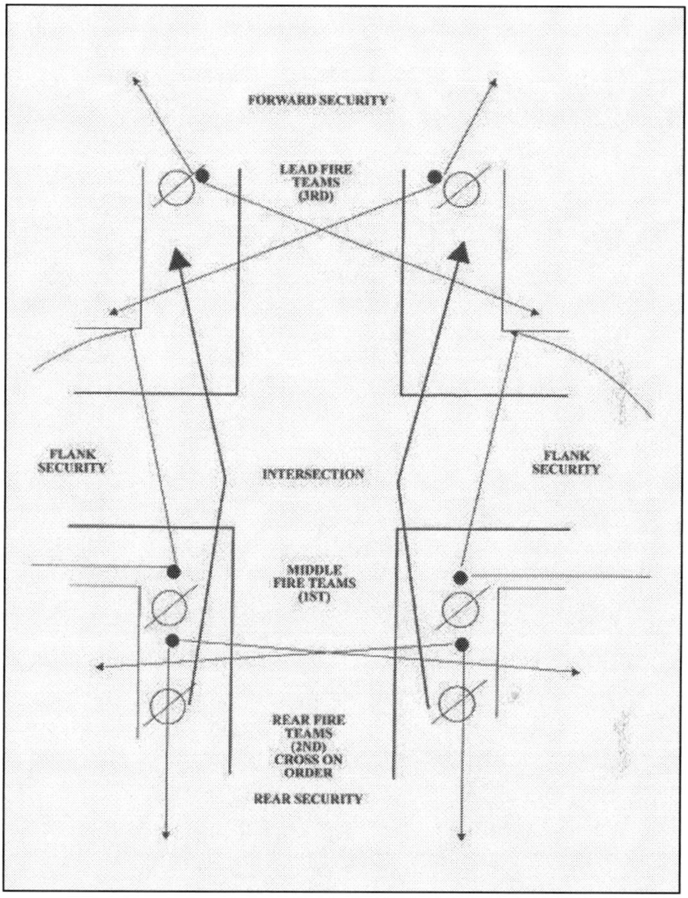

Figure A-73. Crossing an Intersection (Frame 2)

c. Third Fire Team To Cross. The 1st fire teams (now the rear fire teams) provide rear security and cross the intersection on order. The 1st fire teams bypass the 2nd fire teams to become the middle fire teams and provide mutually supporting forward and flank security. The 2nd fire teams become the rear security (Figure A-74).

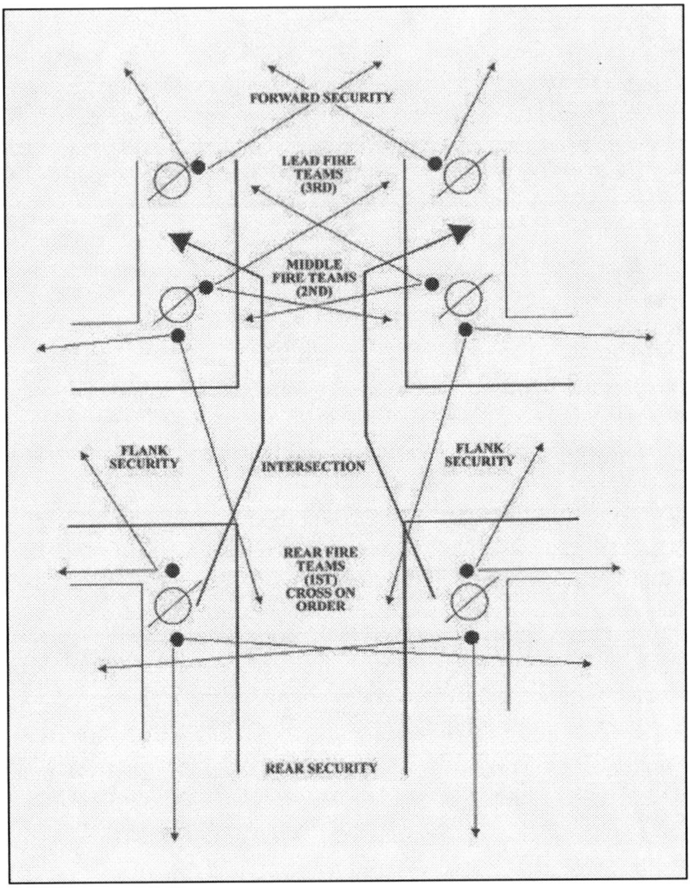

Figure A-74. Crossing an Intersection (Frame 3)

d. The squads and three fire teams have now crossed the intersection, and the fire teams have rotated (Figure A-75). This rotation is convenient because it alternates the lead units, who may quickly become fatigued during MOUT operations. The same procedures are used to cross an intersection with a single column.

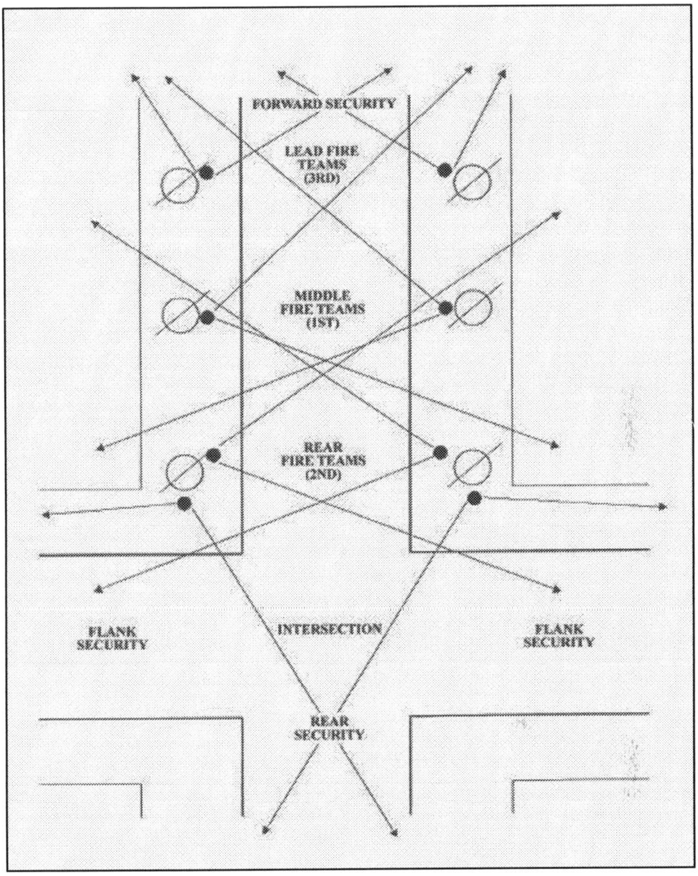

Figure A-75. Crossing an Intersection (Frame 4)

33. Reacting to Enemy Contact. When a platoon or squad is moving along a city street, contact with the enemy could happen at any time, even if the area has been declared secure. If a unit engages or is engaged by the enemy, there are two basic options that Marines may follow. The first and most preferred option is to immediately return fire and conduct a hasty clearing of a structure to seek cover inside of the building to which the unit was adjacent. The second and least preferred method is to remain outside of the building(s) and fight from the street.

a. Hasty Clearing. To conduct a hasty clearing of a structure in order to seek cover, Marines immediately return fire and enter buildings as quickly as possible. The basic principles of entry are applied as discussed earlier in the appendix; however, speed of action is essential. Once inside, Marines quickly scan the area and engage any threat. If no threat is present, the Marines acquire the enemy's positions and deliver carefully aimed shots to achieve fire superiority. At this point, the situation is developed as rapidly as possible and an attack to clear the enemy may be ordered.

b. Remain Outside and Fight. This option should not be conducted unless strict ROE do not allow the occupation of buildings without being fired on from them first or unless buildings cannot be entered because of obstruction, obstacles, or boobytraps. In such cases, the Marines should quickly acquire the target(s) and return fire to gain fire superiority. They should also seek whatever cover may exist. Once fire superiority is achieved, an assault to clear the enemy may be conducted.

Section VIII

Navigation in Built-Up Areas

Navigation in built-up areas present a unique set of challenges. Deep in the city core, the normal terrain features depicted on maps may not apply—buildings become the major terrain features, and units become tied to streets. Fighting in the city destroys buildings, whose rubble then blocks streets. Street and road signs are destroyed during the fighting if they are not switched or removed by the defenders. Operations in subways and sewers present other unique challenges. However, maps and photographs are available to help units overcome these problems. Additionally, the global positioning system (GPS) can help supplement navigational abilities in built-up areas.

34. Military Maps. The military city map is a topographical map of a city that delineates streets and shows street names, important buildings, and other urban elements. The scale of a city map can vary from 1:25,000 to 1:50,000 to 1:100,000 depending on the importance and size of the city, the density of detail, and intelligence information available.

a. Special maps prepared by supporting topographical engineers can assist units in navigating in built-up areas. These maps are designed or modified to provide information that is not covered on a standard military map. This may include maps of road and bridge networks, railroads, built-up areas, and electric power fields. Special maps can be used to supplement military city maps and topographical maps.

b. Once in the built-up area, Marines may use street intersections as reference points as they would use hills and streams in rural terrain. City maps supplement or replace topographical maps as the basis of navigation. These maps enable units moving in the built-up area to know where they are and to move to new locations even though streets have been blocked or a key building has been destroyed.

c. The techniques of compass reading and pace counting can still be used, especially in a blacked-out city where street signs and buildings are not visible. The presence of steel and iron in the MOUT environment may cause inaccurate compass readings. Sewers may be navigated in much the same way. Maps providing the basic layout of the sewer system are maintained by city sewer departments. This information includes directions in which the sewer lines run and distances between manhole covers. Along with basic compass and pace-count techniques, such information enables a unit to move through the city sewers with accuracy.

d. Operations in a built-up area adversely affect the performance of some types of communications-electronic devices such as the GPS, the Position Location Reporting System (PLRS), or other data-distribution systems. These systems function the same as some communications equipment—by LOS. They cannot determine underground locations or positions within a building. These systems should be employed on the tops of buildings, in open areas, and down streets where obstacles will not affect LOS readings.

e. City utility workers are assets to units fighting in built-up areas. They can provide maps of sewers and electrical fields and information about the city. This is especially important with regard to the use of the sewers. Sewers can contain pockets of methane gas that are highly toxic to humans. City sewer workers know the locations of these danger areas and can advise a unit on how to avoid them.

35. Aerial Photographs. Current aerial photographs are also excellent supplements to military city maps and can be substituted for a map. A topographical map or military city map could be obsolete if compiled many years ago. A recent aerial photograph shows changes that have taken place since the map was made. This could include destroyed buildings and streets that have been blocked by rubble as well as enemy defensive preparations. More information can be gained by using aerial photographs and maps together than by using either one alone.

Section IX

Camouflage

To survive and win in combat in built-up areas, a unit should supplement cover and concealment with camouflage. To properly camouflage men, carriers, and equipment, Marines should study the surrounding area and make their fighting positions blend with the local terrain.

36. Application. Only the proper amount of material needed for camouflaging a position should be used. Excess material could reveal the position. Material should be obtained from a wide area around the position. For example, Marines defending a cinder-block building do not strip the front, sides, or rear of the building to camouflage a position.

a. Buildings provide numerous opportunities for establishing concealed positions. Armored vehicles can often find prospective positions under archways or inside small industrial or commercial structures. Thick masonry, stone, or brick walls offer excellent protection from direct fire and may provide concealed routes into or out of the position.

b. After camouflage is completed, the Marine inspects his position from the enemy's viewpoint. He makes routine checks to see whether the camouflage remains natural looking and actually conceals the position. If it does not look natural, the Marine should rearrange or replace it until a natural look is achieved.

c. Positions should be progressively camouflaged as they are prepared. Work should continue until all camouflage is complete. If the enemy has air superiority, work may be possible only at night. Shiny or light-colored objects that attract attention from the air should be hidden.

d. Camouflage face paint is issued in three standard, two-tone sticks. When issue-type face-paint sticks are not available, burnt cork, charcoal, or lampblack can be used to tone down exposed skin. Mud may be used as a last resort because it dries and may peel off, leaving the skin exposed, and may also contain harmful bacteria.

37. Use of Shadows. Buildings throw sharp shadows that can be used to conceal vehicles and equipment (Figure A-76). Marines should avoid movement or positions in areas that are not in shadows. Vehicles may have to be moved periodically as shadows shift during the day. Emplacements inside buildings provide better concealment and cover as well.

38. Color and Texture. Standard camouflage-pattern painting of equipment is not as effective in built-up areas as a solid, dull, dark color hidden in shadows. Because repainting vehicles before entering a built-up area is not always practical, the lighter sand-colored patterns should be subdued with mud or dirt.

Figure A-76. Use of Shadows for Concealment

a. The need to break up the silhouette of helmets and individual equipment exists in built-up areas the same as it does elsewhere. However, burlap or canvas strips are a more effective camouflage than foliage (Figure A-77). Predominant colors are normally browns, tans, and sometimes grays rather than greens, but each camouflage location should be evaluated to determine ideal colors and patterns.

Figure A-77. Helmet Camouflaged With Burlap Strips

b. CPs and logistics emplacements are easier to camouflage and better protected if located underground. Antennas can be remoted to upper stories or to higher buildings based on remote capabilities. Field telephone wire should be laid in conduits, in sewers, or through buildings.

c. Marines should consider the background to ensure that they are not silhouetted or skylined, but rather blend into their surroundings. To defeat enemy urban camouflage, Marines should be alert for common camouflage errors such as:

- Tracks or other evidence of activity
- Shine or shadows
- Unnatural color or texture
- Muzzle flash, smoke, or dust
- Unnatural sounds and smells
- Movement.

d. Dummy positions can be used effectively to distract the enemy and make him reveal his position by firing on them.

e. Built-up areas afford cover, resources for camouflage, and locations for concealment. Following are the basic rules of cover, camouflage, and concealment to which Marines should adhere:

(1) Use the terrain and alter camouflage habits to suit your surroundings.

(2) Employ deceptive camouflage of buildings.

(3) Continue to improve positions. Reinforce fighting positions with sandbags or other fragment- and blast-absorbent material.

(4) Maintain the natural look of the area.

(5) Keep positions hidden by clearing away minimal debris for fields of fire.

(6) Choose firing ports in inconspicuous spots when available.

Note: Remember that a force that *covers* and *conceals* itself has a significant advantage over a force that does not.

Appendix B

Employment and Effects of Weapons

Success against the enemy in urban operations to a large extent depends on the effective employment of weapons. Marines at all levels must understand the effects that their weapons will produce in urban combat.

1. Effectiveness of Weapons and Demolitions. The characteristics and nature of combat in built-up areas affect the results and employment of weapons. Leaders at all levels must consider the following factors in various combinations when choosing weapons to employ.

a. Hard, smooth, flat surfaces are characteristic of urban targets. Rounds rarely impact perpendicular to these flat surfaces, but rather tend to impact at some angle of obliquity. This reduces the effect of a round and increases the threat of ricochets. The tendency of rounds to strike glancing blows against hard surfaces means that up to 25 percent of impact-fuzed explosive rounds may not detonate when fired into rubbled areas.

b. Engagement ranges in MOUT are close. Studies and historical analyses have shown that only 5 percent of all targets are more than 100 meters away. About 90 percent of all targets are located 50 meters or less from the identifying Marine. Few personnel targets will be visible beyond 50 meters, and most occur at 35 meters or less. Minimum arming ranges and unit safety from backblast or fragmentation effects must be considered.

c. Engagement times are short because of the close fight. Enemy personnel present only fleeting targets. Enemy-held buildings or structures are normally covered by mutually supporting fires and often cannot be engaged with deliberate, well-aimed shots without prolonging exposure to their observation and weapons.

d. Depression and elevation limits for some weapons create dead space. Tall buildings form deep canyons that are often safe from indirect fires. Some weapons, such as the M203, can fire rounds to ricochet behind cover and inflict casualties. Target engagement from oblique angles, both horizontal and vertical, demands superior marksmanship skills.

e. Smoke from burning buildings, dust from explosions, shadows from tall buildings, and the lack of light penetrating inner rooms all combine to reduce visibility and to increase a sense of isolation. Added to this is the masking of fires caused by rubble and manmade structures. Targets, even at close range, tend to be indistinct.

f. Urban fighting often leads to confused melees with several small units attacking on converging axes. The risks from friendly fires, ricochets, and fratricide must be considered during the planning phase of operations, and control measures must be continually adjusted to lower these risks. Marines and leaders must maintain a sense of situational awareness and clearly mark their progress in accordance with unit SOPs to avoid fratricide.

g. The firer and target may each be inside or outside the same or separate buildings. The enclosed nature of combat in built-up areas means that a weapon's signature effect, such as a muzzle blast or backblast, must be considered, as well as the round's impact on the target.

h. Usually the manmade structure that is protecting the enemy must be attacked before enemy personnel inside can be engaged. Therefore, weapons and demolitions can be chosen for employment based on their effects against masonry and concrete rather than against enemy personnel.

i. Modern engineering and design improvements mean that most large buildings constructed since World War II are resilient to the blast effects of bomb and artillery attack. Even though modern buildings may burn easily, they often retain their structural integrity and remain standing. Once high-rise buildings burn out, they are still useful to the military and are almost impossible to damage further. A large structure can take 24 to 48 hours to burn out and become cool enough for people to enter.

j. The most common worldwide building type is the 12- to 24-inch brick building. Table B-1 lists the frequency of occurrence of building types worldwide.

Type of Building	Frequency of Occurrence (Percentage)
30-inch stone	1
8- to 10-inch reinforced concrete	6
12- to 24-inch brick	63
6-inch wood	16
14-inch steel and concrete (heavy clad)	2
7-inch steel and concrete (light clad)	12

Table B-1. Types of Buildings and Frequency of Occurrence

2. M16 Rifle and M249 Squad Automatic Weapon. The M16A2 rifle and the M249 squad automatic weapon are the most common weapons used in combat in built-up areas. The M16A2 rifle and the M249 are used to kill enemy personnel, to suppress enemy fire and observation, and to penetrate light cover. Leaders can use 5.56-mm tracer fire to designate targets for other weapons.

a. Employment. Close combat is the predominant characteristic of urban engagements. Marine riflemen must be able to hit small, fleeting targets from bunker apertures, windows, and loopholes. This requires pinpoint accuracy with weapons fired in the semiautomatic mode. Killing an enemy through an 8-inch loophole at a range of 50 meters is a challenge, but one that may be common in combat in built-up areas.

(1) When fighting and engaging targets inside buildings, the weapons handling and firing techniques discussed in Appendix A are used.

(2) Within built-up areas, burning debris, reduced ambient light, strong shadow patterns of varying density, and smoke all limit the effectiveness of night vision and sighting devices. Aiming stakes and noise/light boobytraps coupled with night vision goggles and firing techniques can be used in the defense. During the offense, night vision goggles and illumination munitions can be used. Any Marine using night vision goggles should be teamed with at least one Marine not wearing them.

b. Weapon Penetration. The penetration that can be achieved with a 5.56-mm round depends on the range to the target and the type of material being fired against. Single 5.56-mm rounds are not effective against structural materials (as opposed to partitions) when fired at close range—the closer the range, the less the penetration.

(1) For the 5.56-mm round, maximum penetration occurs at 200 meters. At ranges of less than 25 meters, penetration is greatly reduced. At 10 meters, penetration by the M16 round is poor as a result of the tremendous stress placed on this high-speed round, which causes it to yaw upon striking a target. Stress causes the projectile to break up, and the resulting fragments are often too small to penetrate.

(2) Even with reduced penetration at short ranges, interior walls made of thin wood paneling, sheetrock, or plaster offer no protection against 5.56-mm rounds. Common office furniture such as desks and chairs cannot stop these rounds, but a layer of books 18 to 24 inches thick can.

(3) Wooden-framed buildings and single cinder-block walls offer little protection from 5.56-mm rounds. When clearing such structures, Marines must ensure that friendly casualties do not result from rounds passing through walls, floors, or ceilings.

(4) Armor-piercing rounds are slightly more effective than ball ammunition in penetrating urban targets at all ranges. They are also more likely to ricochet than ball ammunition, especially when the target presents a high degree of obliquity.

c. Protection. The following common barriers in built-up areas stop a 5.56-mm round fired at less than 50 meters:

- One thickness of sandbags
- A 2-inch concrete wall (unreinforced)
- A 55-gallon drum filled with water or sand
- A small ammunition can filled with sand

- A cinder block filled with sand (block will probably shatter)
- A plate-glass windowpane at a 45-degree angle (glass fragments will be thrown behind the glass)
- A brick veneer
- A car body (5.56-mm round will penetrate but normally not exit).

d. Wall Penetration. Although most structural materials repel single 5.56-mm rounds, continued and concentrated firing can breach some typical urban structures (See Table B-2).

Type	Penetration	Rounds Required
8-inch reinforced concrete	Initial Loophole	35 250
14-inch triple brick	Initial Loophole	90 160
12-inch cinder block with single-brick veneer	Loophole Breach hole	60 250
9-inch double brick	Initial Loophole	70 120
16-inch tree trunk or log wall	Initial[1]	1 to 3
12-inch cinder block (filled w/sand)	Loophole	35
24-inch double sandbag wall	Initial[1]	220
3/8-inch mild steel door	Initial[1]	1

[1]Penetration only, no loophole

Table B-2. Structure Penetration Capabilities of the 5.56-mm Round Against Typical Urban Targets (Range 25 to 100 Meters)

(1) The best method for breaching a masonry wall is by firing short bursts (3 - 5 rounds) in a U-shaped pattern. The distance from the gunner to the wall should be minimized for best results—ranges as close as 25 meters are relatively safe from ricochet. Ballistic eye protection, a protective vest, and a helmet should be worn.

(2) Ball ammunition and armor-piercing rounds produce almost the same results, but armor-piercing rounds are more likely to fly back at the firer. The 5.56-mm round can be used to create either a loophole (about 7 inches in diameter) or a breach hole (large enough for a man to enter). When used against reinforced concrete, the M16 rifle and M249 cannot cut the reinforcing bars.

3. Medium/Heavy Machine Guns and Sniper Rifles (7.62 mm and .50 caliber)

a. Machine Guns. In the urban environment, the .50-caliber machine gun and the 7.62-mm M240G machine gun provide high-volume, long-range, automatic fires for the suppression or destruction of targets. They provide final protective fires along fixed lines and can be used to penetrate light structures; the .50-caliber machine gun is most effective in this role. Tracers from both machine guns are likely to start fires, but the .50-caliber tracer is more apt to do so.

(1) Employment. The primary consideration affecting the employment of machine guns within built-up areas is the limited availability of long-range fields of fire. Although machine guns should be emplaced at the lowest level possible, grazing fire at ground level is often obstructed by rubble.

(a) The .50-caliber machine gun is often employed on its vehicular mount during both offensive and defensive operations. If necessary, it can be mounted on the M3 tripod for use in the ground role or in the upper level of buildings. When mounted on a tripod, the .50-caliber machine gun can be used as an accurate, long-range weapon and can supplement sniper fires.

(b) The M240G machine gun is cumbersome, making it difficult to use inside while clearing a building. However, it is useful outside to suppress and isolate enemy defenders. The M240G can be fired from either the shoulder or the hip to provide a high volume of assault and suppressive fires. The use of the long sling to support the weapon and ammunition is preferred.

(c) Because of their reduced penetration power, M240G machine guns are less effective against masonry targets than .50-caliber machine guns. However, their availability and light weight make them well suited to augment heavy machine gun fire, to be used in areas where .50-caliber machine guns cannot be positioned, or as a substitute when heavy machine guns are not available. The M240G machine gun can be employed on its tripod to deliver accurate fire along fixed lines and can then quickly be converted to bipod fire to cover alternate fields of fire.

(2) Penetration. The ability of the 7.62-mm and .50-caliber rounds to penetrate is also affected by the range to the target and type of material against which the rounds are fired. Close ranges affect penetration of the 7.62-mm round less than the 5.56-mm round; the .50-caliber round's penetration is reduced the least.

(a) At 50 meters, the 7.62-mm ball round cannot penetrate a single layer of sandbags. It can penetrate a single layer at 200 meters, but not a double layer. The armor-piercing round does only slightly better against sandbags. It cannot penetrate a double layer but can penetrate up to 10 inches of sand at 600 meters.

(b) The penetration of the 7.62-mm round is best at 600 meters, but most urban targets are closer. The longest engagement range is usually 200 meters or less. Table B-3 lists the penetration capabilities of a single 7.62-mm (ball) round at closer ranges.

Range (meters)	Pine Board (inches)	Dry, Loose Sand (inches)	Cinder Block (inches)	Concrete (inches)
25	13	5	8	2
100	18	4.5	10	2
200	41	7	8	2

Table B-3. Penetration Capabilities of a Single 7.62-mm (Ball) Round

(c) The .50-caliber round is also optimized for penetration at long ranges (about 800 meters). For hard targets, .50-caliber penetration is affected by obliquity and range. Both armor-piercing and ball ammunition penetrate 14 inches of sand or 28 inches of packed earth at 200 meters if the rounds impact perpendicular to the flat face of the target. Table B-4 depicts the effect of a 25-degree obliquity on a .50-caliber penetration.

Thickness (feet)	100 Meters (rounds)	200 Meters (rounds)
2	300	1,200
3	450	1,800
4	600	2,400

Table B-4. Number of Rounds Needed To Penetrate a Reinforced Concrete Wall at a 25-Degree Obliquity

(3) Protection. Barriers that offer protection against 5.56-mm rounds are also effective against 7.62-mm rounds with some exceptions. The 7.62-mm round can penetrate a windowpane at a 45-degree obliquity, a hollow cinder block, or both sides of a car body. It can also easily penetrate wooden-framed buildings. The .50-caliber round can penetrate all of the commonly found urban barriers except a sand-filled 55-gallon drum.

(4) Wall Penetration. Continued and concentrated machine gun fire can breach most typical urban walls. Such fire cannot breach thick reinforced concrete structures or dense natural stone walls. Internal walls, partitions, plaster, floors, ceilings, common office furniture, home appliances, and bedding can be easily penetrated by both 7.62-mm and .50-caliber rounds (Tables B-5 and B-6).

(a) The M240G machine gun can be hard to hold steady enough to repeatedly hit the same point on a wall. The dust created by the bullet strikes also makes precise aiming difficult. Firing with a tripod is usually more effective than without, especially if sandbags are used to steady the weapon. Short bursts of three to five rounds fired in a U-type pattern are best.

(b) Breaching a brick veneer over cinder block presents a special problem for the M240G machine gun. Rounds penetrate the cinder block but leave a net-like structure of unbroken block. Excessive ammunition is required to destroy the net because most rounds only pass through a previously eroded hole. One or two minutes of work with an entrenching tool, crowbar, or ax can remove this web and allow entry through the breach hole.

(c) The .50-caliber machine gun can be fired accurately from the tripod using the single-shot mode. This is the most efficient method for producing a loophole. Automatic fire in three- to five-round bursts in a U-type pattern is more effective in producing a breach.

Type	Thickness (Inches)	Hole Diameter (Inches)	Rounds Required
Reinforced concrete	8	7	100
Triple brick wall	14	7	170
Concrete block with single brick veneer	12	6 and 24	30 and 200
Cinder block (filled)	12	N/A[1]	18
Double brick wall	9	N/A[1]	45
Double sandbag wall	24	N/A[1]	110
Log wall	16	N/A[1]	1
Mild steel door	3/8	N/A[1]	1

[1]Penetration only, no loophole

Table B-5. Structure Penetrating Capabilities of 7.62-mm Round (NATO) Against Typical Urban Targets (Range 25 Meters)

Type	Thickness (Inches)	Hole Diameter (Inches)	Rounds Required
Reinforced concrete	10	12	50
		24	100
	18	7	140
Triple brick wall	12	8 and 26	15 and 50
Concrete block with single brick veneer	12	10 and 33	25 and 45
Armor plate	1	N/A[1]	1
Double sandbag wall	24	N/A[1]	5
Log wall	16	N/A[1]	1

[1]Penetration only, no loophole

Table B-6. Structure Penetrating Capabilities of .50-caliber Ball Against Typical Urban Targets (Range 35 Meters)

b. Sniper Rifles. The 7.62-mm match ball ammunition fired from the M-40A1 sniper rifle will perform similarly to the 7.62-mm ball round. The M-82A1A special application sniper rifle (SASR) can fire the .50-caliber armor-piercing incendiary (API) ammunition, which will have the same performance as when fired from the M-2 HB. The M-82A1A also fires the .50-caliber RAUFOSS ammunition, which contains a tungsten penetrator and a more powerful explosive charge than the API ammunition. When the RAUFOSS penetrator punches through the metal, the explosive detonates inside, acting as an incendiary round. Although the SASR's exact penetration capabilities have yet to be defined, it has penetrated an inch of steel at 2000 yards. Both sniper rifles are organic to the Marine infantry battalion.

4. Grenade Launchers, 40-mm (M203 and MK19). Both the M203 dual-purpose weapon and the MK19 grenade machine gun fire 40-mm HE and HE dual-purpose (HEDP) ammunition. Ammunition for these weapons is not interchangeable, but the grenade and fuze assembly that actually hits the target is identical. Both weapons provide point and area destructive fires as well as suppression. The MK19 has a much higher rate of fire and a longer range. The M203 is much lighter and more maneuverable.

a. Employment. The main consideration affecting the employment of 40-mm grenades within built-up areas is the typically short engagement range. The 40-mm grenade has a *minimum arming range of 14 to 28 meters*. If the round strikes an object before it is armed, it will not detonate. Both the HE and HEDP rounds have 5-meter burst radii against exposed forces, which means that the *minimum safe firing range for combat is 33 meters*. The 40-mm grenades can be used to suppress the enemy in a building or inflict casualties by firing through apertures or windows. The MK19 can use its high rate of fire to concentrate rounds against light structures. This concentrated fire can create extensive damage. The 40-mm HEDP round can penetrate the armor on the flank, rear, and top of Soviet-made BMPs and BTRs. Marines can use the M203 from upper stories to deliver accurate fire against the top decks of armored vehicles. Multiple hits are normally required to achieve a kill.

b. Weapon Penetration. The 40-mm HEDP grenade has a small shaped charge that penetrates better than the HE round. It also has a thin wire wrapping that bursts into a dense fragmentation pattern, creating casualties out to 5 meters. Because they explode on contact, 40-mm rounds achieve the same penetration regardless of range. Table B-7 lists the penetration capabilities of the HEDP round.

Target	Penetration (Inches)
Sandbags	20
Sand-filled cinder block	16
Pine logs	12
Armor plate	2

Table B-7. Penetration Capabilities of the HEDP Round

(1) If projected into an interior room, the 40-mm HEDP can penetrate all interior partition-type walls. It splinters plywood and plaster walls, making a hole large enough to fire a rifle through. It is better to have HEDP rounds pass into a room and explode on a far wall, even though much of the round's energy is wasted penetrating the back wall (Figure B-1). The fragmentation produced in the room causes more casualties than the high-explosive jet formed by the shaped charge.

(2) The fragments from the HEDP round do not reliably penetrate interior walls. They are also stopped by office furniture, sandbags, helmets, and protective vests (flak jackets). The M203 dual-purpose weapon has the inherent accuracy to place grenades into windows at 125 meters and bunker apertures at 50 meters. These ranges are significantly reduced as the angle of obliquity increases. Combat experience shows that M203 gunners cannot consistently hit windows at 50 meters when forced to aim and fire quickly.

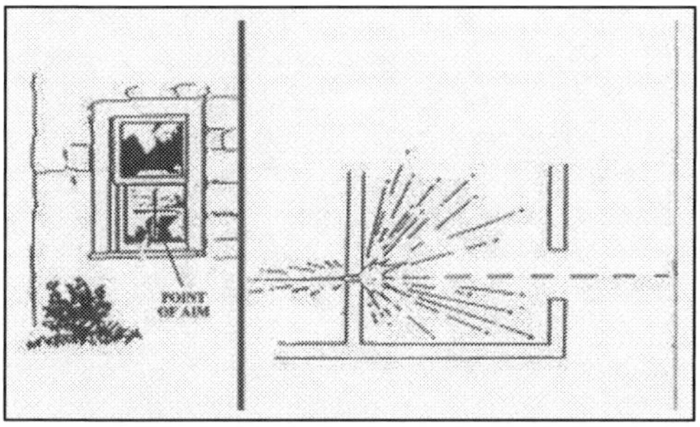

Figure B-1. Aim Point for 40-mm HEDP

c. Wall Penetration. The M203 cannot reasonably deliver the rounds needed to breach a typical exterior wall. The MK19 can concentrate its fire and achieve wall penetration. Firing from a tripod using a locked-down traversing and elevating mechanism is best for this role. Brick, cinder block, and concrete can be breached by using the MK19; individual HEDP rounds can penetrate 6 to 8 inches of brick. The only material that has proven resistant to concentrated 40-mm fire is dense stone such as that used in some European building construction. No precise data exist as to the number of rounds required to produce loopholes or breach holes with the MK19. However, the rounds' explosive effects are dramatic and should exceed the performance of the .50-caliber machine gun.

5. Light and Medium Recoilless Weapons. Light and medium recoilless weapons are used to attack enemy personnel, field fortifications, and LAVs. They have limited capability against main battle tanks, especially those equipped with reactive armor, except when attacking from the top, flanks, or rear. This category of weapons includes the AT4, the M47 Dragon, and the SMAW.

a. Employment. Other than defeating LAVs, the most common task for which light recoilless weapons are used is to neutralize fortified firing positions. Because of the design of the warhead and the narrow blast effect, these weapons are not as effective in this role as heavier weapons such as a tank main gun round. Their light weight allows Marines to carry several AT4s. Light recoilless weapons can be fired from the tops of buildings or from areas with extensive ventilation.

(1) Light and medium recoilless weapons with the exception of the SMAW employ shaped-charge warheads. As a result, the hole they punch in walls is often too small to use as a loophole. The fragmentation and spall that these weapons produce are limited. Normally, shaped-charge warheads do not neutralize enemy forces behind walls unless they are located directly in line with the point of impact.

(2) Against structures, shaped-charge weapons should be aimed about 6 inches below or to the side of a firing aperture (Figure B-2). This enhances the probability of killing the enemy behind the wall. A round that passes through a window wastes much of its energy on the back wall. Because these shaped-charge rounds lack the wire wrapping of the 40-mm HEDP, they burst into few fragments and are often ineffective casualty producers.

(3) Sandbagged emplacements present a different problem (Figure B-3). Because sandbags absorb much of the energy from a shaped charge, the rounds should be aimed at the center of the firing aperture. Even if the round misses the aperture, the bunker wall area near it is usually easiest to penetrate.

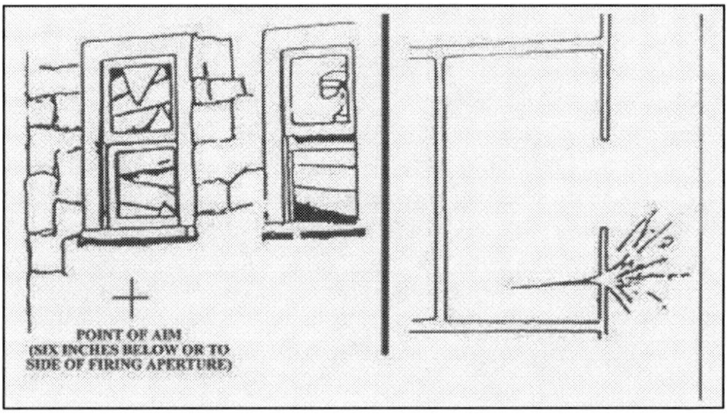

Figure B-2. Point of Aim for a Shaped-Charge Weapon Against a Masonry Structure

Figure B-3. Point of Aim for Sandbagged Emplacement

(4) Light and medium recoilless weapons obtain their most effective short-range antiarmor shots by firing from upper stories or from the flanks and rear. When firing at main battle tanks, these weapons should always be employed against these weaker areas in volley or paired firing. They normally require multiple hits to achieve a kill on a tank. Flank, top, and rear shots hit the most vulnerable parts of armored vehicles. Firing from upper stories protects the firer from tank main gun and coaxial machine gun fire because tanks cannot sharply elevate their cannons. The BMP-2 *can* elevate its 30-mm cannon to engage targets in upper stories. The BTR-series armored vehicles *can* also fire into upper stories with their heavy machine gun.

(5) Modern IFVs such as the BMP-2 and the BTR-80 have significantly improved frontal protection against shaped-charge weapons. Many main battle tanks have some form of reactive armor in addition to their thick armor plate. Head-on, ground-level shots against these vehicles have little probability of obtaining a kill. Even without reactive armor, modern main battle tanks are hard to destroy with a light antiarmor weapon.

(6) The easiest technique to improve the probability of hitting and killing an armored vehicle is to increase the firing depression angle. A 45-degree downward firing angle doubles the probability of a first-round hit as compared to a ground-level shot.

b. Backblast. Backblast characteristics must be considered when employing all recoilless weapons. During combat in built-up areas, the backblast area in the open is more hazardous because of the loose rubble and the channeling effect of the narrow streets and alleys. Figure B-4 shows the backblast areas of Marine light recoilless weapons in the open.

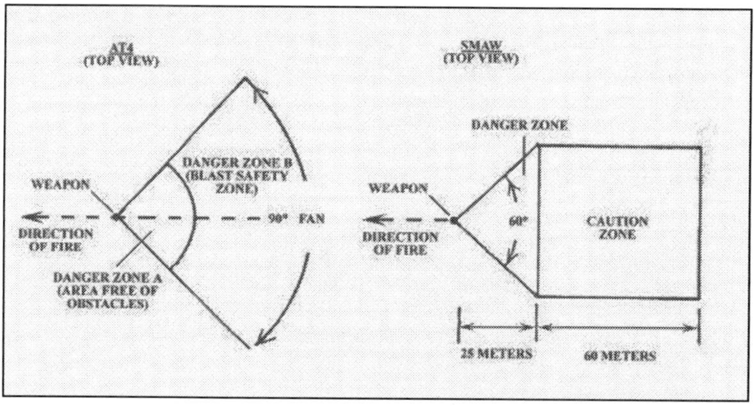

Figure B-4. Backblast Areas of Light Recoilless Weapons in the Open

(1) When firing recoilless weapons in the open, Marines should protect themselves from blast and burn injuries caused by the backblast. All personnel should be out of the danger zone. Anyone not able to vacate the caution zone should be behind cover. Marines in the caution zone should wear helmets, protective vests, and eye protection. The firer and all Marines in the area should wear earplugs.

(2) Extensive testing on the effects of firing recoilless weapons from within enclosures has been conducted since 1948. These tests showed that firing these weapons from enclosures presented no serious hazards, even when the overpressure was enough to produce structural damage to the building. The following were other findings of these tests.

(a) Little hazard exists to the gunnery or crew from any type of flying debris. Loose items were not hurled around the room.

(b) No substantial degradation occurs to the operator's tracking performance as a result of obscuration or blast overpressure.

(c) The most serious hazard that can be expected is hearing loss. This must be evaluated against the advantage gained in combat from firing from cover. To place this hazard in perspective, a gunner wearing earplugs and firing the loudest combination (the Dragon from within a masonry building) is exposed to less noise hazard than if he fired a LAW or AT4 in the open without earplugs.

(d) The safest place for other Marines in the room with the firer is against the wall from which the weapon is fired. Plastic ignition plugs are a hazard to anyone standing directly behind a LAW or TOW when it is fired.

(e) Firers should take advantage of all available sources of ventilation by opening doors and windows. Ventilation does not reduce the noise hazard, but it helps clear the room of smoke and dust and reduces the effective duration of the overpressure.

(f) The only difference between firing these weapons from enclosures and firing them in the open is the duration of the pressure fluctuation.

(g) Frame buildings, especially small ones, can suffer structural damage to the rear walls, windows, and doors. Large rooms suffer slight damage, if any.

(3) Recoilless weapons fired from within enclosures create some obscuration inside the room but almost none from the gunner's position looking out. Inside the room, obscuration can be intense, but the room remains inhabitable. Table B-8 shows the effects of smoke and obscuration.

Building	Weapon	From Gunner's Position Looking Out	Inside the Room	From Outside at a Distance
Masonry	LAW	None	Moderate	Slight smoke
	Dragon	Slight	Moderate	Small flash

Bunker	Dragon TOW	None None	Slight Slight	Moderate flash Moderate smoke
Small frame	LAW Dragon	None None	Moderate Severe	Moderate Moderate
Medium frame	LAW Dragon	None None	Slight Severe	Moderate Slight flash
Large frame	LAW Dragon TOW	None Slight None	Slight Severe Severe	None Slight flash Slight smoke

Table B-8. Smoke and Obscuration

(4) The Dragon causes the most structural damage, but only in frame buildings. There does not seem to be any threat of injury to the gunner because the damage is usually to the walls away from the gunner. The most damage and debris is from flying plaster chips and pieces of wood trim. Large chunks of plasterboard can be dislodged from ceilings. The backblast from a LAW, AT4, Dragon, or TOW rarely displaces furniture. Table B-9 shows the test results of structural damage and debris.

		Damage		
Building	**Weapon**	**Structure**	**Wall Covering**	**Debris Movement**
Masonry	LAW Dragon	None None	Slight Slight	Slight Slight
Bunker	Dragon TOW	None None	None None	None Leaves and dust disturbed
Small frame	LAW Dragon	None Severe	Slight Severe	None None
Medium frame	LAW Dragon	None Slight	None Slight	Slight Lamp and chair overturned
Large frame	LAW Dragon TOW	None Slight Slight	Slight Moderate Severe	Slight None None

Table B-9. Structural Damage and Debris Movement

(5) To fire a LAW, AT4, or SMAW from inside a room, the following safety precautions must be taken (Figure B-5).

(a) The building should be of a sturdy construction.

(b) The ceiling should be at least 7 feet high with loose plaster or ceiling boards removed.

(c) The floor size should be at least 15 feet by 12 feet. (The larger the room, the better.)

(d) At least 20 square feet of ventilation (room openings) should exist to the rear or side of the weapon. An open 7- by 3-foot door would provide minimum ventilation.

(e) All glass should be removed from windows, and small, loose objects should be removed from the room.

(f) Floors should be wet to prevent dust and dirt from blowing around and obscuring the gunner's vision.

(g) All personnel in the room should be forward of the rear of the weapon.

(h) All personnel in the room should wear helmets, protective vests, ballistic eye protection, and earplugs.

(i) If the gunner is firing from the prone position, his lower body must be perpendicular to the bore of the weapon, or the blast could cause injury to his legs.

Figure B-5. Firing an AT4 From Inside a Building

c. Weapon Penetration. The most important tasks to be performed against structures are the neutralization of fortified firing positions, personnel, and weapons behind barriers. Recoilless weapons can be used in this role; none, however, are as effective as heavy, direct-fire weapons or standard demolitions. Each recoilless weapon has different penetrating ability against various targets. Penetration does not always mean the destruction of the integrity of a position. Usually only those enemy soldiers directly in the path of the spall from a high-explosive antitank (HEAT) round become casualties. Other soldiers inside a fortification could be deafened, dazed, or shocked but could eventually return to action.

(1) AT4. The AT4 is heavier than the LAW and has a diameter of 84 millimeters, which gives the warhead much greater penetration. The AT4 can penetrate more than 17.5 inches (450 mm) of armor plate. Its warhead produces highly destructive results behind the armor. Tests against typical urban targets are still ongoing, but the AT4 should penetrate at least as well as the 90-mm recoilless rifle, if not better. The AT4 has a *minimum arming distance of 10 meters*, which allows it to be fired successfully against close targets. Firers should be well covered and protected when firing at close targets.

(2) MK153 83-mm SMAW. The SMAW is a lightweight assault weapon that is easily carried and placed into action by one man. It is used against fortified positions, but it is also effective against LAVs. The SMAW has a 9-mm spotting rifle and a 3.8-power telescope that ensure accuracy over ranges common to combat in built-up areas. The SMAW has excellent incapacitating effects for enemies behind walls and inside bunkers and can arm within 10 meters. Its fuze has the ability to distinguish between armor and soft earth, maximizing its capabilities against buildings, bunkers, or light armor. The warhead detonates immediately against hard targets but delays detonation against soft targets and burrows in to explode inside. The SMAW can destroy most bunkers with a single hit. The 83-mm HE warhead can create an 8-inch mousehole in reinforced concrete but will not cut the steel reinforcing bar. The SMAW is an outstanding urban support weapon that can be decisive during an urban assault. The SMAW allows for quick reduction of obstacles and bunkers as well as creation of entry points. SMAWs fired in a volley can be devastating to a building.

d. Wall Breaching. Wall breaching is a common combat task in built-up areas for which light recoilless weapons can be used. Breaching operations improve mobility by providing access to building interiors without using existing doors or windows. Breaching techniques can also be used to create loopholes for weapons positions or to allow hand grenades to be thrown into defended structures. Breaching holes for unit mobility should be about 24 inches (60 centimeters) in diameter. Loopholes should be about 8 inches (20 centimeters) in diameter (Figure B-6). Only the SMAW provides a one-shot wall breaching ability. However, in breaching walls a number of shots should be planned.

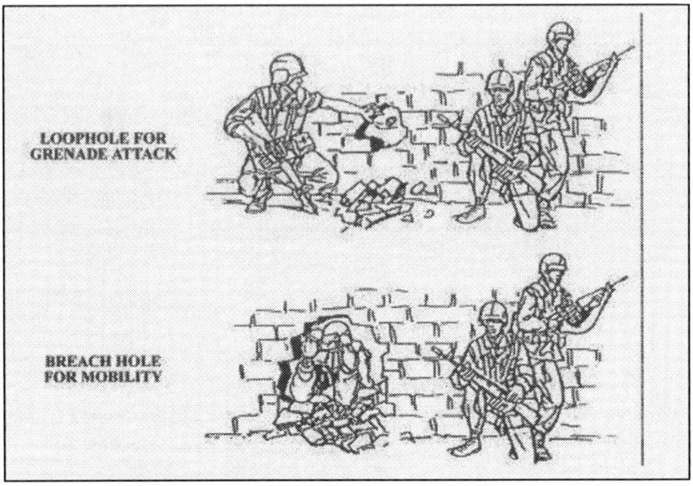

Figure B-6. Tactical Use of Holes in Masonry Walls

(1) Of all the common building materials, heavy stone is the most difficult to penetrate. The AT4 normally will not penetrate a heavy European-style stone wall. Surface cratering is usually the only effect it will have.

(2) Layered brick walls are also difficult to breach with light recoilless weapons. Some brick walls can be penetrated by multiple firings, especially if they are less than three bricks thick. Multiple AT4 rounds fired at the same spot on an 8-inch (double-brick) wall normally produce a loophole. The SMAW produces a hole in brick walls that is often large enough to be a breach hole.

(3) Wooden structural walls offer little resistance to light recoilless weapons. Even heavy timbered walls are penetrated and splintered. Multiple AT4 rounds fired at the same area of a wood-framed wall usually produce a man-sized hole. The SMAW has a devastating effect against wood-framed walls. A single round produces a breach hole as well as significant spall.

(4) Because of its high velocity, the AT4 may penetrate a soft target such as a car body or frame building before exploding.

(5) None of the light recoilless weapons are as effective against structural walls as demolitions or heavier weapons such as tank main guns or field artillery. Of all the light recoilless weapons, the SMAW is the most effective.

6. Antitank Guided Missiles. ATGMs are used mainly to defeat main battle tanks and other armored combat vehicles. They have a moderate capability against bunkers, buildings, and other fortified targets commonly found during combat in built-up areas. This category of weapons includes the TOW and Dragon missiles.

a. Employment. TOWs and Dragons provide overwatch antitank fires during the attack of a built-up area and an extended range capability for the engagement of armor during the defense. Within built-up areas, they are best employed along major thoroughfares and from the upper stories of buildings to attain long-range fields of fire. Their *minimum firing range of 65 meters* could limit firing opportunities in the confines of densely built-up areas.

(1) Obstacles. Rubble or other obstacles could interfere with the flight of missiles when fired from street level. At least 3.5 feet (1 meter) of vertical clearance over such obstacles must be maintained. Figure B-7 shows the most common obstacles to ATGM flight that are found in built-up areas. *Power lines are a special obstacle and present a unique threat to ATGM gunners. If the power in the lines has not been interrupted, the ATGM guidance wires could create a short circuit. This would allow extremely high voltage to pass to the gunner in the brief period before the guidance wires melted. This voltage could either damage the sight and guidance system or injure the gunner. Before any ATGM is fired over a power line, an attempt must be made to determine whether or not the power to that line has been interrupted.*

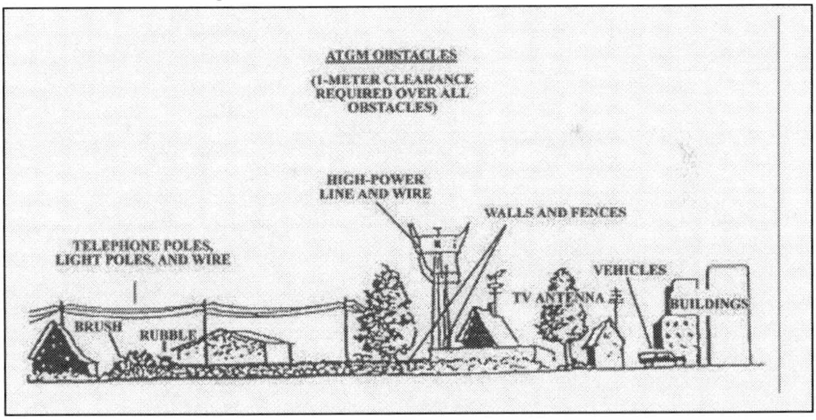

Figure B-7. Common Obstacles to ATGM Flights

(2) Dead Space. Three aspects of dead space that affect ATGM fires are arming distance, maximum depression, and maximum elevation.

(a) Both the Dragon and TOW missiles have a minimum arming distance of 65 meters, which severely limits their use in built-up areas. Few areas in the inner city permit fires much beyond the minimum arming distance. However, ground-level, long-range fires down streets or rail lines and across parks or plazas are possible. ATGMs may be used effectively from upper stories or roofs of buildings to fire into other buildings.

(b) The TOW is limited much more than the Dragon by its maximum depression and elevation. The maximum depression and elevation limits of the TOW mount could result in dead space and preclude the engagement of close targets (Figure B-8). A target located at the minimum arming range (65 meters) cannot be engaged by a TOW crew located any higher than the sixth floor of a building due to maximum depression limits. At 100 meters, the TOW crew can be located as high as the ninth floor and still engage the target.

(3) Backblast. Backblast for ATGMs is of more concern during combat in built-up areas than in open country. Any loose rubble in the caution zone could be picked up and thrown by the backblast, and the channeling effect of walls and narrow streets is even more pronounced due to the greater backblast. If the ATGM backblast strikes a wall at an angle, it can pick up debris or be deflected and cause injury to unprotected personnel (Figure B-9). Both types of ATGM can be fired from inside some buildings. In addition to the helmet and protective vest, eye protection and earplugs should be worn by all personnel in the room.

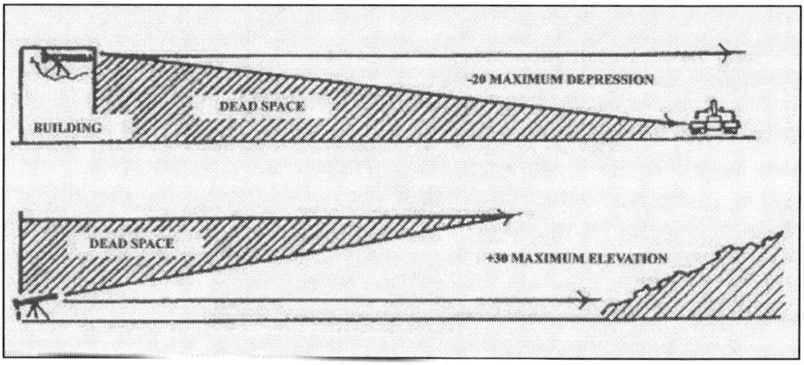

Figure B-8. TOW Maximum Elevation and Depression Limitations

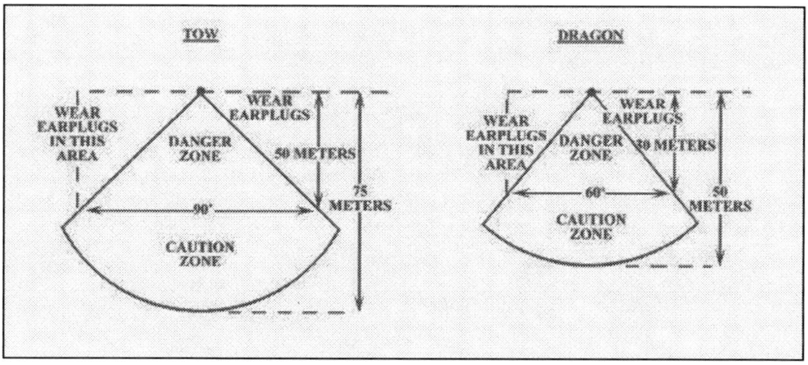

Figure B-9. ATGM Backblast in an Open Street

(a) To fire a TOW from inside a room, the following safety precautions must be taken (Figure B-10):

- The building must be of sturdy construction.
- The ceiling should be at least 7 feet high.
- The floor size of the room should be at least 15 by 15 feet, or larger if possible.
- At least 20 square feet of room ventilation should exist, preferably to the rear of the weapon. An open 7- by 3-foot door is sufficient. Additional ventilation can be created by removing sections of interior partitions.
- All glass must be removed from the windows and all small, loose objects removed from the room.
- All personnel in the room should be forward of the rear of the TOW.
- All personnel in the room should wear ballistic eye protection and earplugs.
- A clearance of 9 inches (23 centimeters) must exist between the launch tube and the aperture from which it is fired. (See FM 23-34, *TOW Weapon System*, for more detailed safety information.)

Figure B-10. TOW Fired From Inside a Room

(b) To fire a Dragon from inside a room, the following safety precautions must be taken.

- The building must be of sturdy construction.
- The ceiling should be at least 7 feet high.
- The floor size should be at least 15 by 15 feet, or larger if possible.
- At least 20 square feet of ventilation should exist (room openings), preferably to the rear of the weapon. An open 7- by 3-foot door would provide minimum ventilation.
- All glass should be removed from windows, and small, loose objects should be removed from the room.
- The room should be clean or the floors must be wet to prevent dust and dirt (kicked up by the backblast) from obscuring the vision of other soldiers in the room.
- All personnel in the room must be forward of the rear of the weapon.
- All personnel in the room must wear ballistic eye protection and earplugs.
- At least a 6-inch clearance must exist between the launch tube and the aperture from which it is fired.

b. Weapon Penetration. ATGMs can penetrate and destroy heavily armored tanks. They have large warheads that contain shaped charges. Because of their size, these warheads can achieve significant penetration against typical urban targets. Penetration, however, does not mean a concurrent destruction of the structural integrity of a position. The shaped-charge warhead produces relatively little spall. Enemy personnel not standing directly behind or near the point of impact of an ATGM may escape injury.

(1) Standard TOW Missiles. The basic TOW missile can penetrate 8 feet of packed earth, 4 feet of reinforced concrete, or 16 inches of steel plate. The improved TOW (ITOW), the TOW 2, and the TOW 2A have all been modified to improve their penetration. They all penetrate better than the basic TOW. All TOW missiles can defeat triple sandbag walls, double layers of earth-filled 55-gallon drums, and 18-inch log walls.

(2) TOW 2B. The TOW 2B uses a different method of defeating enemy armor. It flies over the target and fires an explosively formed penetrator down onto the top armor, which is thinner. *Because of this design feature, the TOW 2B missile cannot be used to attack nonmetallic structural targets. When using the TOW 2B missile against enemy armor, gunners must avoid firing directly over other friendly vehicles, disabled vehicles, or large metal objects such as water or oil tanks.*

(3) Dragon Missile. The Dragon missile can penetrate 8 feet of packed earth, 4 feet of concrete, or 13 inches of steel plate. It can attain effective short-range fire from upper stories or from the rear or flanks of a vehicle. These engagements are targeted against the most vulnerable parts of tanks and can entrap tanks in situations where they are unable to counterfire. Elevated firing positions increase the first-round hit probability. Firing down at an angle of 20 degrees increases the chance of a hit by 67 percent at 200 meters. A 45-degree downward angle doubles the first-round hit probability, compared to a ground-level shot.

c. Breaching Structural Walls. Firing ATGMs is the least efficient means to defeat structures. Because of their small basic load and high cost, ATGMs are better used against tanks or enemy-fortified firing positions. They can be effective against bunkers or other identified enemy firing positions.

7. Flame Weapons. Flame weapons are characterized by both physical and psychological casualty-producing abilities. Flame does not need to be applied with pinpoint accuracy, but it also must not spread to structures needed by friendly forces. Large fires in built-up areas are catastrophic. If they burn out of control, fires can create an impenetrable barrier for hours. The Marine Corps does not possess flame-generating equipment. However, the M34 WP grenade can be used to ignite and destroy flammable objects, especially wooden structures. The Army's M2A1-7 portable flamethrower is stored in war reserve status as a standard "C" item. Its availability is limited; however, Marines could be issued flame-support weapons for urban operations depending on the status of international agreements on the use of flame weapons that are in effect at the time.

a. Employment. Flame weapons used against fortified positions should be aimed directly at the aperture. Even if the round or burst misses, enough flaming material enters the position to cause casualties and disrupt the enemy occupants. The M34 WP grenade is difficult to throw far or into a small opening such as a bunker aperture. However, its effects are dramatic when thrown into a room or building.

b. Effects. The M34 is used to ignite and destroy flammable objects, especially wooden structures. It is also used to create an immediate smoke cloud to conceal movement across a narrow open space such as a street. Its smoke is not toxic but can cause choking in heavy concentrations.

(1) The grenade's explosion, bright flash, smoke, and burning WP particles all combine to make the M34 one of the most effective psychological weapons available. The M34 hand grenade throws WP fragments up to 35 meters from the point of detonation. These fragments can attach to clothing or skin and continue burning. Because of its weight, most infantrymen can throw this grenade only 30 to 40 meters.

(2) The Marine must avoid injury from friendly use of the M34. The M34 can ignite if the WP inside is exposed to the air. Bullets and shell fragments have been known to strike and rupture M34 grenades; therefore, grenades must be protected from enemy fire.

(3) The M34 WP grenade is an effective weapon against enemy armored vehicles when used in the close confines of combat in built-up areas. It can be thrown or dropped from upper stories onto enemy vehicles. The M34 can be combined with flammable liquids, detonating cords, blasting caps, and fuze igniters to create the eagle fireball, a field-expedient antiarmor device. (See FM 21-75, *Combat Skills of the Soldier*, Appendix H.)

(4) The M34 is also excellent as a screening device. A grenade can be thrown from behind cover into an open street or plaza. When it explodes, the enemy's observation is temporarily obscured. Thus, friendly forces can quickly cross the open area—if the enemy fires, it is unaimed and presents less of a danger. If screening smoke is used to cover a squad's movement across short open areas, it will reduce expected casualties from small-arms fire by about 90 percent.

8. Hand Grenades. Hand grenades are used extensively during combat in built-up areas. Smoke grenades are used for screening and signaling. Riot control grenades are used to drive the enemy out of deep fortifications. Fragmentation and concussion grenades are used to clear the enemy out of rooms and basements. Hand grenades are the most-used explosive munition during intense combat in built-up areas. In World War II, it was common for a battalion fighting in a city to use more than 500 hand grenades each day.

a. Employment. Smoke and riot-control grenades have similar employment techniques. Fragmentation grenades are used to produce enemy casualties.

(1) The AN-M8 HC grenade produces a dense white or grey smoke. It burns intensely and cannot be easily extinguished once it ignites. The smoke can be dangerous in heavy concentrations because it makes breathing difficult and causes choking. The M8 grenade is normally used for screening. It produces a slowly building screen of longer duration than the M34 WP grenade without the problem of collateral damage caused by scattered burning particles.

(2) The M18-series smoke grenades produce several different colors of smoke that are used for signaling. Yellow smoke is sometimes difficult to see in built-up areas.

(3) The M7A3 CS riot control grenade can be used to drive enemy troops out of fortifications when civilian casualties or collateral damage constraints are considerations. Built-up areas often create variable and shifting wind patterns. When using CS grenades, Marines must prevent the irritating smoke from affecting friendly forces. The CS grenade burns intensely and can ignite flammable structures. Enemy troops wearing even rudimentary chemical protective masks can withstand intense concentrations of CS gas.

(4) The MK3A2 offensive hand grenade, commonly referred to as the concussion grenade, produces casualties during close combat while minimizing the danger to friendly personnel. For this reason, it is the preferred hand grenade for use during offensive operations in a MOUT environment. The grenade produces severe concussion effects in enclosed areas. It can be used for light blasting and demolitions and for creating breach holes in interior walls. The concussion produced by the MK3A2 is much greater than that of the fragmentation grenade. It is very effective against enemy soldiers within bunkers, buildings, and underground passages.

(5) The M67 fragmentation grenade is the most commonly available grenade during combat in built-up areas. It provides suppression during room-to-room or house-to-house fighting, and it is used while clearing rooms of enemy personnel. The M67 grenade has a 3- to 5-second delay fuze. When used at close ranges, it can be cooked-off for two seconds to deny the enemy the time to throw it back. The fragmentation grenade can be rolled, bounced, or ricocheted into areas that cannot be reached by 40-mm grenade launchers.

b. Effects. Each type of hand grenade has its own specific effect during combat in built-up areas.

(1) Both HC smoke and CS gas *displace* oxygen from enclosed areas. Gas masks will not prevent this from occurring because they only filter air, not provide oxygen. Smoke grenades produce dense clouds of colored or white smoke that remain stationary in the surrounding area. They can also cause fires if used indiscriminately. If trapped and concentrated within a small space, smoke grenades and CS gas *can suffocate Marines*.

(2) The fragmentation grenade has more varied effects in combat in built-up areas. It produces a large amount of small, high-velocity fragments that can penetrate sheetrock partitions and are lethal at short ranges (15 to 20 meters). Fragments lose their velocity quickly and are less effective beyond 25 meters.

The fragments from a fragmentation grenade cannot penetrate a single layer of sandbags, a cinder block, or a brick building, but they can perforate wood-framed and tin buildings if exploded close to their walls.

(3) Inside rooms, fragmentation barriers consisting of common office furniture, mattresses, doors, or books can be effective against the fragmentation grenade. For this reason, a room should never be considered safe just because one or two grenades have been detonated inside. Fragmentation grenades detonated on the floor not only throw fragments laterally, but also send fragments and spall downward to lower floors. Predicting how much spall will occur is difficult because flooring material varies, but wooden floors are usually affected the most.

(4) Some foreign grenades throw fragments that are much larger than those of the U.S.-made M67. Light barriers and interior walls would probably be less effective against these grenades than against the M67. A major problem with the U.S.-made fragmentation grenade is its tendency to bounce back off of hard targets. Grenades are often directed at window openings on the ground floor or second floor. At ranges as close as 20 meters, a thrower's chances of missing a standard 1-meter by 1-meter window are high. The fragmentation grenade normally breaks through standard window glass and enters a room. If the grenade strikes at a sharp angle or the glass is thick plate, the grenade could be deflected without penetrating.

(5) Hand grenades are difficult weapons to use. They involve a high risk of fratricide. Commanders should conduct precombat training with hand grenades as part of normal preparations. Marines must be very careful when throwing hand grenades up stairs to avoid having the grenade roll back down on them.

(6) Once pulled, the pins of both fragmentation and concussion grenades can be replaced if the thrower decides not to use the weapon. This pin replacement must be done carefully (see FM 23-30, *Grenades and Pyrotechnic Signals*) and prior to the spoon being released or detaching.

(7) An estimate of the situation (METT-T) and existing ROE will dictate what type of grenade will be used to clear each room. Because of the high rate of expenditure of grenades, units should use assault packs to carry additional grenades of all types. Additional grenades can also be carried in empty ammunition or canteen pouches.

9. 25-mm Automatic Gun. The 25-mm automatic gun mounted on the LAV-25 is an effective weapon during combat in built-up areas. The primary roles of LAV-25s during combat in built-up areas are to provide suppressive fire and to breach exterior walls and fortifications.

a. Obliquity. The 25-mm gun produces its best urban target results when fired perpendicular to the hard surface (zero obliquity). In combat in built-up areas, however, finding a covered firing position that permits low-obliquity firing is unlikely unless the streets and gaps between buildings are wide. Most shots impact the target at an angle, which normally reduces penetration. With the armor-piercing, discarding, sabot with tracer (APDS-T) round, an angle of obliquity of up to 20 degrees can actually improve breaching. The rounds tend to dislodge more wall material for each shot but do not penetrate as deeply into the structure.

b. Target Types. The 25-mm gun has different effects when fired against different urban targets.

(1) Reinforced Concrete. Reinforced concrete walls, which are 12 to 20 inches thick, present problems for the 25-mm gun when trying to create breach holes. It is relatively easy to penetrate, fracture, and clear away the concrete, but the reinforcing rods remain in place. These create a "jail window" effect by preventing entry but allowing grenades or rifle fire to be placed behind the wall. Steel reinforcing rods are normally 3/4 of an inch thick and 6 to 8 inches apart—there is no quick way of cutting these rods. They can be cut with demolition charges, cutting torches, or special power saws. Firing with either APDS-T or high-explosive, incendiary with tracer (HEI-T) rounds from the 25-mm gun will not always cut these rods.

(2) Brick Walls. Brick walls are more easily defeated by the 25-mm gun regardless of their thickness, and they produce the most spall.

(3) Bunker Walls. The 25-mm gun is devastating when fired against sandbag bunker walls. Obliquity has the least effect on the penetration of bunker walls. Bunkers with earth walls up to 36 inches thick are easily penetrated. At short ranges typical of combat in built-up areas, defeating a bunker should be easy, especially if the 25-mm gun can fire at an aperture.

c. Burst Fire. The 25-mm gun's impact on typical urban targets seems to be magnified if the firing is in short bursts. At close ranges, the gunner might need to shift his point of aim in a spiral pattern to ensure that the second and third bursts enlarge the hole. Even without burst fire, sustained 25-mm gunfire can defeat almost all urban targets.

d. Weapon Penetration. The penetration achieved by the two combat rounds (HEI-T and APDS-T) differs slightly, but both are effective.

(1) APDS-T. The APDS-T round penetrates urban targets by retaining its kinetic energy and blasting a small hole deep into the target. The APDS-T round gives the best effects behind the wall, and the armor-piercing core often breaks into two or three fragments that can create multiple enemy casualties. The APDS-T needs as few as four rounds to achieve lethal results behind walls. Table B-10 lists the number of APDS-T rounds needed to create different-sized holes in common urban walls.

Target	Loophole	Breach Hole
3-inch brick wall at 0-degree obliquity	22 rounds	75 rounds
3-inch brick wall at 45-degree obliquity	22 rounds	35[1] rounds
5-inch brick wall at 0-degree obliquity	32 rounds	50[1] rounds
8-inch reinforced concrete at 0-degree obliquity	22 rounds	75 rounds (Note: Reinforcing rods still in place)
8-inch reinforced concrete at 45-degree obliquity	22 rounds	40[1] rounds (Note: Reinforcing rods still in place)

[1] Obliquity and depth tend to increase the amount of wall material removed.

Table B-10. Breaching Effects of APDS-T Rounds

(a) When firing single rounds, the APDS-T round provides the greatest capability for behind-the-wall incapacitation. The APDS-T round can penetrate more than 16 inches of reinforced concrete with enough energy left to cause enemy casualties. It penetrates through both sides of a wood-framed or brick veneer building. Field fortifications are easily penetrated by APDS-T rounds. Table B-11 lists the number of APDS-T rounds needed to create different-sized holes in commonly found bunkers.

(b) The APDS-T round creates a hazardous situation for exposed personnel because of the pieces of sabot that are thrown off of the round.

Type Bunker	Obliquity	Penetration	Loophole	Small Breach Hole
36-inch sand/timber	0 degree	1 round	25 rounds	40 rounds
36-inch sand/6-inch concrete	0 degree	6 rounds	6 rounds	20 rounds

Table B-11. Number of APDS-T Rounds Needed To Create Different-Sized Holes In Bunkers

WARNING

Personnel not under cover forward of the 25-mm gun's muzzle and within the danger zone could be injured or killed by these sabots, even if the penetrator passes overhead to hit the target.

The danger zone extends at an angle of about 10 degrees below the muzzle level, out to at least 100 meters, and about 17 degrees left and right of the muzzle. Figure B-11 shows the hazard area of the APDS-T round.

Figure B-11. APDS-T Danger Zone

(2) HEI-T. The HEI-T round penetrates urban targets by blasting away chunks of material.

(a) The HEI-T round does not penetrate an urban target as well as the APDS-T, but it does strip away a greater amount of material for each round. The HEI-T does more damage to an urban target when fired in multiple short bursts because the cumulative impact of multiple rounds is greater than the sum of individual rounds. Table B-12 lists the number of HEI-T rounds needed to create different-sized holes.

Target	Loophole	Breach Hole
3-inch brick wall at 0-degree obliquity	10 rounds	20 rounds
3-inch brick wall at 45-degree obliquity	20 rounds	25 rounds
5-inch brick wall at 0-degree obliquity	30 rounds	60 rounds

8-inch reinforced concrete at 0-degree obliquity	15 rounds	25 rounds
8-inch reinforced concrete at 45-degree obliquity	15 rounds	30 rounds

Table B-12. Number of HEI-T Rounds Needed To Create Different-Sized Holes in Bunkers

(b) The HEI-T round does not provide single-round perforation or incapacitating fragments when used against any external masonry structural wall. It can create first-round fragments behind wood-framed and brick veneer walls. HEI-T rounds cannot penetrate a bunker as quickly as APDS-T rounds, but they can create more damage inside the bunker once the external earth has been stripped away. Against a heavy bunker, about 40 rounds of HEI-T ammunition are needed to strip away the external earth shielding and breach the inner lining of concrete or timber. The HEI-T round is also used for suppression against known or suspected firing ports such as doors, windows, and loopholes.

10. Tank Cannon. The powerful, high-velocity cannon mounted on the M1A1 tank provides Marines with a key requirement for victory in built-up areas—heavy direct-fire support. Although the infantry assumes the lead role during combat in built-up areas, tanks and infantry work as a close team. Tanks move down streets after the infantry has cleared them of any suspected ATGM positions and, in turn, support the infantry with fire. The tank is one of the most effective weapons for heavy fire against structures. The primary role of the tank cannon during combat in built-up areas is to provide heavy direct fire against buildings and strongpoints that are identified as targets by the infantry. The wall and fortification breaching effects of the 120-mm tank cannon are major assets to Marines fighting in built-up areas.

a. Obliquity. The tank cannon produces its best urban target effects when fired perpendicular to the hard surface (zero obliquity). During combat in built-up areas, however, finding a covered firing position that permits low-obliquity firing is unlikely. Most shots strike the target at an angle that would normally reduce penetration. With tank cannon APDS rounds, obliquity angles of up to 25 degrees have little effect, but angles greater than 45 degrees greatly reduce penetration.

b. Ammunition. Armor-piercing, fin-stabilized, discarding sabot (APFSDS) rounds are the most commonly carried tank ammunition. These rounds work best against armored vehicles. The 120 mm cannon also carries an effective high-explosive, antitank multipurpose (HEAT-MP) round.

c. Characteristics. The 120-mm tank cannon has two specific characteristics that affect its employment in built-up areas: limited elevation and depression and short arming ranges. In addition, the M1A1 tank has another characteristic that is not involved with its cannon but that affects Marines working with or around the tank—*extremely hot turbine exhaust.*

(1) The M1A1 tank can elevate its cannon 20 degrees and depress it 10 degrees. The lower depression limit creates a 35-foot (10.8-meter) dead space around a tank. On a street 16 meters wide, this dead space extends to the buildings on each side (Figure B-12). Similarly, there is a zone overhead in which the tank cannot fire (Figure B-13). This dead space offers ideal locations for short-range antiarmor weapons and allows hidden enemy gunners to fire at the tank when the tank cannot fire back. It also exposes the tank's most vulnerable areas: the flanks, rear, and top. Infantry must move ahead, alongside, and to the rear of tanks to provide close protection. The extreme heat produced immediately to the rear of the M1A1 tank prevents dismounted infantry from following closely, but protection from small-arms fire and fragments is still provided by the tank's bulk and armor. The M1A1 tank also has a blind spot caused by the zero-degree depression available over part of the back deck. To engage any target in this area, the tank must pivot to convert the rear target to a flank target.

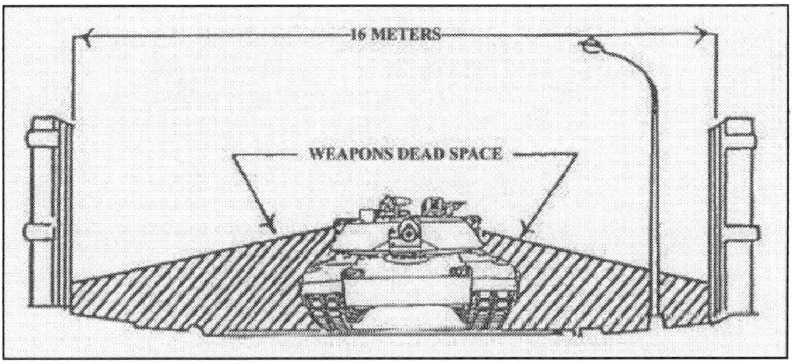

Figure B-12. Tank Cannon Dead Space at Street Level

(2) The 120-mm HEAT-MP round arms at about 36 feet. This arming distance allows the tank to engage targets from short ranges. The armor of the tank protects the crew from both the blowback effects of the round and enemy return fire. The APFSDS round does not need to arm and can, therefore, be fired at almost any range. The discarding portions of the round can be lethal to exposed infantry forward of the tank.

d. Target Effects. HEAT rounds are most effective against masonry walls. The APFSDS round can penetrate deeply into a structure but does not create as large a hole or displace as much spall behind the target. In contrast to lighter HEAT rounds, tank HEAT rounds are large enough to displace enough spall to inflict casualties inside a building. One HEAT round normally creates a breach hole in all but the thickest masonry construction—brick veneer and wood-framed constructions are demolished by a single round. Even the 120-mm HEAT round cannot cut all of the reinforcing rods, which are usually left in place, often hindering entry through the breach hole (Figure B-14). Both HEAT and APFSDS rounds are effective against all field fortifications. Only large earth berms and heavy mass-construction buildings can provide protection against tank fire.

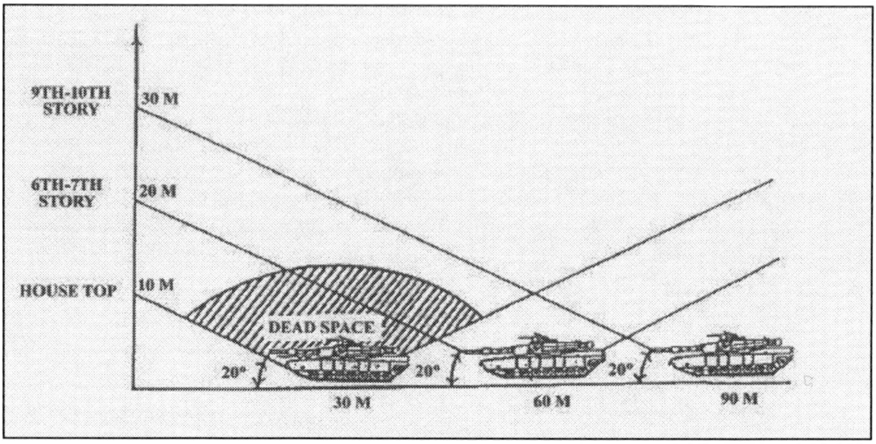

Figure B-13. Tank Cannon Dead Space Above Street Level

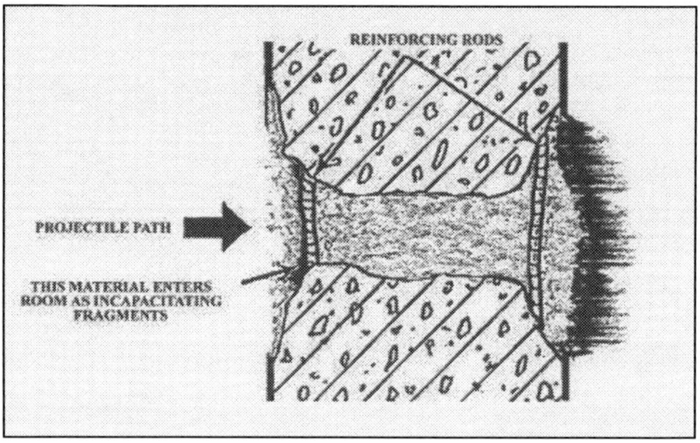

Figure B-14. Tank HEAT Round Effects on Reinforced Concrete Walls

e. Employment. Tank-heavy forces could be at a severe disadvantage during combat in built-up areas, but a few tanks working with the infantry can be most effective, especially if they work well together at the small-unit level. Tank, infantry, and engineer task forces are normally formed to attack a fortified area. The tank platoon or section can work together with rifle squads or platoons.

(1) Tanks need infantry on the ground to provide security in built-up areas and to designate targets. Against targets protected by structures, tanks should be escorted forward to the most covered location that provides a clear shot. On-the-spot instructions by infantry leaders ensure that the tank's fire is accurate and its exposure is limited. The tank commander may have to halt in a covered position, dismount, and reconnoiter his route forward into a firing position.

(2) When the tank main gun fires, it creates a large fireball and smoke cloud. In the confines of a built-up area, dirt and masonry dust are also picked up and added to this cloud. The target is further obscured by the smoke and dust of the explosion. Depending on the local conditions, this obscuration could last as long as two or three minutes. Marines can use this period to reposition or advance unseen by the enemy. Caution must be exercised, however, because the enemy might also move.

(3) A tank cannon creates an overpressure and noise hazard to exposed Marines. All dismounted Marines working near tanks should wear Kevlar helmets and protective vests, as well as ballistic eye protection. If possible, they should also wear earplugs and avoid the tank's frontal 60-degree arc during firing (Figure B-15).

WARNING

- **The overpressure from the tank 120-mm cannon can kill a marine found within a 90-degree arc extending from the muzzle of the gun tube out to 200 meters.**
- **From 200 to 1,000 meters along the line of fire, on a frontage of about 400 meters, dismounted Marines must be aware of the danger from discarding sabot petals, which can kill or seriously injure personnel.**
- **Personnel outside the tank should remain at least 50 meters from the tank in all directions as they may receive damaging effects from firing noise and overpressure.**
- **Personnel must also wear hearing protection when operating within 704 meters of a tank that is firing its main gun.**

TM 9-2350-264-10-2, *Operator's Manual for Operation Under Usual and Unusual Conditions for Tank, Combat, Full-Tracked: 120mm Gun, M1A1 General Abrams*, with Change 4

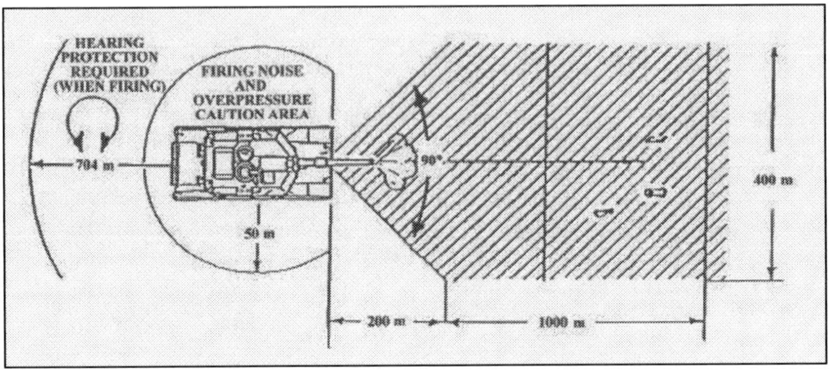

Figure B-15. Danger Areas Around a Tank Firing a 120-mm Gun

(4) Tanks are equipped with powerful thermal sights that can be used to detect enemy personnel and weapons that are hidden in shadows and behind openings. Dust, fires, and thick smoke significantly degrade these sights.

(5) Tanks have turret-mounted grenade launchers that project screening smoke grenades. The grenades use a bursting charge and burning RP particles to create this screen. *Burning particles can easily start uncontrolled fires and are hazardous to dismounted Marines near the tank.* The tank commander and the infantry small-unit leader must coordinate when and under what conditions these launchers can be used. Grenade launchers are a useful feature for protecting the tank but can cause significant problems if unwisely used.

(6) The tank's size and armor can provide dismounted Marines with cover from direct-fire weapons and fragments. With coordination, tanks can provide moving cover for Marines as they advance across small open areas. However, enemy fire that strikes but does not penetrate a tank is a major threat to nearby Marines. Fragmentation generated by antitank rounds and ricochets off of tank armor have historically been a prime cause of casualties while infantry was working with tanks in built-up areas.

11. Artillery. A major source of fire support for infantry forces fighting in built-up areas is field artillery weapons. If the built-up area is near the coast, NGF can be used. Field artillery employment can be in either the indirect- or direct-fire mode.

a. Indirect Fire. Low-angle indirect artillery fire is not effective in attacking targets within walls and masonry structures. It tends to impact on roofs or upper stories rather than on structurally critical wall areas or pillars. High-angle fire can achieve the trajectories necessary to strike targets in urban areas.

Weapons of at least 155 mm are required against thick reinforced concrete, stone, or brick walls. Even with heavy artillery, large expenditures of ammunition are required to knock down buildings of any size. Tall buildings also create areas of indirect-fire dead space, which are areas that cannot be engaged by indirect fire because of a combination of building height and angle of fall of the projectile (Figure B-16). Usually the dead space for low-angle indirect fire is about five times the height of the highest building over which the rounds must pass.

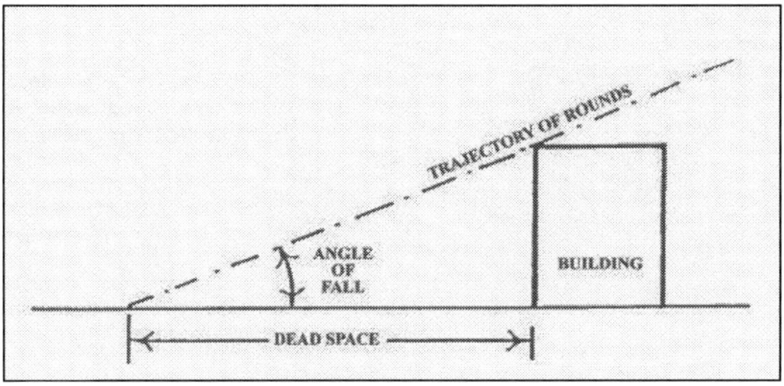

Figure B-16. Indirect-Fire Dead Space (Low Angle)

b. Direct Fire. A likely use of Marine artillery in an urban direct-fire role is to reinforce infantry against tough or important urban targets. Artillery should be used in this role only after an analysis of the need for heavy direct fire and the tradeoff involved in the extreme decentralization of artillery firepower. Artillery has the same need for close security and target designation as tanks.

c. Target Effects. Medium-caliber (155-mm) direct fire has a devastating effect against masonry construction and field fortifications. Smaller artillery (105-mm) pieces in the direct-fire mode are much less destructive than the larger caliber weapons. However, all Marine artillery is towed and, therefore, more difficult to employ in the direct-fire mode.

The 155-mm (M198) towed howitzer is effective because of its rate of fire and penetration. HE rounds can penetrate up to 38 inches of brick and unreinforced concrete. Projectiles can penetrate up to 28 inches of reinforced concrete with considerable damage beyond the wall. HE rounds with concrete-piercing fuzes provide an excellent means of penetrating strong reinforced concrete structures. One round can penetrate up to 46 inches. Five rounds are needed to reliably create a 1.5-meter breach in a wall that is 1 meter thick. About 10 rounds are needed to create the same size breach in a wall that is 1.5 meters thick. Superquick fuzing causes the rubble to be blown into the building, whereas delay fuzing tends to blow the rubble outward into the street. (See Table B-13 for employment considerations for artillery.)

Organization for Combat	Movement/Positioning	Delivery of Fire	Security	Command and Control
Centralized control is required during initial phases; decentralized control is required during later phases to support semi-independent actions of small units.	Movement should occur during night or periods of reduced visibility when possible. There are few displacements, often by platoon or section. Positions should be selected that minimize masking, provide several routes of escape, and afford as much cover and concealment as possible. Use of existing structures (garages, office buildings, highway overpasses) is recommended. Special techniques for emplacing howitzers, such as spades against a curb when the ground is not suitable for emplacement, may be required. Explosives may be required to soften emplacement of howitzers. Reconnaissance, selection, and occupation of position (RSOP) elements should be well armed because they may have to clear areas to be occupied. Extensive route reconnaissance is required. Target acquisition devices are somewhat degraded. Radars should be emplaced to cover likely areas of enemy indirect-fire weapon employment. Radars should not be placed in the midst of an urban area because of masking.	Both direct and indirect fires are delivered for supported units. Destruction of fortifications may require assault fire techniques. High-angle fires may be required. Need for accurate meteorological (MET) and survey data increases because most targets are point targets. Improved conventional munition and variable time (fuze) effects are reduced by structures, although they are effective against personnel on rooftops. HE delay is used for penetration effects. Illumination, chemical incendiary ammunition, and smoke are effective. Ammunition expenditures will be heavy. Lasers and PGMs permit destruction of targets with minimal rubbling of adjacent buildings. Tall buildings may hamper laser use. Batteries must be prepared for hasty survey techniques. Magnetic instruments are impaired.	Positions must be fortified.	Radio communications are impaired by buildings. Wire can usually be run overhead. Make use of civilian communications. A greater use of messengers and prearranged audio and visual signals is required.

Table B-13. Artillery Employment Considerations in Built-Up Areas

12. Mortars. The urban environment greatly restricts low-angle indirect fires because of overhead masking. While all indirect-fire weapons are subject to overhead masking, mortars are less affected than field artillery weapons because of the mortar's higher trajectory. For low-angle artillery fire, dead space is about five times the height of the building behind which the target sits. For mortar fire, dead space is only about one-half the height of the building. Because of these advantages, mortars are even more important than field artillery to the infantry during combat in built-up areas.

a. Employment. Not only can mortars fire into the deep defilade created by tall buildings, but they can also fire out of it. Mortars emplaced behind buildings are difficult for the enemy to accurately locate and even harder to hit with counterfire. Because of their light weight, even heavy mortars can be hand-carried to firing positions that may not be accessible to vehicles.

(1) The 60-mm and 81-mm mortars have limited effects on structural targets. Even with delay fuzes, they seldom penetrate more than the upper stories of light buildings. However, their wide area coverage and multioption fuzes make them useful against an enemy force advancing through streets, through other open areas, or over rubble.

(2) Mortar platoons often operate as separate firing sections during combat in built-up areas. The lack of large open areas can preclude establishing a platoon firing position. Figure B-17 shows how two mortar sections that are separated by only one street can be effective in massing fires and be protected from countermortar fire by employing defilade and dispersion.

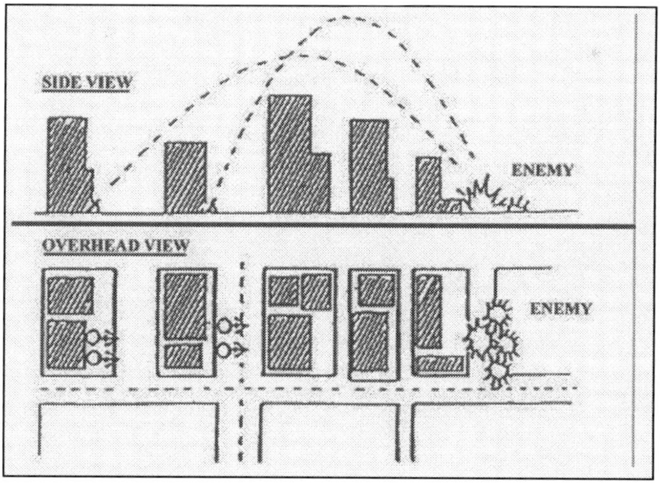

Figure B-17. Split-Section Mortar Operations on Adjacent Streets

(3) All three of the standard mortar projectiles are useful during combat in built-up areas. HE fragmentation rounds are most commonly used. WP is effective for starting fires in buildings and forcing the enemy out of cellars and light-framed buildings, and it is also the most effective mortar round against dug-in enemy tanks. Even near-misses blind and suppress the tank crew, forcing them to button up. Hits are difficult to achieve but are effective.

(4) Because the artificial roughness of urbanized terrain reduces wind speed and increases atmosphere mixing, mortar smoke tends to persist longer and give greater coverage in built-up areas than in open terrain.

(5) Urban masking has an impact on the use of illumination. In built-up areas, it is often necessary to plan illumination behind friendly positions, which places friendly troops in shadows and enemy troops in the light. Illumination rounds are difficult to adjust and are often of limited use because of the deep canyon nature of the urbanized area. Rapidly shifting wind currents in built-up areas also affect mortar illumination, making it less effective.

b. Effects of Mortar Fire. The multioption fuze on newer U.S. mortar rounds makes them effective weapons on urbanized terrain. Delay settings can increase penetration slightly, and proximity bursts can increase the lethal area covered by fragments. Tall buildings can cause proximity-fuzed mortar rounds to detonate prematurely if they pass too closely to the buildings.

(1) 60-mm Mortar. The 60-mm mortar round cannot penetrate most rooftops, even with a delay setting. Small explosive rounds are effective, however, in suppressing snipers on rooftops and preventing roofs from being used by enemy observers. The 60-mm WP round is not normally a good screening round because of its small area of coverage. In combat in built-up areas, however, the tendency of smoke to linger and the small areas to be screened make it more effective. During the battle for Hue in South Vietnam, 60-mm WP rounds were used to create small, short-term smoke screens to conceal movement across open areas such as parks, plazas, and bridges. Fragments from 60-mm HE rounds landing as close as 10 feet cannot penetrate a single sandbag layer or a single-layer brick wall. The effect of a 60-mm mortar HE round that achieves a direct hit on a bunker or fighting position is equivalent to 1 - 2 pounds of TNT. Normally, the blast will not collapse a properly constructed bunker, but it can cause structural damage. The 60-mm mortar will not normally create craters in a hard-surfaced road.

(2) 81-mm Mortar. The 81-mm mortar has much the same effect against urban targets as the 60-mm mortar. It has a slightly greater lethal area, and its smoke rounds (WP and RP) are more effective. A direct hit is equivalent to about 2 pounds of TNT. The 81-mm round cannot cause significant craters in a hard-surfaced road. With a delay setting, the 81-mm round can penetrate the roofs of light buildings.

13. Naval Gunfire. The primary NGF weapon, the 5-inch/54, has a high rate of fire and is roughly equivalent to the 155-mm howitzer in target effect. Target engagement considerations are the same as for artillery; however, the 5-inch/54's flat trajectory is even more affected by terrain masking and a constantly changing gun-target line. NGF can achieve greater target engagement by firing reduced charges at high angles; however, range will be reduced.

14. Aerial Weapons. Both rotary- and fixed-wing aircraft can quickly deliver large volumes of firepower over large built-up areas. Specific targets are hard to distinguish from the air. Good ground-to-air communications are vital in successfully employing aerial firepower. Planners have historically tended to overestimate the effects of HE on defenders. Modern, large buildings are remarkably resistant to damage from bombs and rocket fire.

a. Rotary-Wing Aircraft. Attack helicopters can be used to engage targets in built-up areas. Enemy armored vehicles in small parks, boulevards, or other open areas are good targets for attack helicopters.

(1) The HELLFIRE missile has a larger warhead and greater range than the TOW, but it too is a shaped-charge warhead and is not specifically designed for use against masonry targets. Laser target designation for the HELLFIRE may be difficult as a result of laser reflections off of glass and shiny metal surfaces. The use of attack helicopters to deliver ATGMs against targets in the upper stories of high buildings is sometimes desirable.

(2) The 2.75-inch folding-fin aerial rocket and the 20-mm cannon on Marine attack helicopters are good area weapons to use against enemy forces in the open or under light cover. They are usually ineffective against a large masonry target. The 20-mm cannon produces many ricochets, especially if armor-piercing ammunition is fired into built-up areas.

b. Fixed-Wing Aircraft. Close air support to ground forces fighting in built-up areas is a difficult mission for fixed-wing aircraft. Targets are hard to locate and identify, enemy and friendly forces could be intermingled, and enemy short-range air defense weapons are hard to suppress.

(1) Because enemy and friendly forces may be separated by only one building, accurate delivery of ordnance is critical. Marking panels, lights, electronic beacons, smoke, or some other positive identification of friendly forces is needed.

(2) General-purpose bombs that weigh from 500 to 2,000 pounds are moderately effective in creating casualties among enemy troops located in large buildings. High-dive-angle bomb runs increase accuracy and penetration but also increase the aircraft's exposure to antiaircraft weapons. Low-dive-angle bomb runs using high-drag (retarded) bombs can be used to get bombs into upper stories. Penetration is not good with high-drag bombs. Sometimes aerial bombs pass completely through light-clad buildings and explode on the outside.

(3) Aerial rockets and 20-mm cannons are only moderately effective against enemy soldiers in built-up areas because rockets lack the accuracy to concentrate their effects. The 20-mm cannon rounds penetrate only slightly better than the .50-caliber round, 20-mm armor-piercing rounds can ricochet badly, and tracers can start fires.

(4) The AC-130 aircraft has weapons that can be most effective during combat in built-up areas. This aircraft can deliver accurate fire from a 20-mm Vulcan cannon, 40-mm rapid-fire cannon, and 105-mm howitzer. The 105-mm howitzer round is effective against the roofs and upper floors of buildings. The AC-130 is accurate enough to concentrate its 40-mm cannon and 105-mm howitzer fire onto a single spot to create a rooftop breach that allows fires to be directed deep into the building.

(5) Laser and optically guided munitions can be effective against high-value targets. Specially developed, heavy, laser-guided bombs can penetrate hardened targets. However, problems associated with dense smoke and dust clouds hanging over the built-up area and laser scatter can restrict their use. If the launching aircraft can achieve a successful laser designation and lock-on, these weapons have devastating effects, penetrating deep into reinforced concrete before exploding with great force. If launched without a lock-on, or if the laser spot is lost, these weapons are unpredictable and can travel long distances before they impact.

15. Demolitions. Combat in built-up areas requires the extensive use of demolitions. All Marines, not just engineers, should be trained to employ demolitions. See FM 5-250, *Explosives and Demolitions*, for specific information on the safe use of demolitions.

a. Demolitions. Demolitions come in two types: TNT and C4. Exposed Marines must take cover or move at least 300 meters away from bulk explosives that are being used to breach walls.

(1) TNT comes in 1/4-, 1/2-, and 1-pound blocks. About 5 pounds of TNT are needed to breach a nonreinforced concrete wall that is 12 inches thick if the explosives are laid next to the wall and are not tamped. If the explosives are tamped, about 2 pounds are sufficient.

(2) C4 comes in many different-sized blocks. Normally, it is found in 1 1/4- and 2 1/2-pound blocks within prepackaged satchel charges. About 10 pounds of C4 placed between waist and chest high will blow a hole in the average masonry wall large enough for a man to walk through.

b. Shaped Charges. There are two sizes of shaped charges: a 15-pound M2A3 and a 40-pound M3A3. The M3A3 is the shaped charge that is most likely to be used in built-up areas. It can penetrate 5 feet of reinforced concrete. The hole tapers from 5 inches down to 2 inches. The amount of spall thrown behind the target wall is considerable. There is also a large safety hazard area for friendly forces.

c. Satchel Charges. There are two standard satchel charges: the M183 and the M37. Both come in their own carrying satchel with detonators and blasting cords. Each weighs 20 pounds. The M183 has 16 individual 1 1/4-pound blocks that can be used separately. When used untamped, a satchel breaches a 3-foot-thick concrete wall. Satchel charges are very powerful. Debris is thrown great distances. Friendly forces must move away and take cover before detonation.

d. Cratering Charges. The standard cratering charge is a 43-pound cylinder of ammonium nitrate. This explosive does not have the shattering effect of bulk TNT or C4. It is more useful in deliberate demolitions than in hasty ones.

Appendix C

Attacking and Clearing Buildings

Offensive operations on urbanized terrain will require Marines to attack and clear buildings.

1. Introduction. At the rifle platoon and squad level, the major offensive tasks for combat in built-up areas are attacking and clearing buildings. This involves:

- Suppressing defensive fires
- Advancing Marine assault forces
- Assaulting buildings
- Clearing buildings
- Reorganizing assault forces.

To accomplish these tasks, the assaulting unit task organizes into assault, security, and support forces or elements. (See Appendix A)

2. Requirements. There are five interrelated requirements for attacking a defended building: *fire support, movement, assault, clearing,* and *reorganization.* Proper application and integration of these requirements reduce casualties and hasten accomplishment of the mission. The proper integration of these requirements are determined by METT-T. For example, larger cities may have high-rise apartment buildings and industrial complexes that are separated by parking areas or parks. Increased fire support may be required to suppress and obscure enemy gunners covering the open terrain and spaces between buildings. Conversely, a built-up area with few open areas can provide numerous covered routes that may decrease fire support requirements.

3. Fire Support. Fire support necessary to advance the assault force may include:

- Suppressing and obscuring enemy gunners within the objective building(s) and adjacent structures
- Isolating the objective building(s) with fires to prevent enemy withdrawal, reinforcement, or counterattack
- Breaching walls en route to and in the objective structure
- Destroying enemy positions with direct-fire weapons

- Securing cleared portions of the objective
- Providing replacements for the assault force
- Providing resupply of ammunition and explosives
- Evacuating casualties and prisoners.

a. The size of the support force is determined by the type and size of the objective building(s), whether the adjacent terrain provides open or covered approaches, and the organization and strength of enemy defenses.

b. The support force for a Marine rifle squad may consist of one fire team that could be reinforced with heavy machine guns, antitank weapons, and/or armored vehicles. In situations involving a larger assault force, a rifle platoon or company reinforced with armored vehicles, engineers, and/or artillery may be required to support the movement and assault of an adjacent platoon or company.

c. After seizing objective buildings, the assault force reorganizes and may be required to provide supporting fires for a subsequent assault. Each weapon is assigned a target or area to cover. Individual small-arms weapons place fires on likely enemy weapon positions—loopholes, windows, and roof areas. Snipers are best employed in placing accurate fire through loopholes or engaging long-range targets. M203 fires are directed through windows or loopholes.

d. AT4s and demolitions are employed to breach walls, doors, barricades, and window barriers on the ground level of structures. The tank cannon or LAV-mounted 25-mm gun engages first-floor targets and breaches walls for attacking Marines. Armored vehicle- or wheeled vehicle-mounted machine guns can engage suspected enemy positions on upper floors and in adjacent structures. In addition to destroying or weakening structures, the tank's main gun rounds cause casualties by their explosive effects and by hurling debris throughout the interior of structures.

e. NGF, artillery, and mortars should initially use time fuzes to initially clear exposed enemy personnel, weapons, OPs, and radio sites from rooftops. They should then use delayed fuze action to cause casualties among the defenders inside the structure as a result of shrapnel and falling debris. They may also provide smoke to cover the movement and assault phases of the attack. Artillery can also be used in the direct-fire mode.

f. In support of the movement and assault phases of the attack, buildings can be attacked by fixed-wing aircraft or by attack helicopters.

4. Movement. The assault force minimizes the effects of enemy defensive fires during movement by:

- Using covered routes
- Moving after defensive fires have been suppressed or obscured
- Moving at night or during other periods of reduced visibility
- Selecting routes that will not mask friendly suppressive fires
- Crossing open areas (e.g., streets and spaces between buildings) quickly under the concealment of smoke and suppressive fires
- Moving on rooftops that are not covered by enemy direct fires.

a. The requirement for speed may dictate movement through streets and alleys without clearing all buildings. When moving in this manner, the maneuver force should employ infantry to lead the column, and supported by AAVs, LAVs, wheeled vehicle-mounted machine guns, or tanks following.

b. Marines should move along each side of a street with the leading squad staying abreast of armored vehicles. When not accompanied by armored vehicles, Marines move in a single file along one side of the street under cover of fires from supporting weapons. Individual Marines should be detailed to observe and cover a specific area, such as second-floor windows on the opposite side of the street.

5. Assault. Assault forces quickly and violently execute their assault and subsequent clearing operations. Momentum is maintained to prevent the enemy from organizing a more determined resistance on other floors or in other rooms. Small-unit leaders should keep the assault force moving without allowing the operation to become disorganized.

a. An assault in a built-up area involves the elementary skills of close combat. Leaders must:

- Be trained in the required techniques to defeat the enemy in a face-to-face encounter
- Keep themselves in excellent physical condition
- Have confidence in their abilities and those of their unit.

b. The composition of the assault force will vary depending on METT-T; however, its basic organization at the squad level consists of two assault elements carrying a fighting load of equipment and as much ammunition as possible, especially grenades.

6. Clearing. Entering at the top and fighting downward is the preferred method of clearing a building. Clearing a building is easier from an upper story because gravity and building construction become assets to the assault elements when throwing hand grenades and moving from floor to floor. However, this method is only feasible when access to an upper floor or rooftop can be gained from the windows or roofs of adjoining buildings and assault elements can be transported to the rooftops by helicopter (Figure C-1). Helicopters should land only on those buildings that have special heliports or parking garages on the roofs. When these are unavailable, Marines can rappel or fastrope onto the roof or dismount as the helicopter hovers a few feet above the roof. Assault elements then breach the roof or common walls with explosives and use ropes to enter the lower floors. Stairs are guarded by a covering team or security force when not in use to protect the assault force from a flank attack.

Figure C-1. Helicopters Used To Clear Buildings

a. Although the top-to-bottom method is preferred for clearing a building, assaulting the bottom floor and clearing upward is a common method. When using this technique, the assault force closes on the flank(s) or rear of the building. It then clears each room on the ground floor and then, moving upward, begin a systematic clearance of the remaining floors.

b. Entry can be gained through walls breached by explosives or gunfire. Assault elements should avoid windows and doors because they are usually covered by fire or boobytrapped. If tanks are attached to the company, they can breach the wall by firing their main gun (Figure C-2).

Figure C-2. Main Gun Used To Breach Exterior

c. Shortly before the assault, suppressive fires on the objective should be increased until masked by the advancing assault force. Once masked, fires are shifted to upper windows and continued until the assault force has entered the building. At that time, fires are shifted to adjacent buildings to prevent enemy withdrawal or reinforcement.

d. Hand grenades should be used to clear the space before assault element members enter through a breached wall. Immediately after the explosion, the assault element clearing team enters and uses three round bursts to fire into the interior of the room, concentrating on areas that are possible enemy positions.

e. Once inside the building, the assault force must cover staircases and seize rooms that overlook approaches to the building. These actions are required to isolate enemy forces within the building and to prevent reinforcement. Assault elements should clear each ground floor room and then the basement. Sequence for the assault is that:

(1) The assault force leader determines which room(s) to clear first.

(2) The support force provides suppressive fire while the assault force systematically clears the building. It also provides suppressive fire on adjacent buildings to prevent enemy reinforcements or withdrawal. The support force destroys any enemy trying to exit the building.

(3) The security force isolates the immediate area.

(4) After assault elements establish a foothold in the building, the assault force leader proceeds with the systematic clearing of the building.

f. When clearing a building from the top floor down, assault forces should gain entrance through a common wall or the roof of an adjoining building (Figure C-3). Demolition charges are used to breach the wall and gain entrance to the top floor. Access to lower floors and rooms may be gained by breaching holes in the floor and having the Marines jump or slide down ropes to the lower floors. Stairs can be used if they are first cleared.

g. When using the top-to-bottom method of clearing, requirements remain the same as for other methods. After the floor is breached to gain access to a lower floor, a grenade is allowed to cook-off and is dropped to the lower room. A Marine then sprays the lower room with gunfire using three-round bursts and drops through the mousehole.

Figure C-3. Upper Floors Secured

h. Marines should avoid predictability when entering and clearing rooms. They should vary entry through doors and mouseholes to keep the enemy off guard (Figure C-4). As rooms are cleared, doors should be left open, and a predetermined mark (cloth, tape, spray paint) should be placed on the doorjamb or over the door, as established by unit SOPs.

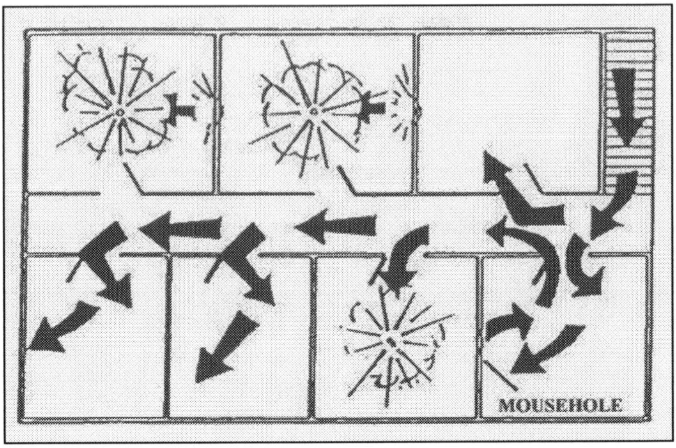

Figure C-4. Varying Techniques for Clearing Rooms

i. If there is a basement, it should be cleared as soon as possible, preferably at the same time as the ground floor. The procedures for clearing a basement are the same as for any room or floor. However, basements often contain entrances to tunnels such as sewers and communications cable tunnels. These should be cleared and secured to prevent the enemy from infiltrating back into cleared areas.

j. Types of buildings that may be cleared include brick buildings and houses, box-wall buildings, heavy-clad framed buildings, and light-clad framed buildings.

(1) Brick Buildings and Houses. The best way to enter a brick building is to blow a breaching hole in the side by using a tank firing HEAT ammunition (Figure C-5). If tanks are not available, a door or window in the rear of the building usually provides better cover and concealment for entry than one in the front. If there is enough cover and concealment, the assault force should enter the rear of the building at an upper level by using a fire escape or grappling hook.

Brick houses usually have similar floor plans on each floor (Figure C-6); therefore, ground floors are cleared the same way as upper floors.

Converely, the floor plans in larger brick buildings typically are different on ground-floor levels than on upper levels (Figure C-7).

Figure C-5. Entering a Brick Building

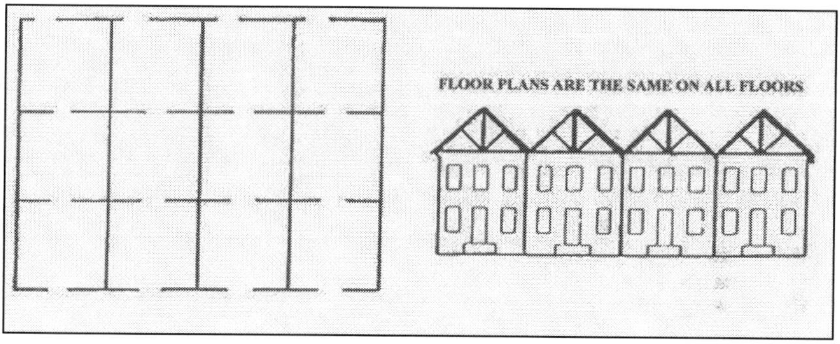

Figure C-6. Brick Houses Have Similar Floor Plans

To clear from building to building, the best method is to move from rooftop to rooftop because the roofs of brick buildings are usually easy to breach. The walls between buildings are at least three bricks thick (total of six bricks between buildings) and require large quantities of demolitions to breach. Walls are normally easier to breach on an upper floor than on a lower floor because the walls are thinner on upper floors. If rooftops are covered by fire and there are not enough demolitions to breach walls between buildings, clearing from rear to rear of buildings is safer than clearing from front to front.

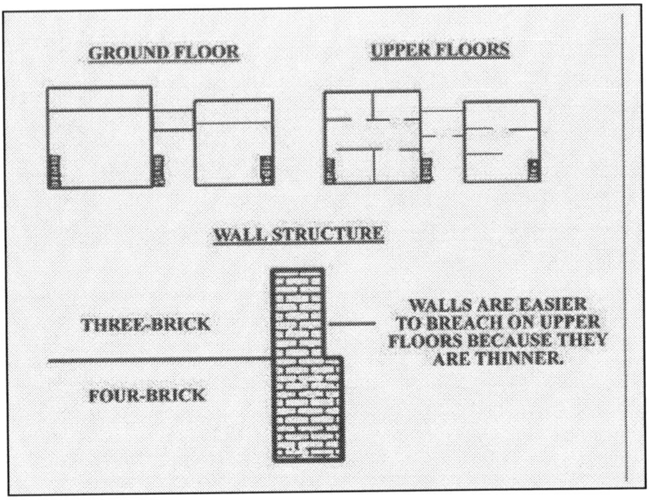

Figure C-7. Floor Plans of Brick Buildings

(2) Box-Wall Buildings. Box-wall buildings often have reinforced concrete walls (Figure C-8), which are difficult to breach because of reinforcing bars. Therefore, the best way to enter is to blow down the door or to blow in one of the side windows. The floor plans of these buildings are predictable; clearing rooms is usually done from one main hallway. Interior walls are also constructed of reinforced concrete and are difficult to breach. The stairways at the ends of the building must also be secured during clearing.

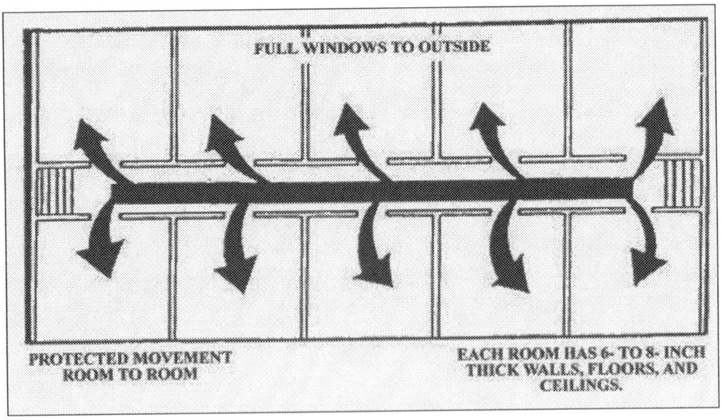

Figure C-8. Box-Wall Buildings

(3) **Heavy-Clad Framed Buildings.** Tanks can be used to breach a hole in heavy-clad framed buildings. Their floor plans are normally oriented around a stairway or elevator. this must be secured during clearing (Figure C-9). The interior walls of these buildings can be breached, although they may require use of demolitions.

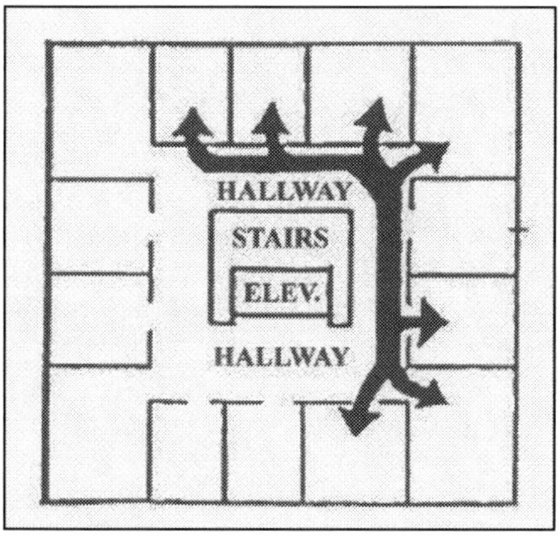

Figure C-9. Heavy-Clad Framed Buildings

(4) **Light-Clad Framed Buildings.** Light-clad framed buildings (Figure C-10), can be cleared in the same manner as heavy-clad buildings. Secure the central stairway and clear in a circular pattern. Walls are usually thin enough to be breached with an ax.

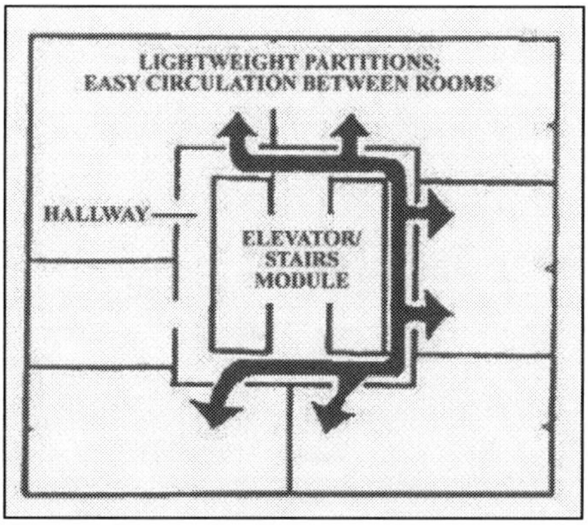

Figure C-10. Light-Clad Framed Buildings

7. **Consolidation and Reorganization.** Consolidation and reorganization of the assault force in a cleared building must be accomplished quickly to repel enemy counterattacks and to prevent the

enemy from infiltrating back into the cleared building. After securing a floor (bottom, middle, or top), selected members of the assault force are assigned to cover potential enemy counterattack routes to the building. Those sentinels alert the assault force and place a heavy volume of fire on enemy forces approaching the building. They should cover:

- Enemy mouseholes between adjacent buildings
- Covered routes to the building
- Underground routes into the basement
- Approaches over adjoining roofs.

As the remainder of the assault force completes search requirements, defensive positions are assigned and tasks are accomplished as established by the priorities of work in the defense. The following actions may then be taken:

- Resupplying and redistributing ammunition
- Marking the building to indicate to friendly forces that the building has been cleared
- Assuming an overwatch mission and supporting an assault on another building
- Treating and evacuating wounded personnel
- Further developing a defensive position if the building is to be occupied for any period.

Appendix D

Fighting Positions

A critical defensive task in combat in built-up areas is the selection and preparation of fighting positions.

1. Considerations. Leaders should consider the following factors when establishing fighting positions:

a. Protection. Leaders should select buildings that provide protection from direct and indirect fires. Reinforced concrete buildings with three or more floors provide suitable protection, while buildings constructed of wood, paneling, or other light materials require reinforcement to gain sufficient protection. One- to two-story buildings without strongly constructed cellars are vulnerable to indirect fires and require construction of overhead protection for each firing position.

b. Dispersion. Normally, a position should not be established in a single building when it is possible to occupy two or more buildings that permit mutually supporting fires. A position in one building without mutual support is vulnerable to bypass, isolation, and subsequent destruction from any direction.

c. Concealment. The fighting position should be concealed. Buildings that are obvious defensive positions should not be selected. The occupation of exposed buildings may be necessary because of requirements for security and fields of fire. Therefore, reinforcement of the structure must be accomplished to provide suitable protection within the building.

d. Fields of Fire. Positions should be mutually supporting and have interlocking fields of fire in all directions. Clearing fields of fire may require the destruction of adjacent buildings by using explosives, engineer equipment, and field expedients. Care should be taken to avoid highlighting the building as a defensive position.

e. Covered Routes. Defensive positions should have at least one covered route that permits resupply, MEDEVAC, reinforcement, or withdrawal from the building. The route can be established by one of the following means:

- Through walls to adjacent buildings
- Through underground systems
- Through communications trenches
- Behind protective buildings.

f. Observation. The building should permit observation of enemy avenues of approach and adjacent defensive sectors.

g. Fire Hazard. Avoid selecting positions in buildings that are a fire hazard. If flammable structures must be occupied, the danger of fire can be reduced by wetting the structure, laying an inch of sand on the floors, and providing fire extinguishers and firefighting equipment. Routes of escape should also be prepared in case of fire.

h. Time. The time available to prepare the defense could be the most critical factor. If enough time is insufficient, buildings that require extensive preparation should not be used. Conversely, buildings located in less desirable areas that require little improvement may be a better choice for a defensive position.

2. Preparation. Preparation of fighting positions depends on proper selection and construction.

a. Selecting Positions. Each weapon should be assigned a primary sector of fire to cover enemy approaches. Alternate positions that overwatch the primary sector should also be selected. These positions are usually located in an adjacent room on the same floor (Figures D-1A, D-1B, and D-1C). Each weapon must be assigned a supplementary position (to engage attacks from other directions) and an FPL.

b. Building Positions. There are many ways to establish a fighting position in a building.

(1) Window Position. Marines should kneel or stand on either side of a window. To fire downward from upper floors, tables or similar objects can be placed against the wall to provide additional elevation, but they must be positioned to prevent the weapon from protruding through the window (Figure D-2). Leaders should inspect positions to determine the width of sector that each position can engage.

(2) Loopholes. To avoid establishing a pattern of always firing from windows, loopholes should be prepared in walls. Marines should avoid firing directly through loopholes to enhance individual protection.

(a) Several loopholes are usually required for each weapon (primary, alternate, and supplementary positions). The number of loopholes should be carefully considered because they can weaken walls and reduce protection. Engineers should be consulted before numerous loopholes are made. Loopholes should be made by punching or drilling holes in walls and should be placed where they are concealed. Blasting loopholes can result in a large hole that can easily be seen by the enemy.

(b) Loopholes should be cone-shaped to obtain a wide arc of fire, to facilitate engagement of high and low targets, and to reduce the size of the exterior aperture (Figure D-3). The edges of a loophole splinter when hit by bullets; therefore, protective linings such as an empty sandbag held in place by wire mesh will reduce spalling effects. When not in use, loopholes should be covered with sandbags to prevent the enemy from firing into or observing through them.

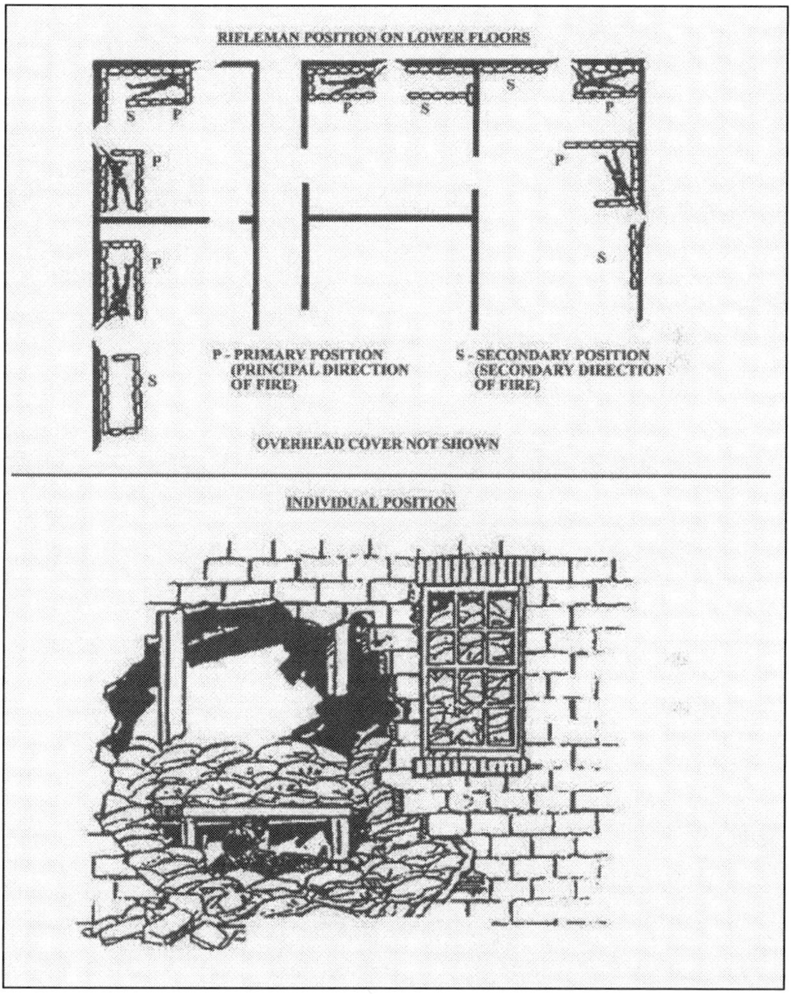

Figure D-1A. Weapon Positions

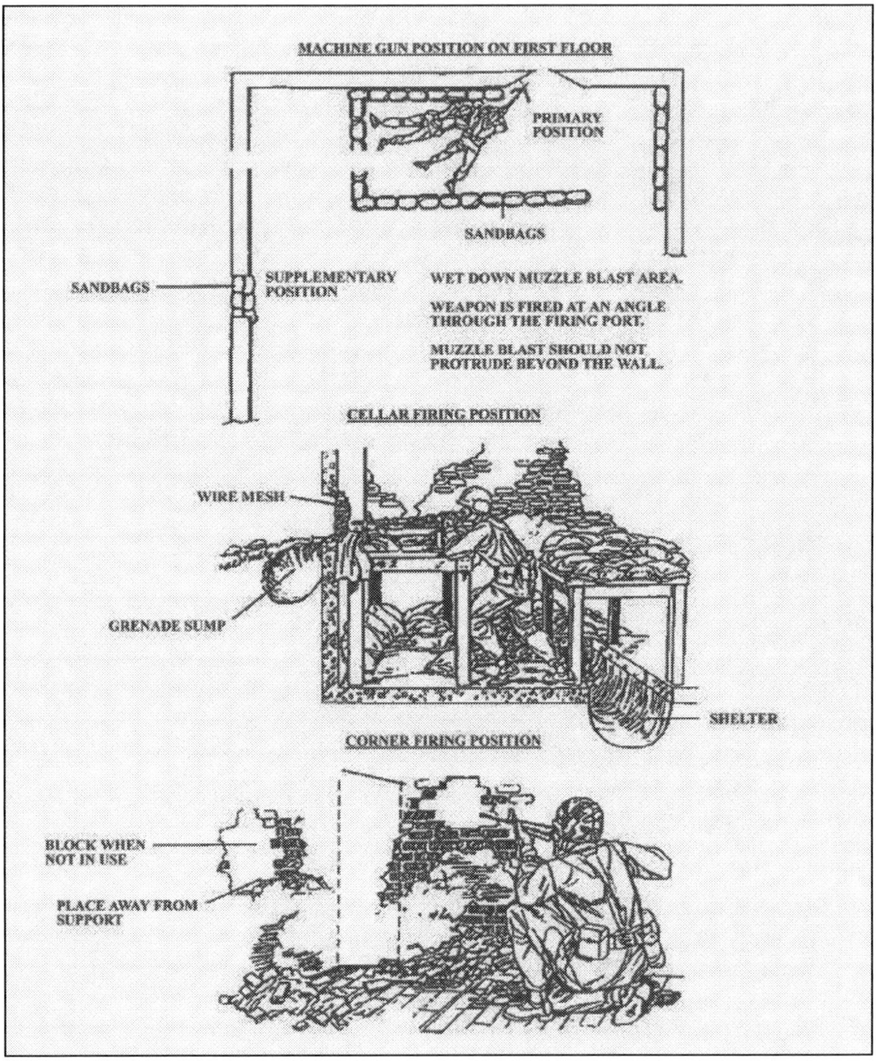

Figure D-1B. Weapon Positions (Continued)

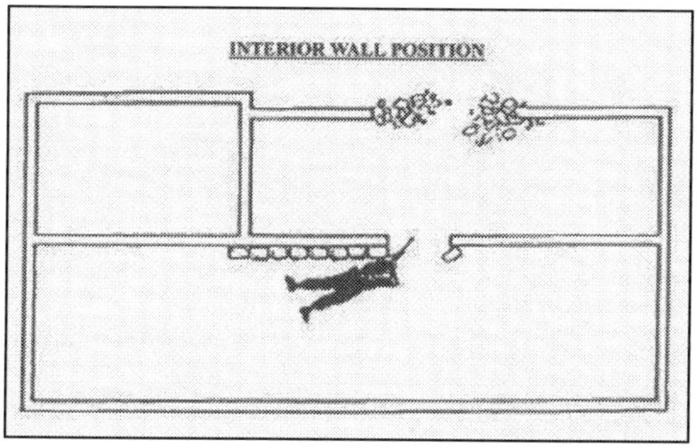

Figure D-1C. Interior Wall Position

Figure D-2. Window Position

(c) Loopholes should also be prepared in interior walls and ceilings of buildings to permit fighting within the position. Interior loopholes should enable overwatch of stairs, halls, and unoccupied rooms and be concealed by pictures, drapes, or furniture. Loopholes in floors permit the defender to engage enemy personnel on lower floors with small-arms fire and grenades.

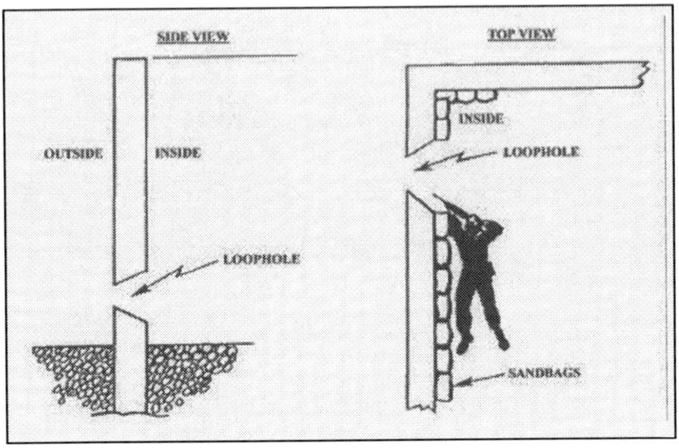

Figure D-3. Cone-Shaped Loopholes

(d) Although walls provide some frontal protection, they should be reinforced with sandbags, furniture filled with dirt, or other expedients. Each position should have overhead and all-around protection (Figure D-4).

Figure D-4. Position With Overhead and All-Around Protection

c. Other Construction Tasks. Other construction tasks may need to be performed. These include work on:

(1) Basements and Ground Floors. Basements require preparation similar to that of the ground floor. Any underground system that is not used by the defender but that could provide enemy access to the position must be blocked.

(a) Doors. Unused doors should be locked, nailed shut, blocked, and reinforced with furniture, sandbags, or other field expedients. Outside doors can be boobytrapped by engineers or other personnel.

(b) Hallways. If not required for the defender's movement, hallways should be blocked with furniture and tactical wire (Figure D-5). If authorized, boobytraps should be employed to impede enemy movement.

(c) Stairs. Defenders should block stairs not used by the defense with furniture and tactical wire (Figure D-5) or remove them. If possible, all stairs should be blocked, and ladders should be used to move from floor to floor and then removed when not being used. Boobytraps should also be employed on stairs.

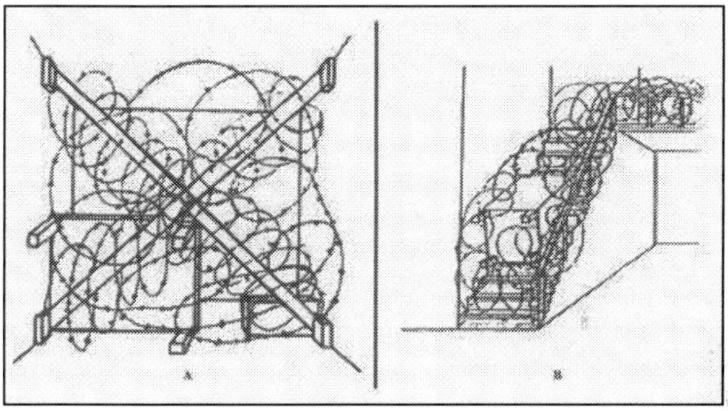

Figure D-5. Blocking Hallways and Stairs

(d) Windows. All glass should be removed. Windows not being used should be blocked with boards or sandbags.

(e) Floors. Fighting positions should be constructed on the ground or basement floors. If there is no basement, fighting positions on the first floor must be constructed to provide additional protection from heavy direct-fire weapons.

(f) Ceilings. Supports that can withstand the weight of rubble from upper floors should be placed under ceilings to provide overhead cover (Figure D-6).

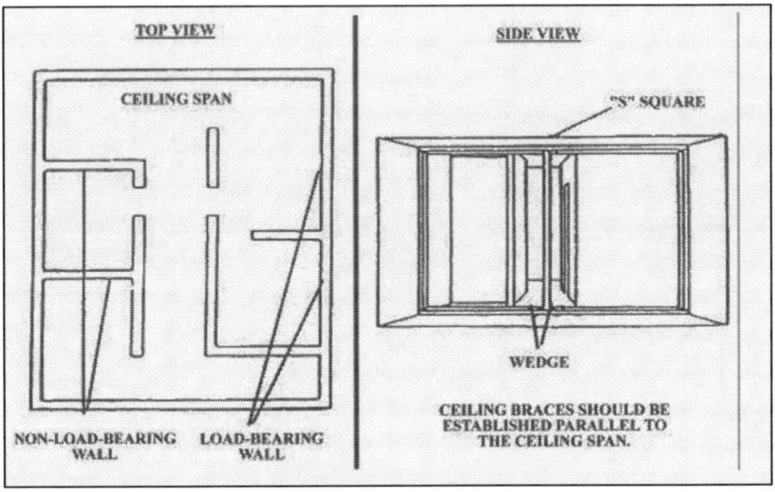

Figure D-6. Ceiling Reinforcement

(g) Unoccupied Rooms. Rooms not required for defense should be blocked with tactical wire, boobytrapped, or both.

(2) Upper Floors. Upper floors require the same preparation as ground floors. Windows need not be blocked, but they should be covered with wire mesh to block grenades thrown from the outside. The wire should be loose at the bottom to permit the defender to drop grenades on enemy forces below.

(3) Interior Routes. Routes are required that permit defending forces to move within the building to engage enemy forces from any direction or reinforce other forces (Figure D-7). Escape routes should also be planned and constructed to permit rapid evacuation of a room or the building. Small holes (called mouseholes) should be made through interior walls to permit movement between rooms. Once the defender has withdrawn to another level, such holes should be clearly marked for both day and night identification. All personnel must be briefed as to where the various routes are located. Rehearsals should be conducted so that everyone becomes familiar with the routes.

(4) Fire Prevention. Buildings that have wooden floors and raftered ceilings require extensive fire prevention measures. The attic and other wooden floors should be covered with about 1 inch of sand or dirt, and buckets of water should be positioned for immediate use. Firefighting materials (dirt, sand, fire extinguishers, and blankets) should be placed on each floor for immediate use. Water basins and bathtubs should be filled as a reserve for firefighting. All electricity and gas should be turned off. Fire breaks can be created by destroying buildings adjacent to the defensive position.

Figure D-7. Movement Between Floors

(5) Communications. Telephone lines should be laid through adjacent buildings or underground systems or buried in shallow trenches. Radio antennas can be concealed by placing them among civilian television antennas, along the sides of chimneys and steeples, or out windows that direct FM communications away from enemy electronic warfare sources and ground observation. Telephone lines within the building should be laid through walls and floors.

(6) Rubbling. Rubbling parts of the building may provide useful cover and concealment for weapons emplacements. Rubbling should be performed only by trained engineers.

(7) Rooftops. Positions in flat-roofed buildings require obstacles that restrict helicopter landings (See Figure F-2). Rooftops that are accessible from adjacent structures should be covered with tactical wire or other expedients and must be guarded. Entrances to buildings from rooftops can be blocked if compatible with the overall defensive plan. Any structure on the outside of a building that could assist in scaling the buildings to gain access to upper floors or to the rooftop should be removed or blocked.

(8) Obstacles. Obstacles should be positioned adjacent to buildings to stop or delay tanks and infantry.

(9) Fields of Fire. Fields of fire should be improved around the defensive position. Selected buildings may be destroyed to enlarge fields of fire. Obstacles to ATGMs, such as telephone or electrical wires, should be cleared. Dead space should be covered with mines and obstacles.

3. Tank and Other Armored Vehicle Positions. Fighting positions for tanks and other armored vehicles are essential to a complete and effective defensive plan in built-up areas.

a. Armored Vehicle Positions. Armored vehicle positions are selected and developed to obtain the best cover, concealment, observation, and fields of fire while retaining the vehicle's ability to move.

(1) If fields of fire are restricted to streets, hull-down positions should be used to gain cover and to fire directly down streets (Figure D-8). From those positions, armored vehicles are protected and can rapidly move to alternate positions. Buildings collapsing from enemy fires are a minimal hazard to armored crews.

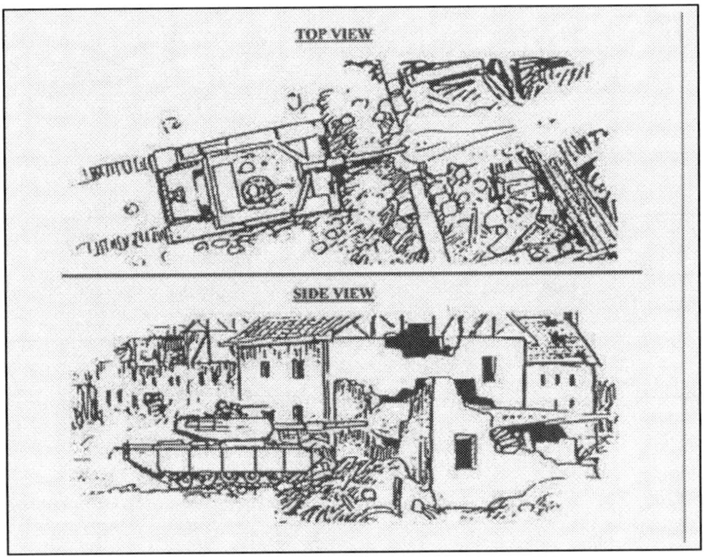

Figure D-8. Hull-Down Position

(2) The hide position (Figure D-9) covers and conceals the vehicle until time to move into position for engagement of targets. Because the crew will not be able to see advancing enemy forces, an observer from the vehicle or a nearby infantry unit must be concealed in an adjacent building to alert the crew. The observer acquires the target and signals the armored vehicle to move to the firing position and fire. After firing, the vehicle moves to an alternate position to avoid compromising one location.

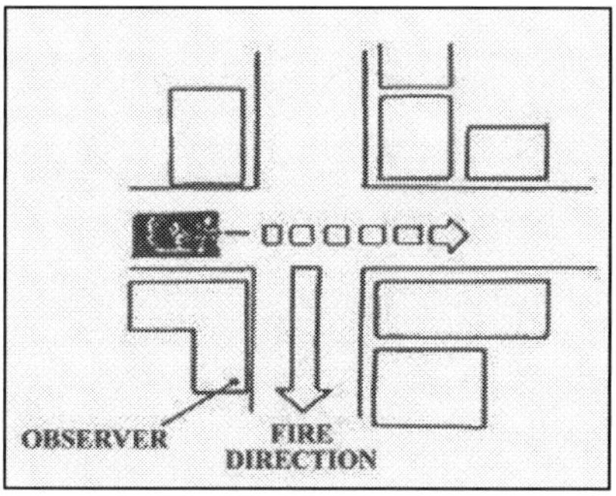

Figure D-9. Hide Position

(3) The building hide position (Figure D-10) conceals the vehicle inside a building. If basement hide positions are inaccessible, engineers can evaluate the building's floor strength and prepare for the vehicle. Once the position is detected, it should be evacuated to avoid enemy fires.

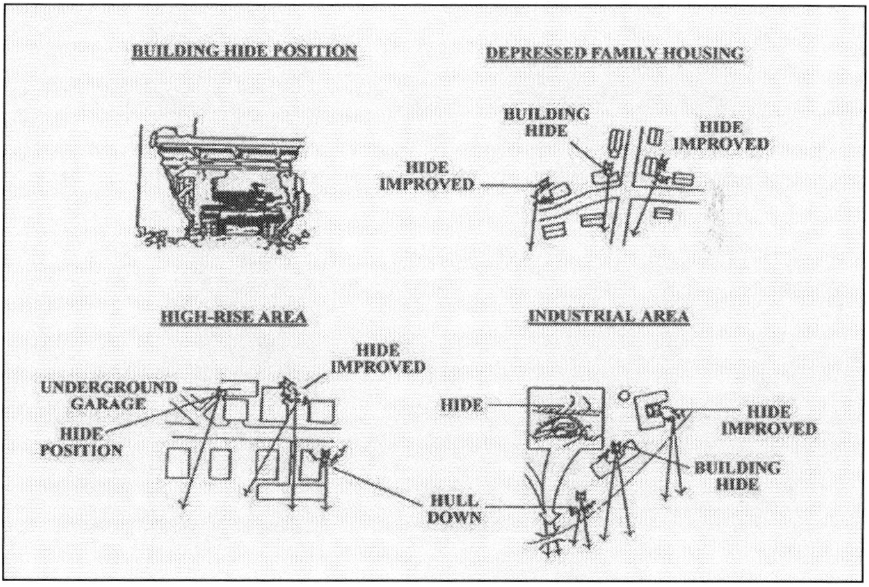

Figure D-10. Building Hide Position

4. Antitank Guided Missile Positions. HMMWVs with ATGMs are employed in areas that maximize their capabilities in the built-up area. Figure D-11 shows mounted and dismounted firing positions.

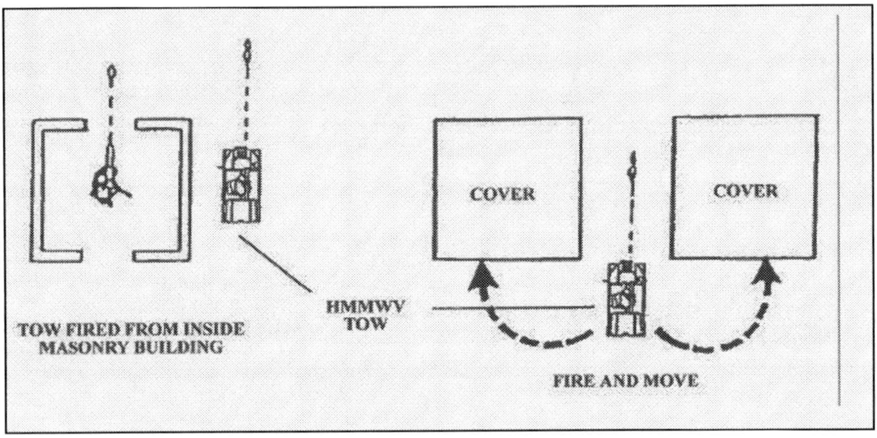

Figure D-11. ATGM Positions

a. Rubble may interfere with missile flight when ATGMs are fired from the HMMWV or from street-level or bottom-floor fighting positions. When firing down streets, missiles must have at least 1 meter of clearance over rubble. Other obstacles to missile flight include trees and brush, vehicles, television antennas, buildings, power lines and wires, walls, and fences (See Figure B-7).

b. An AT4 is best suited for built-up areas because its short arming distance allows employment at close range. AT4s and other light and medium antitank weapons are not effective against the front of modern battle tanks. Because tanks have the least armor protection on the top and rear deck and the tank presents a larger target when engaged from above, AT4s should fire down onto tanks.

5. Sniper Positions. Snipers contribute to combat in built-up areas by firing on selected enemy targets. Effective sniper employment can trouble the enemy far more than its relative personnel numbers reflect.

a. General areas (a building or group of buildings) are designated as sniper positions (Figure D-12), but the sniper selects the best position for engagement. Masonry buildings that offer the best protection, long-range fields of fire, and all-around observation are preferred. The sniper also selects several secondary and supplementary positions to cover his areas of responsibility.

b. Engagement priorities for snipers are determined by the relative importance of the targets to the effective operations of the enemy. Sniper targets may include tank commanders, crewmen of direct fire support weapons, crewmen of crew-served weapons, officers, FOs, and radiotelephone operators.

c. The urban environment often limits snipers to firing down or across streets. Snipers can be employed to cover rooftops, obstacles, dead space, and gaps in final protective fires.

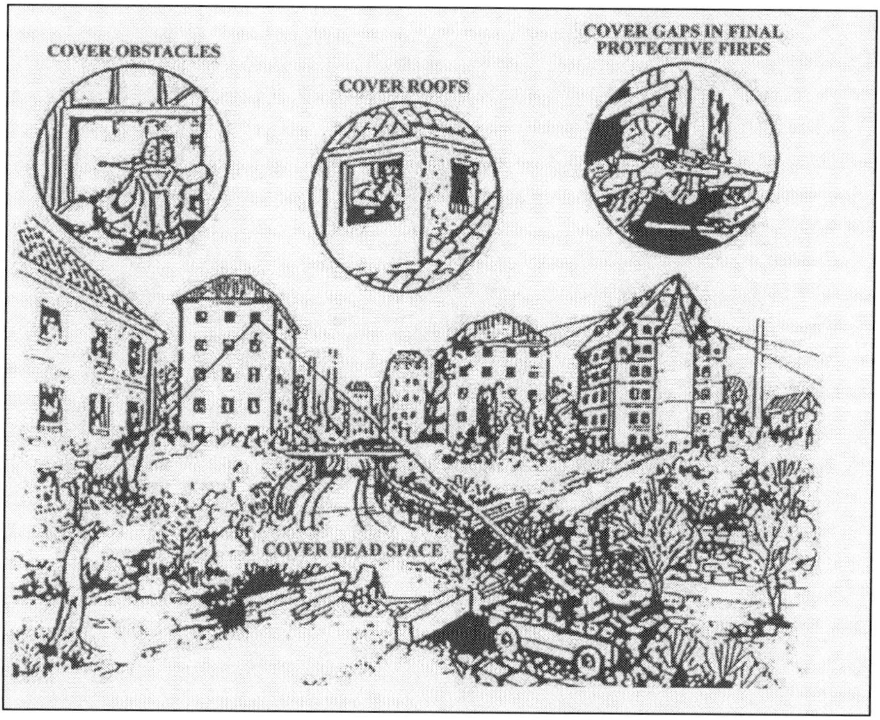

Figure D-12. Sniper Positions

Appendix E

Subterranean Operations

Knowledge of the nature and location of underground facilities is valuable to both the urban attacker and defender.

1. Tactical Value. Fighting in MOUT is multidimensional. Larger cities may have subterranean features that include sunken garages, underground passages, subway lines, utility tunnels, sewers, and storm drains (Figure E-1). Most of these features allow for the movement of units. Even in smaller European towns, sewers and storm drains permit Marines to move beneath the fighting and surface behind the enemy.

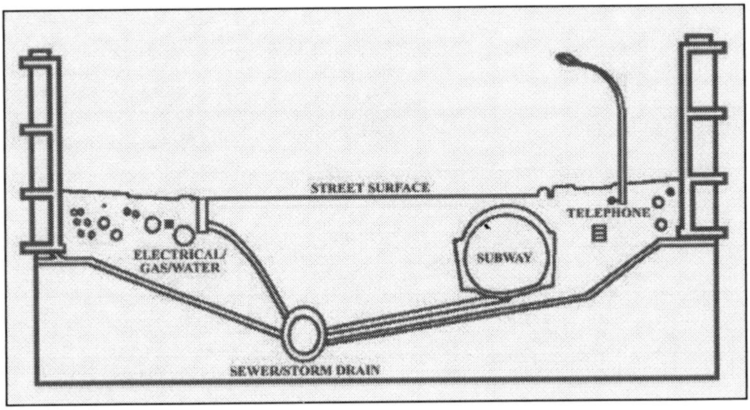

Figure E-1. Potential Underground Passages

a. Subterranean passages may provide the attacker with covered and concealed routes into and through built-up areas. This may allow an enemy to launch his attack along roads that lead into the city while infiltrating a smaller force into the defender's rear. An enemy unit in the defender's rear area may disrupt his defense and obstruct the avenues of withdrawal for his forward defense.

b. An attack along a subterranean avenue of approach could become the main effort. Even if the subterranean effort is not immediately successful, it forces the defender to fight on atleast two levels and to extend resources to more than just street-level fighting.

c. The existence of subterranean passages forces the defender to defend the built-up area above and below ground. While these passages are usually more of a disadvantage to the defender than to the attacker they may offer some advantages in the defense. When thoroughly reconnoitered and controlled by the defender, subterranean passages provide excellent covered and concealed routes to move reinforcements or to launch counterattacks. They also provide ready-made covered lines of communication for the movement of supplies and evacuation of casualties and provide places to cache supplies for forward companies. Subterranean passages also offer the defender a ready-made conduit for communications wire, with protection from armored vehicles and indirect fires.

2. Denial to the Enemy. Subterranean passages are useful to the defender only to the extent that the attacker can be denied their use. The defender has an advantage in that, given the confining, dark environment of these passages, a small group of determined Marines in a prepared position can defeat a numerically superior force.

a. Tunnels afford the attacker little cover and concealment except for the darkness and any manmade barriers. The passageways provide tight fields of fire and amplify the effects of grenades. Obstacles at intersections in tunnels support excellent ambush sites and may turn the subterranean passages into a maze. These obstacles can be created quickly by using chunks of rubble, furniture, and parts of abandoned vehicles interspersed with M18 claymore mines.

b. A thorough reconnaissance of the subterranean or sewer system is required. To be effective, obstacles must be located at critical intersections in the passage network so that they trap attackers in a kill zone but allow defenders freedom of movement (Figure E-2).

3. Subterranean Reconnaissance Techniques. The mission of local reconnaissance for a platoon- or company-sized area of responsibility should be given to a squad-sized element.

a. The squad leader tasks one fire team with providing forward security. One fire team follows in support, and one fire team provides rear security (Figure E-3). The squad leader moves directly behind the pointman and records data collected by the patrol. If available, an engineer should be with the lead fire team. The rear security fire team protects the ground- and subterranean-level point of entry. This fire team is responsible for detecting enemy movement on the patrol's rear and serves as the communications link between the squad leader and higher headquarters. Organization of the patrol may vary based on METT-T.

b. The squad leader should carry a map, compass, street plan, and notebook in which he has detailed the information that he must gather. One Marine is assigned to carry the tools needed to open manhole covers. If the patrol is to move more than 200 to 300 meters, or if the platoon leader directs, the squad may carry a sound-powered telephone (TA-1) and wire dispenser (XM-306A) for communications. Radios are often unreliable in this environment. Marines (especially the pointman, squad leader, engineer, and fire team leaders) should be equipped with night vision devices to maintain surveillance within the sewer. Thermal night devices are also useful as there may not be enough ambient light for accurate visibility with night vision devices.

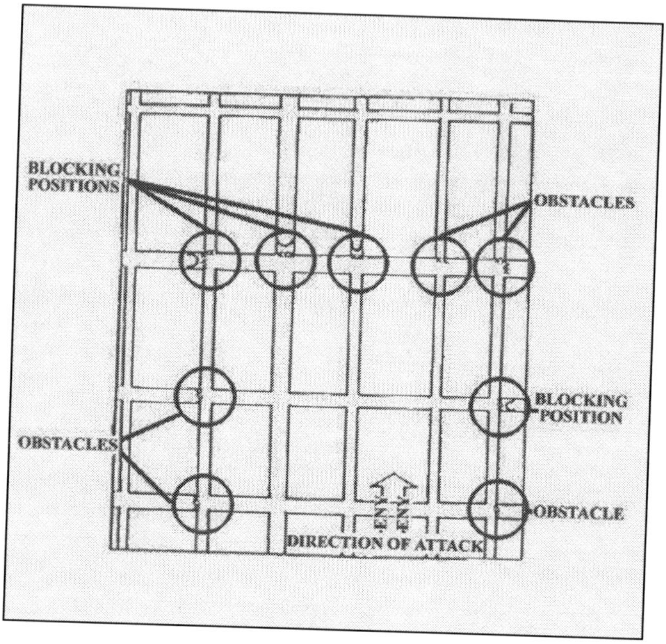

Figure E-2. Defense of a Sewer System

Figure E-3. Organization of the Patrol Unit

c. Marines entering the sewer should carry a sketch of the sewer system that includes magnetic north, azimuths, distances, and manholes. They should also carry protective masks, flashlights, gloves, and chalk (or chemiluminescent lights) for marking features along the route. The patrol should also be equipped with a safety rope to which each man is tied. To improve their footing in slippery sewers and storm drains, the squad members may wrap chicken wire or screen wire around their boots.

d. A constant concern to units conducting a subterranean patrol is chemical defense. Enemy chemical agents used in tunnels may be suddenly encountered in dense concentrations with little dispersion. The pointman should carry an automatic chemical agent alarm system to provide instant warning of the presence of chemical agents. M8 detector paper can also be used to detect chemical agents. At the first indication that harmful gases are present, the squad should mask.

e. In addition to enemy chemical agents, noxious gases from decomposing sewage can also pose a threat. These gases are not detected by chemical agent alarm systems, nor are they completely filtered by the protective mask. Physical signs that indicate their presence in harmful quantities are nausea and dizziness. The squad leader should be constantly alert to these signs and know the shortest route to the surface for fresh air. Some sewer gases deplete oxygen, a situation against which the field protective mask provides no protection.

f. Once the squad is organized and equipped, it moves to the entrance of the tunnel. This may be a manhole. With the manhole cover removed, the squad waits 15 minutes before entry to allow any gases that may be present to dissipate. The pointman then descends into the tunnel to determine whether the air is safe to breathe and whether movement is restricted. The pointman should remain in the tunnel for 10 minutes before the rest of the squad follows. If he becomes ill or is exposed to danger, he can be pulled out by the safety rope.

g. When the squad is moving through the tunnel, the pointman moves about 10 meters in front of the squad leader. Other squad members maintain 5-meter intervals. If water in the tunnel is flowing faster than 2.5 meters per second or the sewer contains slippery obstacles, those intervals should be increased to prevent all patrol members from falling if one man slips. All squad members should stay tied into the safety rope so that they can easily be retrieved from danger. The last Marine in the patrol marks the route with chalk or chemiluminescent lights so other units can find the squad.

h. The squad leader should record the azimuth and pace count of each turn he takes in the tunnel. When he encounters a manhole to the surface, the pointman should open it and determine the location, which the patrol leader then records. The use of recognition signals (Figure E-4) prevents friendly forces from accidentally shooting the pointman as he appears at a manhole.

Figure E-4. Recognition Signals

i. Once the patrol has returned and submitted its report, the platoon leader must decide how, or whether, to use the tunnel. In the offense, a tunnel might provide a covered route to move behind the enemy's defenses. In the defense, a tunnel might provide a covered passage between positions. In any case, the squad members that provided the reconnaissance should act as guides along the route.

j. If the tunnel is to be blocked, the platoon may emplace concertina wire, early warning devices, and antipersonnel mines. A security force should be established at the entrance of the sewer (Figure E-5) to provide warning against any enemy trying to approach the platoon's defense. It should be equipped with command-detonated illumination. The overpressure from grenades, mines, and boobytraps exploding in a sewer or tunnel can have adverse effects on friendly forces, including ruptured eardrums and wounds from flying debris. Also, gases found in sewers can be ignited by the blast effects of these munitions. For these reasons, small-arms weapons should be employed in tunnels and sewers. Friendly personnel should be outside of tunnels or out of range of the effects when mines or demolitions are detonated.

4. Psychological Considerations. Combat operations in subterranean passages are similar to night combat operations. The psychological factors of night operations reduce confidence, cause fear, and increase a feeling of isolation. This feeling of isolation is further magnified by the tight confines of the tunnels. The layout of tunnels could require greater dispersion between positions.

Figure E-5. Security Position Established at the Entrance to a Sewer

a. Leaders must enforce measures to help dispel the feelings of fear and isolation experienced when operating in tunnels. These measures include leadership training, physical and mental fitness, sleep discipline, and stress management.

b. Leaders maintain communication with Marines manning positions in tunnels either through personal visits, messengers, or field telephone. Communication increases situational awareness and provides insight into the mental state of Marines. Marines manning positions below ground should be given as much information as possible on the organization of the tunnels and the importance of the mission. They should be briefed on contingency plans and alternate positions should their primary positions become untenable.

c. Physical and mental fitness can be maintained by periodically rotating Marines out of tunnels so they can stand and walk in fresh air and sunlight. Stress management is also a factor of operations in tunnels. Historically, combat in built-up areas has been one of the most stressful forms of combat. Fighting in continuous darkness and restricted maneuver space causes even more stress to Marines than street fighting.

Appendix F

Obstacles, Mines, and Demolitions

Obstacles are used extensively in combat in built-up areas to allow the defender to canalize the enemy, impede his movement, and disrupt his attack.

Section I

Obstacles

1. Introduction. Obstacles are designed to prevent movement by personnel, to separate infantry from tanks, and to slow or stop vehicles. (See MCRP 3-17A, *MAGTF Breaching Operations.*)

2. Types. Antipersonnel mines, barbed wire, boobytraps, and exploding flame devices are used to construct antipersonnel obstacles (Figure F-1). (See FMFRP 13-5/FM 5-34, *Engineer Field Data*; FM 5-250, *Explosives and Demolitions*; and FM 3-11, *Flame, Riot Control Agent, and Herbicide Operations,* for more detailed information.) These obstacles are used to block the following infantry approaches:

- Streets
- Buildings
- Roofs
- Open spaces
- Dead space
- Underground systems.

a. The approval authority to boobytrap buildings is normally the GCE commander. This authority may be delegated to subordinate commanders.

b. The three types of wire obstacles used in defensive operations are protective, tactical, and supplementary. Wire obstacles should be integrated with other obstacles and covered by fire and observation.

(1) Protective wire is usually located beyond hand-grenade range (40 to 100 meters) from the defensive position.

(2) Tactical wire is positioned to increase the effectiveness of friendly weapons fire. Tactical wire is usually positioned on the friendly side of machine gun FPLs.

(3) Supplementary wire is used to break up the pattern of tactical and protective wire to prevent the enemy from easily locating friendly weapons.

c. Obstacles may be placed in dead spaces to restrict infantry movement in areas that cannot be observed and that are protected from direct fires.

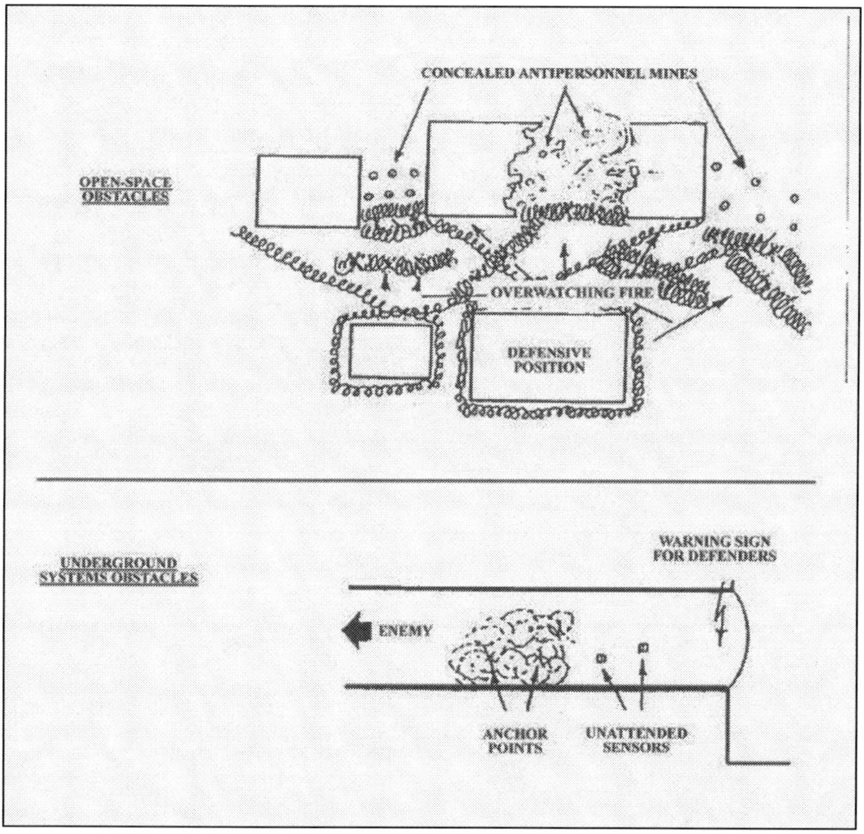

Figure F-1. Antipersonnel Obstacles

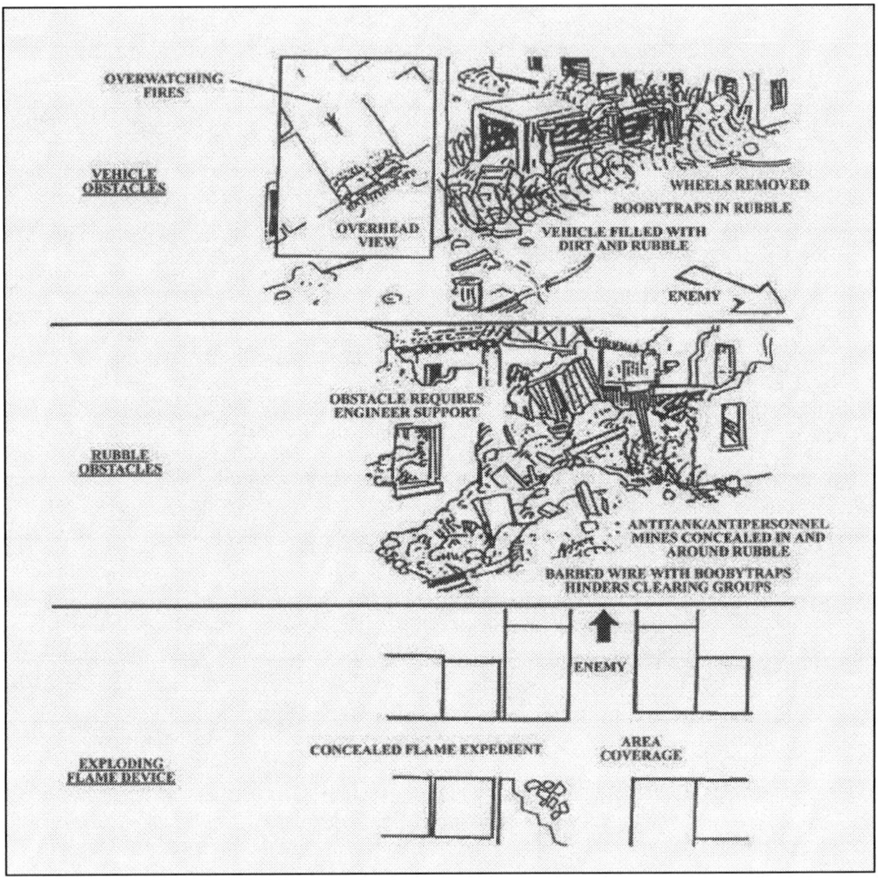

Figure F-1. Antipersonnel Obstacles (Continued)

d. Antiarmor obstacles are mainly restricted to streets and other armored avenues of approach (Figure F-2).

e. Some antitank mines may be sited and employed as antihelicopter mines to cover probable HLZs (Figure F-2).

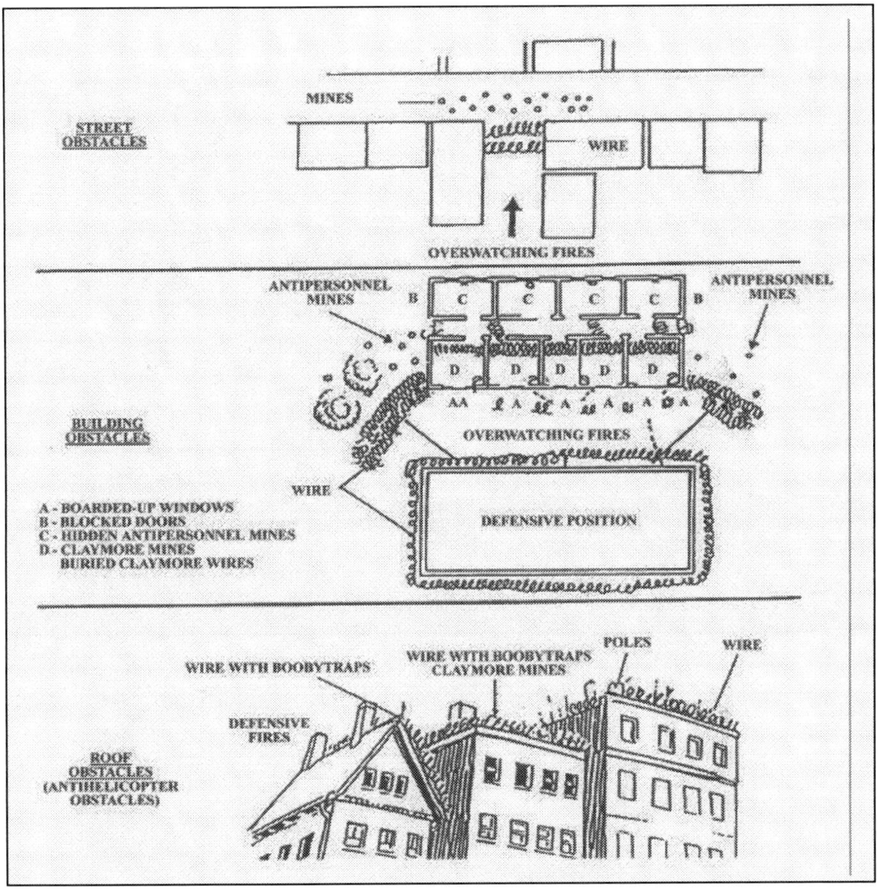

Figure F-2. Antiarmor Obstacles

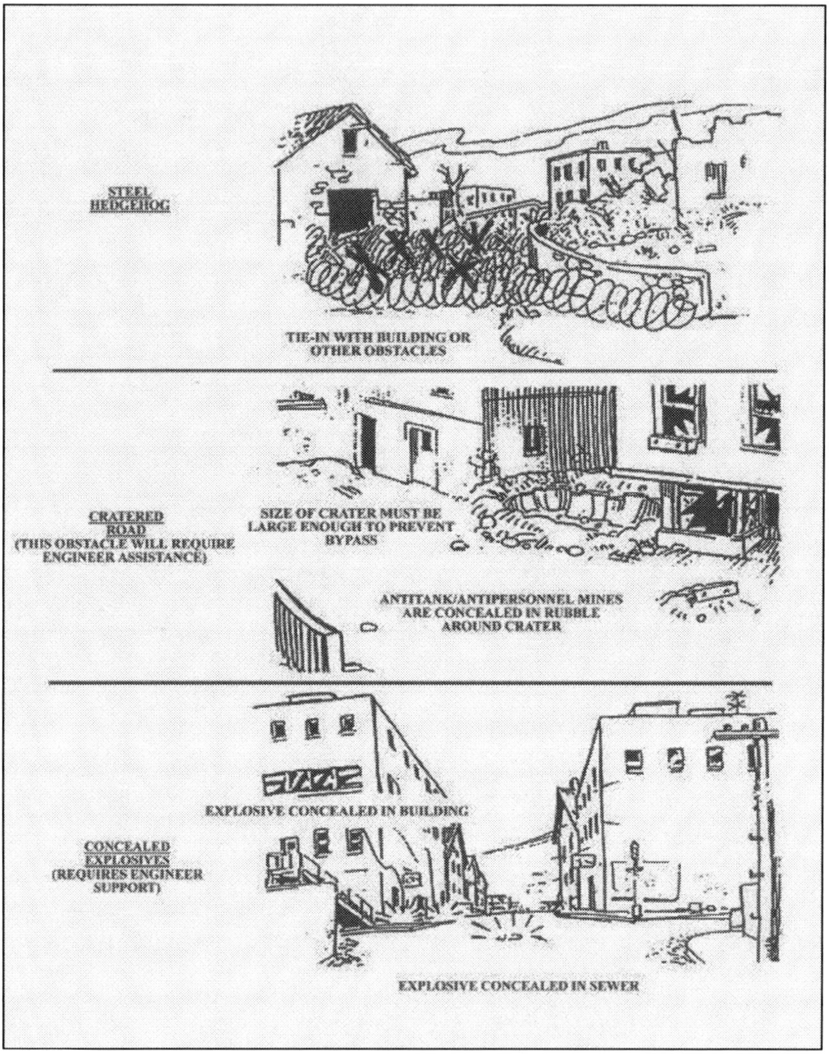

Figure F-2. Antiarmor Obstacles (Continued)

3. Construction of Obstacles. Obstacles are constructed in buildings to deny enemy infantry covered routes and weapons positions that are close to friendly defensive positions. They can be constructed of explosives, by using wire, or by using boobytraps within buildings. The building may be prepared as an explosive or flame trap for command detonation after enemy occupation.

Section II

Mines

4. Introduction. The employment or discovery of mines in built-up areas should be recorded on a building sketch (Figure F-3). (See FM 20-32.) The sketch should include the number of the building (taken from a city map, if available) and all floor plans. It should also include the type of mine and firing device. When possible, mined buildings should be marked on the friendly side (Figure F-4). Clearing areas or buildings that have been mined is extremely difficult and dangerous. Therefore, such areas should be considered "no go" areas for infantry. This factor must be carefully considered when planning and authorizing the placement of mines. (See Table F-1 for notional approving authority for minefields, but with the understanding that information in that table is subject to change with international agreements and ROE.)

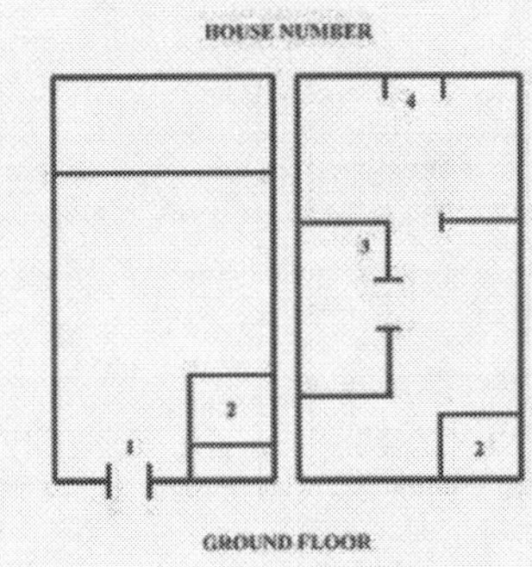

MINE	LOCATION	TYPE	FIRING DEVICE
1	FRONT DOOR	CLAYMORE	COMMAND
2	STAIRS	CLAYMORE	COMMAND
3	CLOSET	CLAYMORE	COMMAND
4	DRESSER	CLAYMORE	COMMAND

Figure F-3. Building Sketch Showing Mines

Figure F-4. Marking Mined Buildings

	Minefields (Type)		FASCAM (Duration)		Other Obstacles	
Level	Tactical	Protective	Long	Short	Tactical	Protective
MEF	X	X	X	X	X	X
Division	X	X	X	X	X	X
Regiment	X	X	X	X	X	X
Battalion	X	X		X	X	X
Company		X				X

Table F-1. Notional Minefield Employment Authority Matrix

5. Types. Several types of mines may be encountered or employed in built-up areas.

a. The M14 mine (a plastic type of antipersonnel landmine) cannot currently be used by American forces due to the Leahy Amendment of 1995 and the Certain Conventional Weapons Convention, of which the United States is a signatory nation. It may, however, be encountered in MOUT or be employed by coalition or enemy forces. For this reason, the following information is given as a guide for what U.S. forces may encounter. This type of antipersonnel landmine is normally used in conjunction with metallic antipersonnel, antitank, or chemical mines to confuse and hinder enemy breaching attempts. When encountered, it must be carefully handled because its light weight makes it easy to displace (Figure F-5). However, its size makes it ideal for covering obscure places such as stairs and cellars.

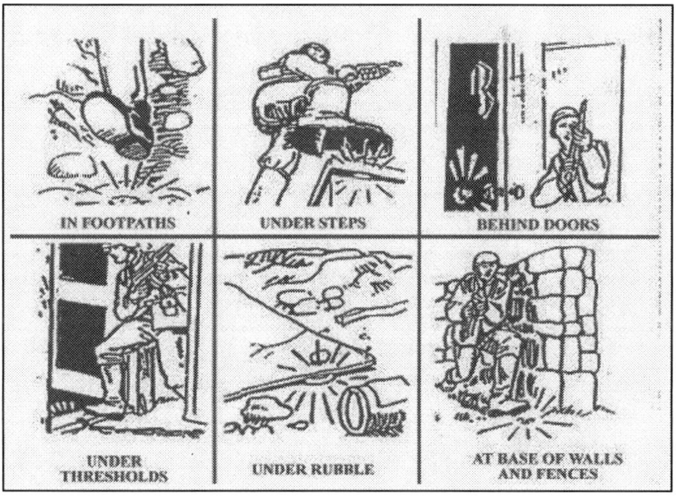

Figure F-5. Typical Employment of the M14 Antipersonnel Mine

b. The M16A2 antipersonnel landmine is subject to the same legal limitations as the M14. It is a boundary type of antipersonnel landmine, which makes it ideal for covering large areas such as rooftops, backyards, parks, and cellars. It can be expediently rigged for command detonation by attaching a rope or piece of communications wire to the pull ring (Figure F-6).

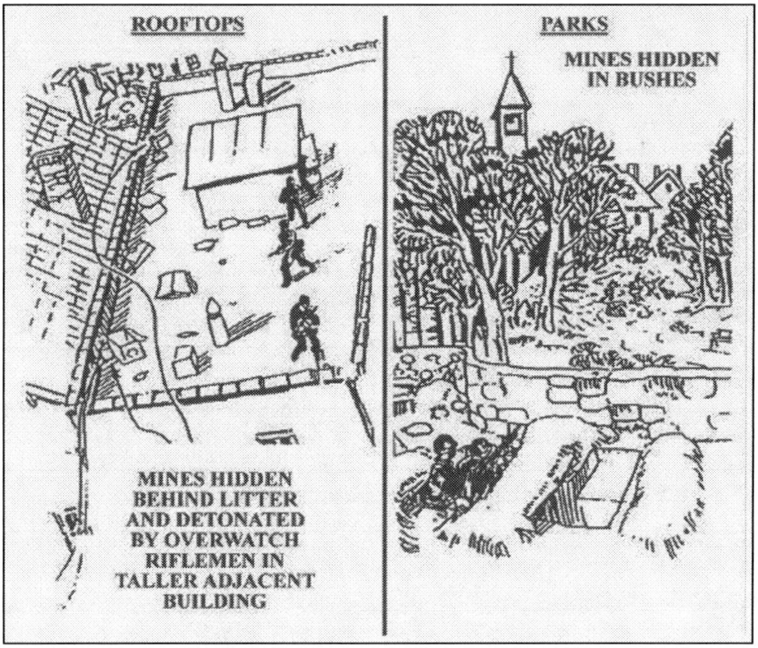

Figure F-6. Emplacements of M16A2 Antipersonnel Mine

c. The M18A1 claymore mine can be employed during the reorganization and consolidation phase on likely enemy avenues of approach. It does not have to be installed in the street, but can be employed on the sides of buildings or any other sturdy structure. The claymore mine, in command-detonated mode, is currently legal for employment by U.S. forces.

(1) Claymore mines can also be used for demolition against thin-skinned buildings and walls.

(2) Claymore mines may be configured for detonation by tripwire. They can help fill the dead space in the final protective fires of automatic weapons (Figure F-7).

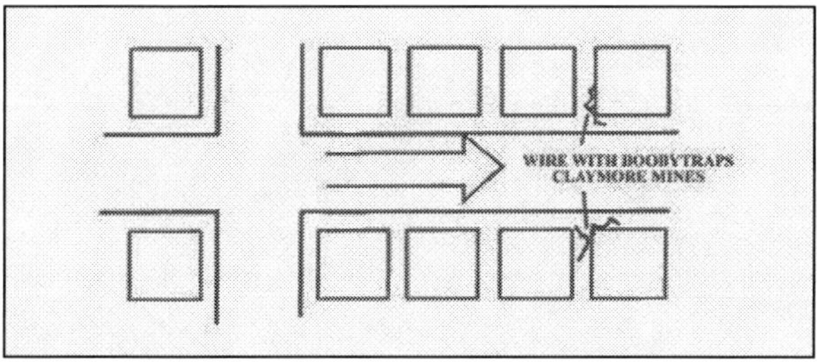

Figure F-7. Claymore Mine Used To Cover Dead Space of Automatic Weapons

(3) Claymore mines can be used in several ways in the offense. For example, if friendly forces are advancing on a city, claymore mines can be used in conjunction with blocking positions to cut off enemy avenues of escape (Figure F-8).

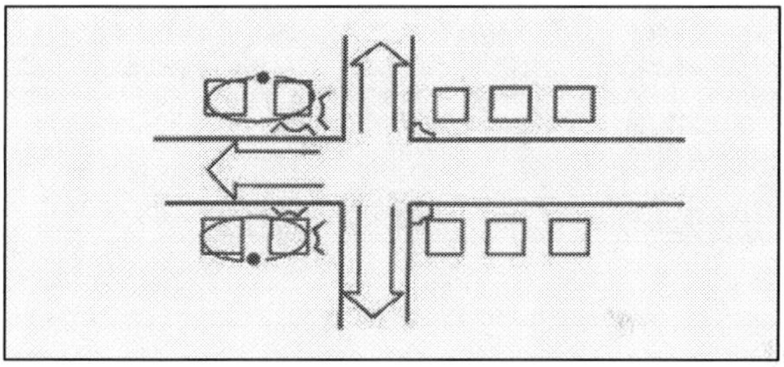

Figure F-8. Claymore Mines Used To Block Enemy Avenues of Escape

d. The M15, M19, and M21 antitank mines are used to stop or turn enemy armor and should be employed (Figure F-9):

- In conjunction with other manmade obstacles and covered by observation and fire
- In streets or alleys to block routes of advance in narrow defiles
- As command-detonated mines with other demolitions.

6. Enemy Mines and Boobytraps. Buildings contain many areas and items that are potential hiding places for boobytraps, for example, doors, windows, telephones, stairs, books, canteens, and so on.

Figure F-9. Emplacement of Antitank Mines

When moving through a building, Marines must not pick up anything—food, valuables, enemy weapons, or other items. Such items could be rigged with pressure devices that explode when they are moved. Marines must be well-dispersed so that if a boobytrap explodes, the number of casualties will be few. Many different types of mines and boobytraps could be encountered during combat in built-up areas (Figure F-10).

a. The equipment commonly used to detect mines in clearing operations includes:

- Mine detectors
- Probes
- Grappling hooks
- Ropes
- Explosives
- Flak vests
- Eye protection
- Engineer tape.

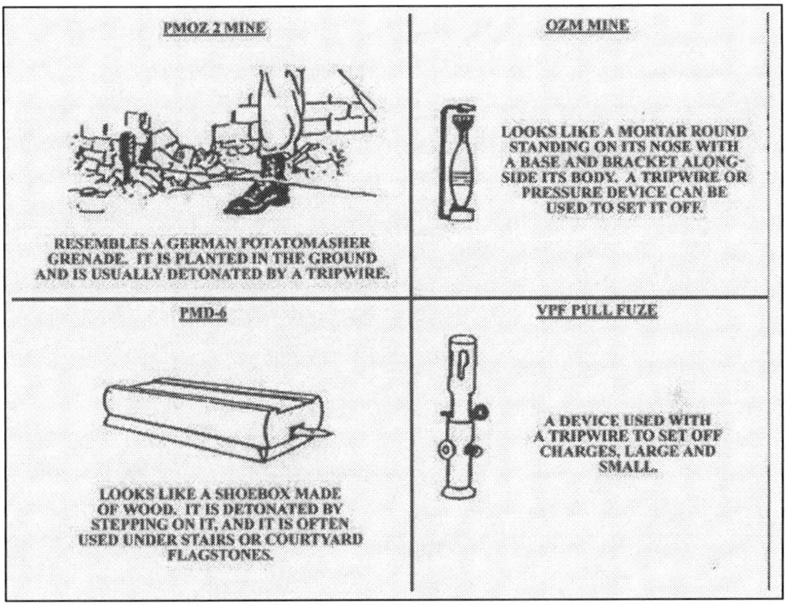

Figure F-10. Common Threat Antipersonnel Mines and Boobytraps

b. To detect tripwires, a Marine can use a 10-foot pole with 5 feet of string tied on one end. A small weight attached to the loose end of the string will snag on the tripwire. This allows the lead Marine to easily detect a tripwire (Figure F-11). Marines can also use a flashlight to cause reflections and shadows to find tripwires. A wand with a dangling thread can be swept over the ground ahead with the thread being watched for any disturbance by a tripwire.

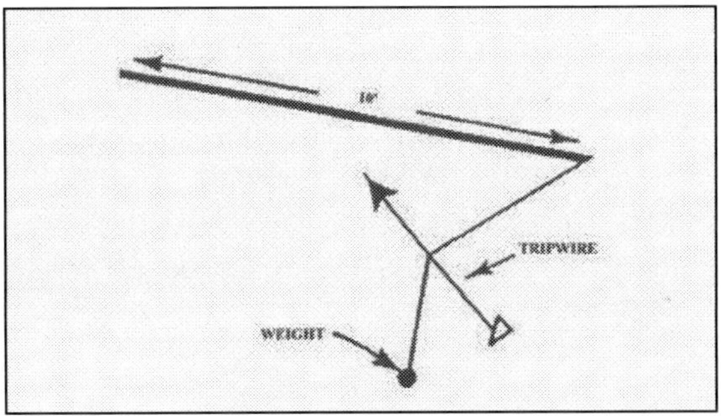

Figure F-11. Tripwire Detection

c. Many standard antipersonnel mines are packed in boxes and crates. If a Marine discovers explosive storage boxes, he should sketch them and turn the sketch over to the platoon commander for forwarding up the chain of command.

d. Most boobytraps should be neutralized by explosive ordnance disposal (EOD) personnel. If EOD teams are not available, boobytraps may be blown in place. Personnel should be protected by adequate cover. If the boobytrap is in a building, all personnel should go outside before the boobytrap is destroyed. Engineer tape placed around the danger area is one method of marking boobytraps. If tape is not available, strips ripped from bedsheets can be used. If possible, a guide should lead personnel through boobytrapped areas. EPWs and civilians may be a good source of information on where and how boobytraps are employed.

Section III

Demolitions

7. Introduction. Demolitions are used more often during combat in built-up areas than during operations in open natural terrain. Demolition operations should be directed by the engineers that support the GCE. However, if engineers are involved in the preparation and execution of the barrier plan, other trained Marines can prepare mouseholes, breach walls, and rubble buildings themselves if assisted and advised by attached engineers.

8. Offensive Use. When assaulting or clearing a built-up area, demolitions may enable the GCE commander to create an avenue of approach through buildings.

a. Every other man in an assault force should carry demolitions, and other selected personnel should carry blasting caps. The same man should *not* carry both the explosives and the blasting caps because of the danger of sympathetic detonation. As the demolitions are expended by the assault force, they should be replaced by explosives carried by the support force.

b. One of the most difficult breaching operations faced by the assault force is the breaching of masonry and reinforced concrete walls. When demolitions must be used, composition C4 is the ideal explosive to use. Normally, building walls are 15 inches thick or less. Assuming that all outer walls are constructed of reinforced concrete, a rule of thumb for breaching is to place 10 pounds of C4 against the target between waist and chest height. When detonated, this normally blows a hole large enough for a man to go through. The amounts of TNT and C4 required to breach concrete are shown in Table F-2.

c. Mouseholes provide the safest method of moving between rooms and floors. They can be created with C4. Because C4 comes packaged with an adhesive backing or can be emplaced using pressure-sensitive tape, it is ideal for this purpose. When using C4 to blow a mousehole in a lath and plaster wall, one block or a strip of blocks should be placed on the wall from neck to knee height. Charges should be primed with detonating cord or electrical blasting caps to obtain simultaneous detonation, which will blow a hole large enough for a man to fit through.

9. Defensive Use. The use of demolitions in defensive operations is similar to their use in offensive operations. When defending a built-up area, demolitions are used to create covered and concealed routes through walls and buildings that can be used for withdrawals, reinforcements, or counterattacks. Demolitions are also used to create obstacles and clear fields of fire.

a. Marines can use demolitions for creating mouseholes and constructing command-detonated mines. Expedient C4 satchel charges can be concealed in firing positions and along movement routes for detonation in the vicinity of the enemy.

Thickness of Material	TNT	C4	Size of Opening
	Reinforced Concrete		
Up to 10 cm (4 inches)	.792 lbs	.591 lbs	10 to 15 cm (4 to 6 inches)
10 to 20 cm (4 to 8 inches)	3.4 lbs	2.54 lbs	15 to 30 cm (6 to 12 inches)
	Nonreinforced Concrete Masonry		
Thickness of Material	TNT	C4	Size of Opening
Up to 35 cm (14 inches)	3.66 lbs	2.73 lbs	35 cm (14 inches)
35 to 45 cm (14 to 18 inches)	7.5 lbs	5.6 lbs	45 cm (18 inches)
45 to 50 cm (18 to 20 inches)	10.2 lbs	7.61 lbs	50 cm (20 inches)

Table F-2. Amounts of TNT and C4 Required To Breach Concrete

b. The engineers should furnish technical assistance for selective rubbling. Normally, buildings can be rubbled by using shaped charges or C4 on the supports and major beams.

c. Charges should be placed directly against the surface to be breached unless a shaped charge is used. Whenever possible, demolitions should be tamped to increase their effectiveness. Tamping materials could be sandbags, rubble, or desks and chairs (Figure F-12).

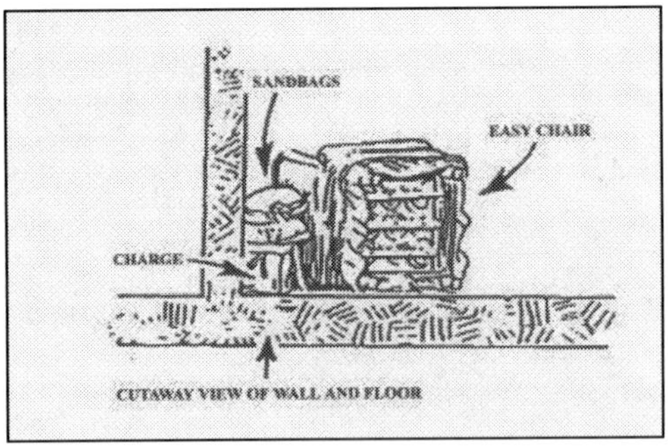

Figure F-12. Chair Used To Tamp Breaching Charge

d. For exterior walls, tamping of breaching charges may be impossible because of enemy fire. Thus, the untamped charge normally requires twice the explosive charge to produce the same effect as a tamped charge (Figure F-13).

METHODS OF PLACEMENT				
THICKNESS OF CONCRETE				
FEET	POUNDS OF TNT	POUNDS OF C4	POUNDS OF TNT	POUNDS OF C4
2	14	11	28	21
2 1/2	27	21	54	41
3	39	30	78	59
3 1/2	62	47	124	93
4	93	70	185	138
4 1/2	132	99	263	196
5	147	106	284	211
5 1/2	189	141	376	282
6	245	186	490	366

Figure F-13. Breaching Reinforced Concrete

e. When enemy fire prevents an approach to the wall, the breaching charge can be attached to a pole and slid into position for untamped detonation at the base of the wall (Figure F-14).

Figure F-14. Charge Placement When Small-Arms Fire Cannot Be Suppressed

f. The internal walls of most buildings function as partitions rather than as load-bearing walls. Therefore, smaller explosive charges can be used to breach them. In the absence of C4 or other military explosives, internal walls can be breached by using a claymore mine (Figure F-15). The claymore should be tamped to increase its effectiveness and to reduce the amount of explosive force directed to the rear.

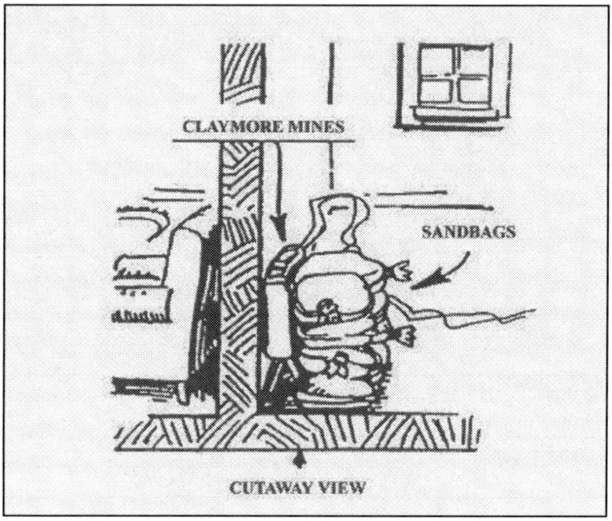

Figure F-15. Tamping a Claymore Mine To Breach Internal Walls

g. The Molotov cocktail (Figure F-16) is an expedient device for disabling both wheeled and tracked vehicles. It is easy to make because the materials are readily available. The results are most effective because of the close engagement in built-up areas. The objective is to ignite a flammable portion of the vehicle such as the fuel or ammunition that it is transporting. The following materials are needed to make a Molotov cocktail:

- Bottle or other glass container
- Gas (60 percent)
- Oil (40 percent)
- Rag for use as a wick.

The gas and oil are mixed thoroughly (60 percent gas to 40 percent oil). The rag is soaked with the mixture, then the mixture is placed into the bottle. The rag is then inserted in the opening of the bottle as a wick. When a target is sighted, the wick is lit and the bottle is thrown hard enough to break.

Figure F-16. Molotov Cocktail

WARNING

Ensure that a safe distance is maintained when throwing a Molotov cocktail.

h. The bunker bomb (Figure F-17) is an expedient, explosive flame weapon that is best used against fortified positions or rooms. This expedient munition should be used with a nonelectrical firing system. The following materials are required to make a bunker bomb:

- 1 metal small-arms ammunition container
- 1 gallon of gasoline
- 50 feet of detonating cord
- 1 nonelectrical blasting cap
- 1 M60 fuze igniter
- 7 1/2 feet of M700 time fuze
- 1 M49 trip flare or M34 WP grenade.

(1) Step 1. Fill the ammunition can 3/4 full with fuel and secure the lid.

(2) Step 2. "Hasty whip" the device with 15 turns around the center of the container using 44 feet of detonating cord. Leave 2-foot "pigtails" for attaching the igniter and fuze igniter.

(3) Step 3. Tape the igniter (M49 trip flare or M34 WP grenade) to the container handle.

(4) Step 4. Place one detonating cord pigtail end under the igniter spoon handle. Tape it in place.

(5) Step 5. Attach the M60 fuze igniter and the nonelectrical blasting cap to the M700 time fuze.

(6) Step 6. Attach the nonelectrical firing system to the other pigtail by making a loop in the detonating cord and attaching the blasting cap to it.

(7) Step 7. Remove the safety pin from the igniter (M49 trip flare or M34 WP grenade). The device is ready to be fired.

WARNING

Never carry the device by the handle or igniter. Remove the igniter safety pin only when it is time to use the device. Use extreme care when handling or carrying nonelectrical firing systems. Protect blasting caps from shock and extreme heat. Do not allow the time fuze to kink or become knotted. Doing so may disrupt the powder train and may cause a misfire. Prime detonating cord and remove the time fuze igniter safety pin only when it is time to use the device.

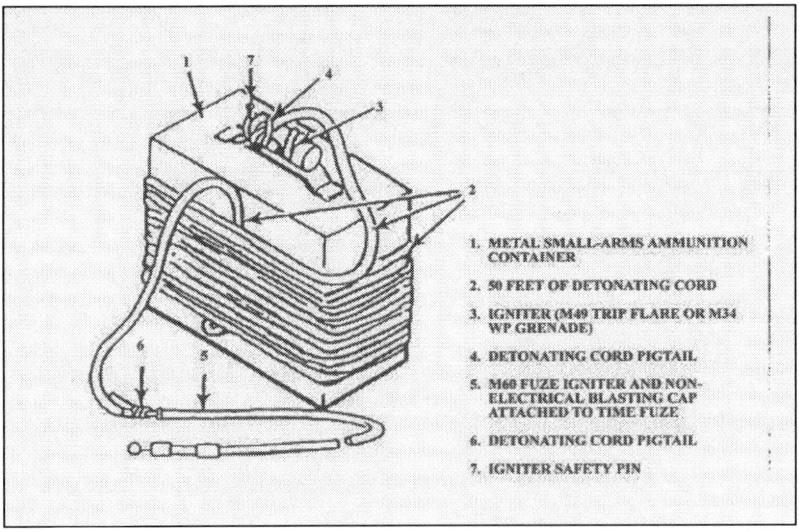

Figure F-17. Bunker Bomb Made From an Ammunition Can

10. Safety. The greatest danger to friendly personnel from demolitions is the debris thrown by the explosion. Leaders must ensure that protective measures are enforced. The safe distance listed in Table F-3 indicates the danger of demolition effects.

Pounds of Explosive	Safe Distance in Meters[1]	Pounds of Explosive	Safe Distance in Meters[1]
1 to 29	300	150	534
30	311	175	560
35	327	200	585
40	342	225	609
45	356	250	630
50	369	275	651
60	392	300	670
70	413	325	688
80	431	350	705
90	449	375	722
100	465	400	737
125	500	425	750

[1]These distances will be modified in combat when troops are in other buildings, around corners, or behind intervening walls.

Table F-3. Minimum Safe Distances for Personnel in the Open

a. The following are the guidelines for using demolitions:

- Keep the blasting machine under the control of a noncommissioned officer.
- Wear helmets and flack jackets at all times while firing explosives.
- Handle misfires with extreme care.
- Clear the room and protect personnel when blowing interior walls.

b. Some charges should be prepared beforehand, without detonators, to save time. Examples include 10- or 20-pound breaching charges of C4 and expedient-shaped charges in No. 10 cans.

- Use C4 to breach hard targets (masonry construction).

- Do not take chances.
- Do not divide responsibility for demolition work.
- Do not mix demolitions and detonators until ready to use.
- Do not carry demolitions and caps together.

Appendix G

Nuclear, Biological, and Chemical Considerations

Current U.S. policy regarding lethal or incapacitating agents is that their use against an armed enemy requires approval at the NCA level. Potential enemies may not operate under the same restrictions. Field commanders must be prepared to assume an adequate NBC defensive posture when engaged in urban fighting.

1. Protection From Nuclear, Biological, and Chemical Weapons. Generally, the lowest floor or basement of a reinforced concrete or steel-formed building offers good protection from nuclear hazards and initial liquid chemical contamination although some chemical agents tend to collect in lower areas. Tanks and LAVs/AAVs also provide protection. (See FMFM 11/MCWP 3-37 series manuals for greater detail.)

a. Biological attacks are difficult to detect and recognize. Biological agents can be disseminated by using aerosols, vectors, and covert methods. Because biological agents can be sprayed or dropped in bomblets, personnel who observe such indicators should promptly report them. Prompt reporting and treatment of the sick speed the employment of medical countermeasures. Although buildings and shelters provide some protection against spraying, they provide little protection against biological agents.

b. Chemical agents cause casualties by being inhaled or absorbed through the skin. They may afford Marines a few seconds to mask. Buildings have a channeling effect and tend to contain the effects of an agent, causing great variations in chemical concentration from room to room or from building to building. Chemical agents usually settle in low places, making sewers and subways hazardous hiding places. A prepared defender should include some collective protective measures in the defensive network. Personnel using fans may be able to put enough overpressure into tunnels to keep some chemical agents from entering. The protective mask and battle dress overgarment provide the best protection against chemical agents.

c. Personal hygiene is a critical defensive measure against infection and disease. Unfortunately, built-up areas are characterized by sophisticated sanitation systems. When those systems are destroyed, the resulting sanitary conditions become much worse than those in areas where sanitation facilities do not exist. Commanders must ensure that personnel employ sanitation measures and that their immunizations are current.

d. Commanders should plan their mission-oriented protective posture (MOPP) with the realization that increased logistics demands in built-up areas also apply to NBC equipment. Protective clothing, detection and decontamination equipment, and sealed containers of food and water must be stockpiled the same as other critical supplies. When operating in protective clothing, commanders must make allowances for the strenuous activities normally associated with combat in built-up areas.

(1) Detection. After an NBC attack, battalions should dispatch their detection and survey teams. Detection in built-up areas is complicated by the numerous rooms within buildings. Teams should conduct tests and surveys of major streets, intersections, and buildings in their area for inclusion in initial NBC reports. A systematic survey of all buildings, rooms, and underground facilities must be accomplished before occupation by unmasked personnel. All data should be forwarded using the appropriate NBC report.

(2) Decontamination. Personnel begin decontamination operations as soon after an NBC attack as the mission allows. Personnel should conduct individual decontamination of themselves and their personal equipment. Unit commanders determine the need for MOPP gear exchange and the requirements for a hasty or deliberate decontamination operation.

(a) Nuclear. Personnel should wear wet-weather gear for certain decontamination operations (hosing down buildings) to prevent radioactive material from touching the skin.

(b) Chemical and Biological. Roads, sidewalks, and other hard surfaces are best decontaminated by weathering, if time permits. Agents can also be covered with several inches of dirt or sand to provide protection. Fragment testing should be conducted periodically to ensure that the agent has not seeped through the covering. For critical sections of roads, a truck-mounted M12A1 (power-driven decontamination apparatus (PDDA)) can be used to spray supertropical bleach (STB) slurry; this aids rapid decontamination. Buildings are difficult to decontaminate, especially wooden ones. Some techniques for their decontamination are scrubbing with STB slurry; washing with hot, soapy water; washing or spraying with a soda solution; and airing.

2. Smoke Operations. The use of smoke is an integral part of either offensive or defensive operations in built-up areas. In the offense, smoke can support the maneuver of combat units and deception operations. Smoke employed in the defense obscures enemy air and ground observation, limiting the accuracy of enemy fires and intelligence collection.

a. Smoke should not be used when it degrades the effectiveness of friendly forces. Likewise, an extremely dense concentration of smoke in a closed area displaces the oxygen, suffocating Marines even when they are wearing protective masks.

b. Smoke pots, generators, or artillery smoke munitions should be used to cover the withdrawal of defending forces or the movement of attacking forces. Artillery-delivered WP can be effective against enemy forces by screening, in addition to causing casualties and fires. Consideration should be given to the incendiary effects of WP, smoke (HC and M825), and illumination munitions on the litter and debris of built-up areas.

c. Smoke grenades can be massed to provide a hasty screen for concealing personnel moving across streets and alleys. Smoke grenades can also be used for signaling; those launched by an M203 can be used to mark targets for fixed and rotary-wing air support.

d. The use of smoke in built-up areas is affected by the complex wind patterns caused by buildings. When covering a built-up area with a smoke haze or blanket, personnel must include all buildings. Failure to obscure tall buildings, towers, and steeples provides enemy observers with reference points for fire placement within the built-up area.

3. Riot Control Agents. When authorized, riot control agents such as CS and CN can be used to drive enemy forces from positions or to deny them areas for occupation. Riot control agents are incapacitating but generally have no lasting effects. They may be appropriate when preventing civilian casualties is a planning consideration. However, riot control agents may not be effective against an enemy that is well trained in chemical defense.

Appendix H

MOUT Under Limited-Visibility Conditions

To be successful, leaders must use limited-visibility conditions to their advantage.

1. Advantages. When fighting in built-up areas during night or periods of limited visibility, attacking or defending forces have several advantages.

a. In many cases, Marines may have a technological advantage in thermal imagery and light intensification over their opponents. This enables Marines to identify, engage, and destroy enemy targets before being detected by the enemy.

b. Possessing the capability to conduct operations as well at night or during periods of reduced visibility as during daylight gives Marines a tremendous advantage over an enemy with reduced capability. It also aids the generation of tempo and allows Marines to gain or retain the initiative.

c. Generally, ranges of direct-fire engagements are greatly reduced in the MOUT environment. During periods of limited visibility, effective unaided target acquisition ranges are further reduced. This enables attacking forces to close to shorter ranges, thus increasing the lethality and accuracy of weapons. Attacking forces can also take advantage of the enemy's reduced ability to see and can more effectively engage the enemy before being detected with thermal imagery or light intensification devices.

d. Helicopterborne assaults are best conducted during periods of limited visibility because the enemy's air defenses are degraded.

e. Attacking during periods of limited visibility gives the attacker a greater chance of achieving surprise.

2. Disadvantages. When fighting in built-up areas during periods of limited visibility, attacking and defending forces also face some disadvantages.

a. Command and control is difficult in any operation in a built-up area; periods of limited visibility increase this difficulty.

b. Marines have an instinctive tendency to move close together during periods of limited visibility. Constant attention must be given to prevent Marines from "bunching up."

c. Marines may become disoriented easily because of the combined effects of low visibility and the characteristics of built-up areas.

d. Target identification becomes more difficult in limited-visibility conditions. Depending on the training of the individual, the Marine may fire at anything seen or may hesitate too long before firing. Improper target identification is one of the leading causes of fratricide, so leaders must pay close attention to individual target engagement.

3. Fratricide Avoidance. The risk of fratricide is much greater during periods of limited visibility. The key to avoiding fratricide is increased situational awareness by leaders and individuals, coupled with realistic training in target identification. (See *Fratricide: Reducing Self-Inflicted Losses*, Newsletter, No. 92-4, April 92, Center for Army Lessons Learned, U.S. Army Combined Arms Command, Fort Leavenworth, Kansas.) Other considerations include:

a. Graphic control measures should be clearly defined and obvious. Examples include distinctive buildings, large boulevards, rivers, and so forth.

b. Leaders must exercise firm control when engaging targets. Movements should also be coordinated and controlled.

c. Cleared rooms and buildings should be distinctly marked to identify cleared areas and friendly forces to base-of-fire elements supporting the maneuver.

d. Visible markers (for example, glint tape or thermal strips) should be attached to individual Marines for rapid identification.

e. Far and near recognition symbols should be coordinated and used properly.

f. Units employing close air support must exercise firm control. Failure to do so may lead to the pilot becoming disoriented and engaging friend and foe alike.

4. Urban Environmental Effects on Night Vision Devices. Built-up areas affect standard night vision devices and sights differently than do open areas. This may cause some confusion because the images Marines receive through their night vision devices are unusual compared to those they may be used to.

a. Most built-up areas have electric power; therefore, street lights and/or building lights may "white out" some light intensification devices unless the power is disrupted or lights are turned off.

b. Fires may be burning in the urban area. Open flames cause problems not only for light intensification devices, but also for thermal devices.

c. Subterranean areas and the interiors of buildings will not have ambient light if the power is off. Passive night vision devices must then use an artificial light source, such as infrared radiation, to provide enough ambient light for the devices to work effectively.

d. The many reflective surfaces found in built-up areas may cause false images, particularly for laser range finders and laser target designators.

e. Thermal imaging devices may not be able to see through large amounts of dust particles suspended in the air.

f. Smoke and dust degrade the effectiveness of night vision devices.

g. Fog degrades long-range target acquisition from thermal sights.

h. Weapons flashes within enclosed areas appear to be much brighter. This causes Marines to lose their night vision and washes out light intensification devices.

5. Considerations. The environment of built-up areas presents special challenges and considerations during periods of limited visibility.

a. The use of glint tape, thermal tape, or chemiluminescent lights is an important consideration. These can be used to mark the FLOT, casualties, cleared buildings and rooms, weapons positions, and individual Marines. Their use must be clearly addressed in the unit's SOP. When markers are used for extended periods, their meanings should change because the enemy may be able to capture or manufacture and use these marking devices to their advantage.

b. The use of tracer and incendiary ammunition may be restricted to prevent fires. The light of fires whites out some night vision devices and may interfere with or confuse thermal devices.

c. The control of power stations may be essential to operations during periods of limited visibility. Control of power stations enables friendly forces to control, to a degree, background illumination. Shutting off the power to the street lights is much easier than shooting out the lights. Commanders must balance the tradeoff between force protection and maintaining law and order after the battle is over. During cold weather, the control of power stations may be critical to the welfare of the civilian population.

d. Identifying friendly units, noncombatant civilians, and enemy troops becomes more difficult during limited-visibility operations.

e. Locating the source of sounds becomes more difficult because of the natural echoing in built-up areas and the tendency of sounds to carry farther at night.

f. Locating boobytraps and obstacles also becomes more difficult at night. Movement rates are normally slower than during periods of normal visibility.

6. Special Equipment. Fighting during periods of limited visibility requires some specialized equipment to maximize advantages.

a. As a rule, thermal imaging devices such as the AN/PAS-7 infrared viewer and the AN/TAS-5 Dragon infrared sight are better for limited-visibility operations than are light intensification devices such as the AN/PVS-7B. Light intensification devices are more easily washed out by background light, weapons flashes in enclosed areas, and fires. Thermal devices, while also affected by fires, are not as easily washed out.

b. The AN/PAQ-4A infrared aiming device is similar to its off-the-shelf laser aiming sight counterparts, except it is not visible to the naked eye. Infrared pen lights can also be attached to weapons and, when used in conjunction with night vision devices, can provide a quick sight picture, illuminate rooms and hallways, identify obstacles and boobytraps, and identify friendly forces.

c. Other night sights for weapons include the AN/TVS-5 crew-served weapon night vision sight, the AN/PVS-4 individual weapons night vision sight, and the AN/UAS-12 night vision sight, which mounts to the modular universal laser equipment (MULE) and the TOW missile system.

d. Trip flares, flares, illumination from mortars, and artillery and spotlights (visible light or infrared radiation) can be used to blind enemy night vision devices or to artificially illuminate the battlefield (Figure H-1). (See FMFM 6-8, *Supporting Arms Observer, Spotter, and Controller*, for more information on illumination from mortars and artillery.)

e. Spare batteries for the night vision devices should be carried to keep the devices operational. Soft, clean rags should be used to clean the lenses.

7. Combat Support. Coordinating employment of forces and fire support in limited-visibility conditions is a major concern to leaders during MOUT. The concentration of forces and fires at the point of decision is facilitated by the technological edge that night vision devices provide.

a. Any degradation of accuracy in artillery fire will likely be a result of the limitations of target acquisition assets. While FOs and FACs may have thermal sights and laser range finders, most Marines on the battlefield do not yet have devices that will enable them to acquire targets accurately. The following are some devices and techniques to improve target acquisition for indirect fires:

(1) If the target is within LOS, tanks and LAVs can rapidly identify the target and provide an accurate range.

(2) Preregistered targets are effective if the target reference point can be observed and the observer has clear communications with the firing unit.

(3) Fixed-wing and rotary-wing aircraft can be used to identify targets and adjust supporting arms.

Figure H-1. Use of Indirect-Fire Illumination During MOUT

8. Combat Service Support. Maneuver unit commanders and their Marines are not the only individuals who must adjust to combat under limited-visibility conditions in MOUT. Logisticians at every level must anticipate requirements for these conditions.

a. Units conducting resupply operations during periods of limited visibility should consider:

- Issuing drivers and vehicle commanders night vision devices so that the vehicles going to and from logistic release points do not require illumination (This also prevents the enemy from acquiring resupply locations by following vehicles with blackout lights on.)

- Maintaining strict noise and light discipline

- Following a clearly marked route to avoid obstacles and prevent the resupply vehicle(s) from becoming disoriented

- Providing radios to resupply vehicles

- Issuing each vehicle a map of the AO (preferably a city map with the street names).

b. Combat units operating for an extended time under conditions of limited visibility should have enough batteries to keep the night vision devices functioning at optimum power and sensitivity.

c. A large operational readiness float of night vision devices should be maintained.

d. Casualty collection during periods of limited visibility is more difficult. Clear methods for marking any casualties must be established before the operation begins.

e. CSS operations located in existing structures should not be visible from a distance. This includes limiting vehicle traffic to an absolute minimum, sealing doors and windows to prevent light leakage, and dispersing assets as much as possible.

9. Operational Considerations. Marine units conduct attacks during periods of limited visibility to gain or sustain the momentum of the attack. Before conducting a limited-visibility attack, the commander must balance the risks and ensure that every Marine understands the mission, intent, and control measures. Rehearsals and strict command and control reduce casualties and greatly enhance the chances for mission accomplishment.

a. To reduce confusion, Marines should clear buildings and rooms using the same techniques they use during the day.

b. Movement rates are slower in the dark. Each Marine must remain alert for mines, boobytraps, and enemy positions.

c. Rifle squads and fire teams should be equipped with a mixture of both thermal imaging and light intensification devices whenever possible. This enables the squads and fire teams to obtain a better picture of the night environment and enables Marines to balance the strengths and weaknesses of each type of night vision device for maximum results.

d. When moving through buildings, assault forces mark cleared rooms and buildings and communicate with the support forces. Marking cleared rooms is especially critical if more than one assault force or element is in the same building. Communications with supporting forces are imperative to avoid fratricide.

e. If flashlights or chemiluminescent lights are used, they should be held away from the head or chest area. This will make it harder for enemy soldiers firing at the light to kill the Marine holding the flashlight or chemiluminescent light.

f. Units must know where everyone is during offensive operations. This not only reduces the risk of fratricide, but also decreases the time spent identifying, locating, and treating casualties. Also, it greatly reduces the chance of Marines becoming disoriented and separated from the unit.

g. Assault units should be aware of adjacent fires that may diminish the effectiveness of night vision devices. Weapons flashes within small rooms cause Marines to lose their night vision and can wash out light intensification devices. Also, enemy soldiers may use flares inside and outside of buildings to deliberately render night vision devices ineffective.

h. Leaders must ensure that all Marines follow the ROE and the laws of land warfare. This is critical if the enemy is intermixed with the local civilian population. Also, leaders must follow all control measures.

i. Enemy forces can be expected to use periods of limited visibility to the same advantages as Marine forces do.

j. Enemy forces may have access to sophisticated night vision devices manufactured in Europe, the United States, Japan, Korea, and the former Soviet Union.

Appendix I

Urban Building Analysis

As in other types of operations, success in urban combat depends largely on the ability to analyze the military aspects of terrain. Marines must be able to recognize certain terrain features when evaluating urbanized terrain.

1. Types of Mass-Construction Buildings. Mass-construction buildings are those in which the outside walls support the weight of the building and its contents. Additional support, especially in wide buildings, comes from using load-bearing interior walls, strongpoints (called pilasters) on the exterior walls, cast-iron interior columns, and arches or braces over the windows and doors (Figure I-1). Modern types of mass-construction buildings are wall and slab structures, such as many modern apartments and hotels, and "tilt-up" structures commonly used for industry or storage. Mass-construction buildings are built in many ways:

- The walls can be built in place by using brick, block, or poured-in-place concrete.
- The walls can be prefabricated and then tilted up, or they can be reinforced-concrete panels.
- The walls can be prefabricated and assembled like boxes.

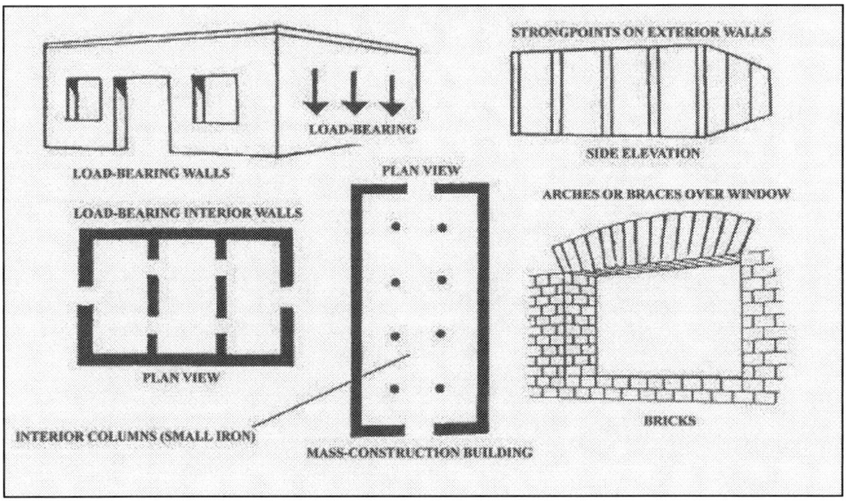

Figure I-1. Mass-Construction Buildings

.3

a. Brick buildings are, in some regions, the most common and most important of the mass-construction buildings. In Europe, brick buildings are commonly covered with a stucco veneer so that bricks do not show (Figure I-2). One of the most common uses of brick buildings is the small store. These buildings are found in all built-up areas but are most common in the core periphery (Figure I-3).

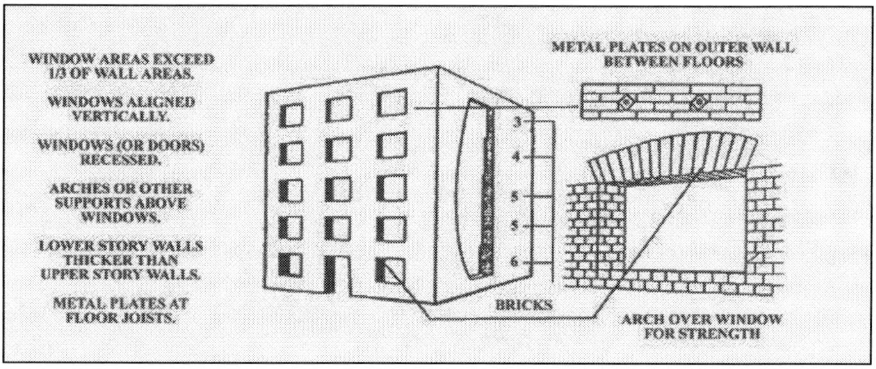

Figure I-2. Brick Buildings

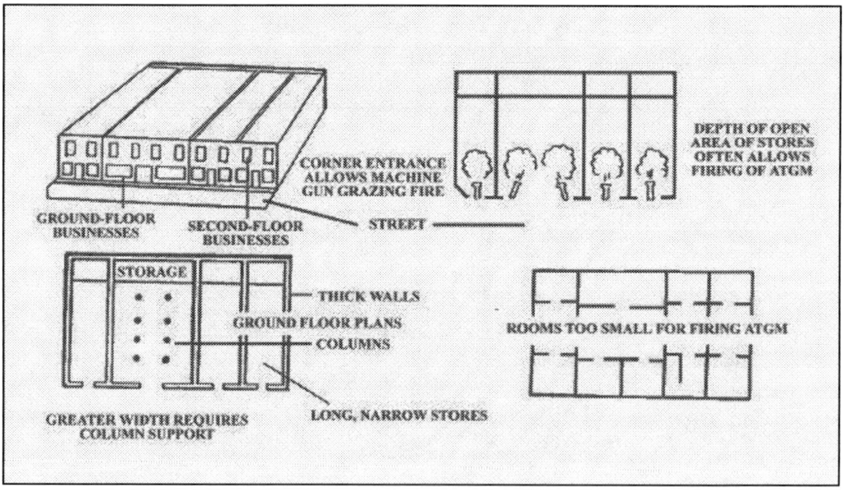

Figure I-3. Brick Store

b. Another common mass-construction building found mainly in industrial areas and along commercial ribbons is the warehouse. It is built of poured-in-place concrete reinforced with steel bars or of prefabricated tilt-up walls. The walls of warehouses provide good cover, although the roofs are generally light and vulnerable. The warehouses' large open bays permit firing of ATGMs or surface-to-air missiles and, because they are normally found in outlying areas, often afford adequate fields of fire for ATGMs. These buildings are built on slabs that can normally support the weight of vehicles and can provide excellent cover and concealment for tanks (Figure I-4).

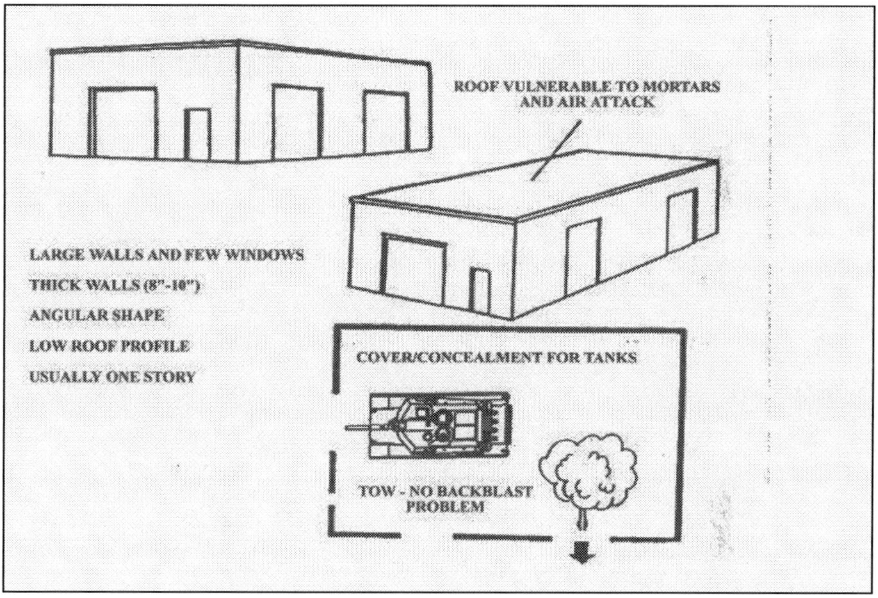

Figure I-4. Warehouse

c. Another type of mass-construction building is the box-wall-principle type. It is made from prefabricated panels of 6- to 8-inch-thick reinforced concrete. The outside wall is often glass. The box-wall-principle building provides good cover, except at the glass wall. The rooms are normally too small for ATGMs to be fired. A good circulation pattern exists from room to room and from floor to floor. These buildings are commonly used as hotels or apartments and are located mainly in residential and outlying areas (Figure I-5).

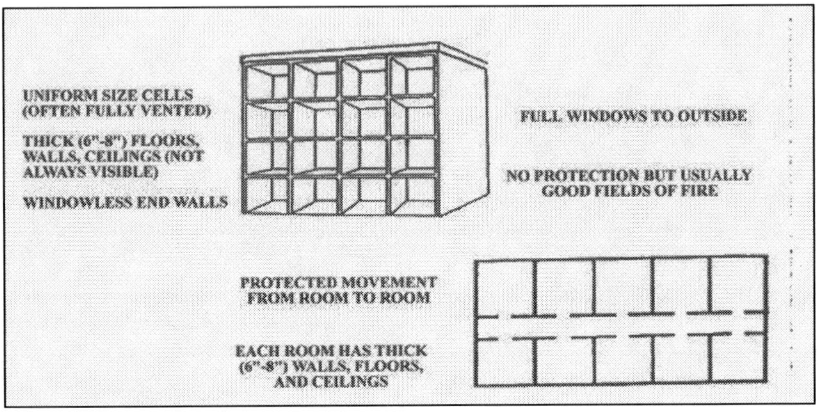

Figure I-5. Box-Wall-Principle Building

d. Public gathering places (churches, theaters) are mass-construction buildings with large, open interiors. The walls provide good cover, but the roof does not. The interior walls are not load-bearing and are normally easy to breach or remove. These buildings have adequate interior space for firing ATGMs or surface-to-air missiles. They are often located next to parks or other open areas and, therefore, have long fields of fire. Public gathering places are most common in core, core periphery, residential, and outlying high-rise areas (Figure I-6).

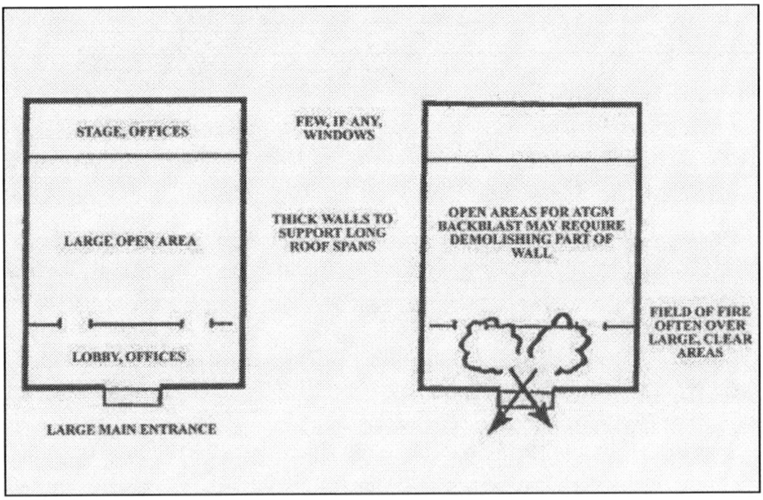

Figure I-6. Public Gathering Places

2. Types of Framed Buildings. Framed buildings are supported by a skeleton of columns and beams and are usually taller than frameless buildings (Figure I-7). The exterior walls are not load-bearing and are referred to as either heavy clad or light clad. Another type of framed building often found in cities is the garage, which generally has no cladding.

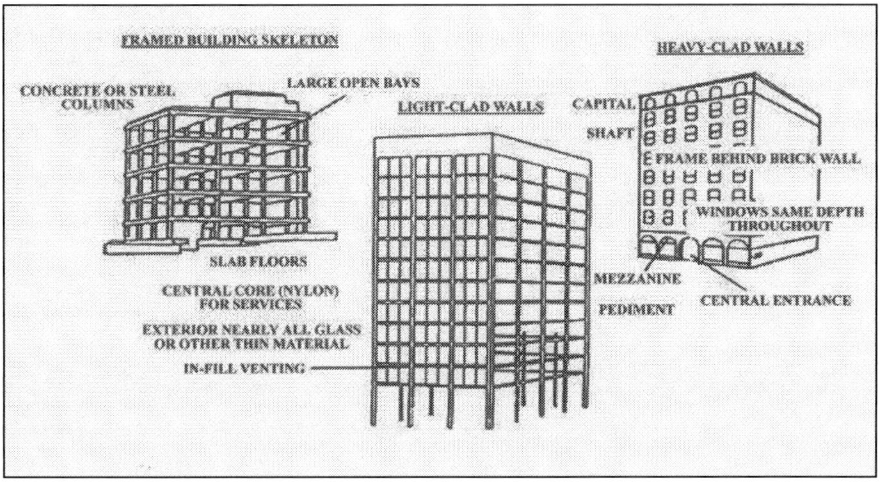

Figure I-7. Framed Buildings

a. Heavy-clad buildings were common when framed buildings were first introduced. Their walls are made of brick and block that are sometimes almost as thick as frameless brick walls, although not as protective. Heavy-clad framed buildings are found in core and core periphery areas. They can be recognized by a classic style or architecture in which each building is designed with three sections: the pediment, shaft, and capital. Unlike the brick building, the walls are the same thickness on all floors, and the windows are set at the same depth throughout. Often the frame members (the columns) can be seen, especially at the ground floor. The cladding, consisting of layers of terra cotta blocks, brick, and stone veneer, does not provide as good a cover as the walls of brick buildings. It protects against small-arms fire and light shrapnel but does not provide much cover against heavy weapons (Figure I-8).

(1) The floor plans of these buildings depend on their functions. Office buildings normally have small offices surrounding an interior hall. These offices have the same dimensions as the distance between columns (some large offices are as large as two times the distance between columns). These rooms are too small to permit firing of ATGMs but do provide some cover for snipers or machine gunners (Figures I-9 and I-10).

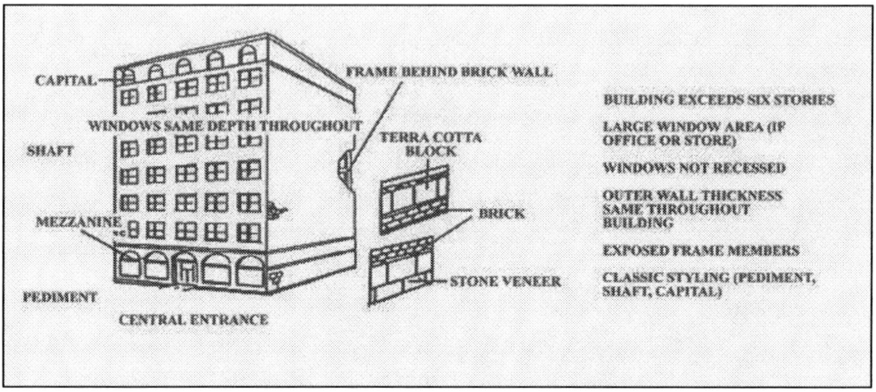

Figure I-8. Heavy-Clad Framed Building

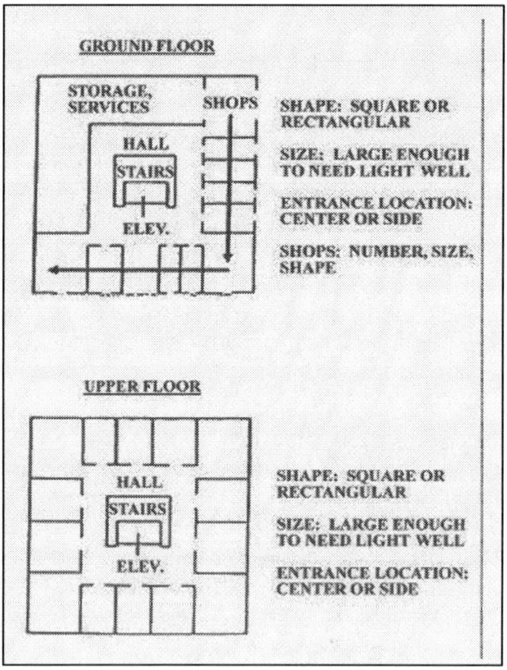

Figure I-9. Floor Plan of Heavy-Clad Framed Office Building

(2) Department stores normally have large, open interiors (Figure I-11). Such areas permit firing of ATGMs or man-portable surface-to-air missiles (if there are adequate fields of fire). Often a mezzanine level with a large backblast area permits firing down onto tanks. Steel fire doors often exist between sections of the store. The steel fire doors are activated by heat. Once closed, they are difficult to breach or force open, but they effectively divide the store into sections (Figure I-12).

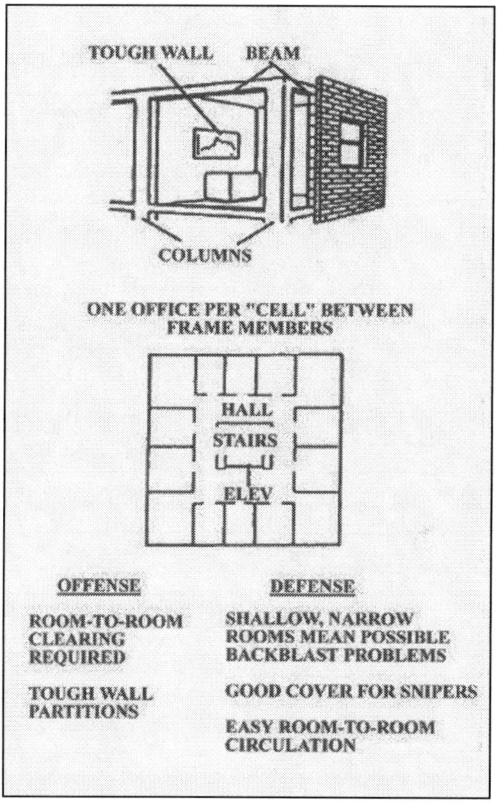

Figure I-10. Heavy-Clad Framed Office

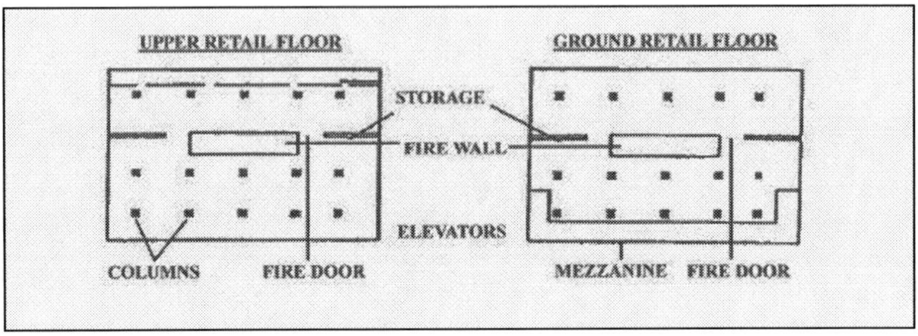

Figure I-11. Heavy-Clad Framed Department Store

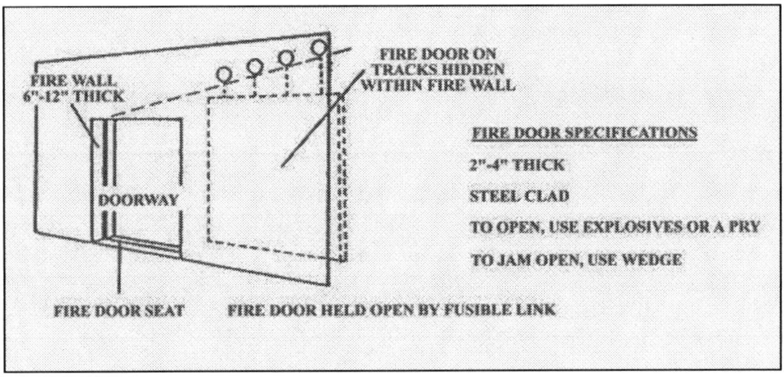

Figure I-12. Fire Wall and Fire Door

(3) Another type of heavy-clad framed building is used as a high-rise factory (Figure I-13). Such buildings are normally easily recognized because the concrete beams and columns are visible from the outside. They are usually located in older industrial areas. The large windows and open interior favor the use of ATGMs. Because the floors are often made to support heavy machinery, these buildings provide good overhead cover.

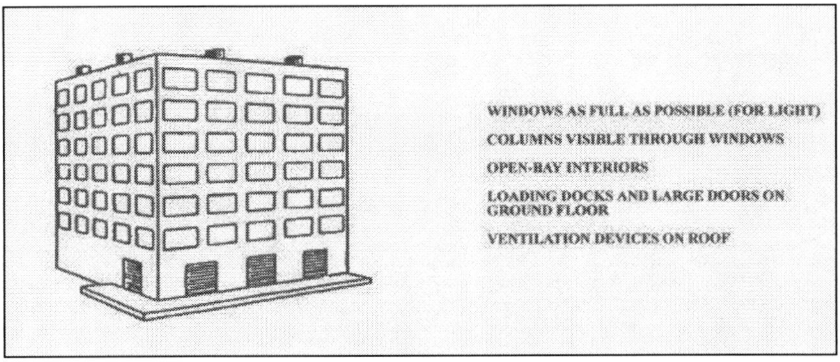

Figure I-13. High-Rise Factory

b. Light-clad buildings are more modern and may be constructed mostly of glass (Figure I-14). Most framed buildings built since World War II are light-clad buildings. They are found in both core and outlying high-rise regions. Their walls consist of a thin layer of brick, lightweight concrete, or glass. Such materials provide minimal protection against any weapon. However, the floors of the buildings are much heavier and provide moderate overhead cover. The rooms in light-clad framed buildings are much bigger than those in heavy-clad buildings. This feature, along with the fact that the buildings usually stand detached from other buildings, favors the employment of ATGMs. The interior partitions are thin, light, and easy to breach (Figure I-15).

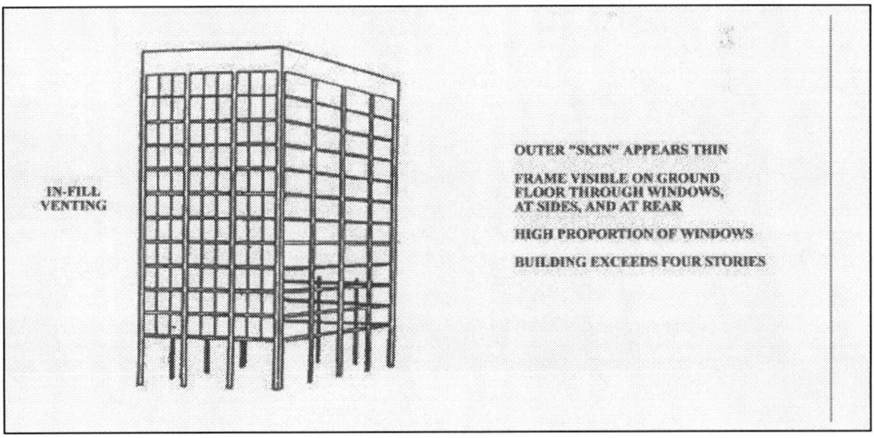

Figure I-14. Light-Clad Framed Building

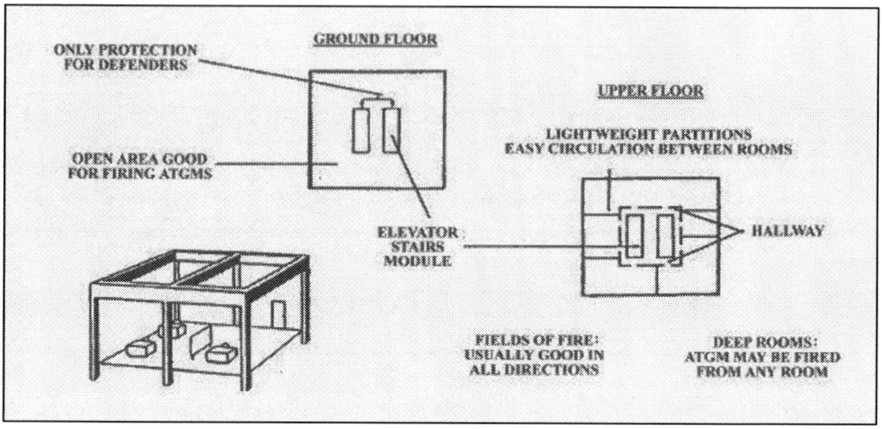

Figure I-15. Light-Clad Framed Room

c. The garage is one of the few buildings in an urban area in which all floors support vehicles. It provides a means to elevate vehicle-mounted TOWs, and the open interiors permit firing of ATGMs. Garages are normally high enough to provide a 360-degree field of fire for antiaircraft weapons. For example, a Stinger could hide under the top floor of the garage, come out to engage an aircraft, and then take cover again (Figure I-16).

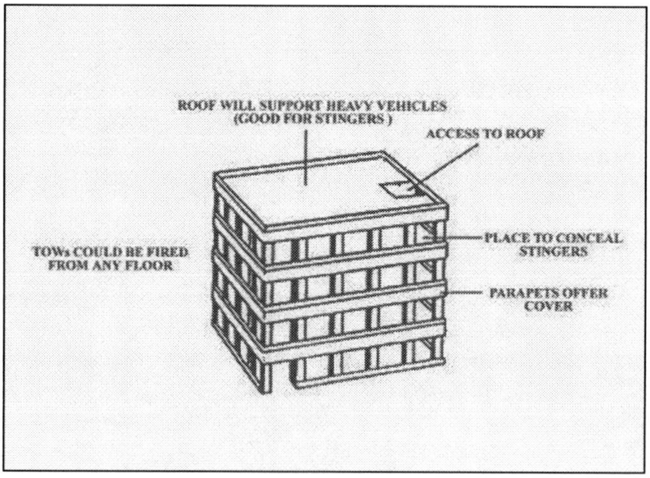

Figure I-16. Garage

3. Floor Plans. Floor plans in buildings follow predictable patterns. One of the factors that determines floor plans is building shape (Figure I-17). The basic principle governing building shape is that rooms normally have access to outside light. This principle helps Marines to analyze and determine the floor plans of large buildings.

4. Residential Areas. The two basic types of houses in the western world are located in and around cities and in rural areas. City-type houses are normally mass-construction brick buildings. Rural-type buildings in the continental United States, South America, and Southeast Asia are commonly made of wood, while in continental Europe, Southwest Asia, and sub-Saharan Africa, where wood is extremely scarce, rural buildings are normally constructed of concrete blocks. Rural-type buildings can also be found within cities (Figure I-18).

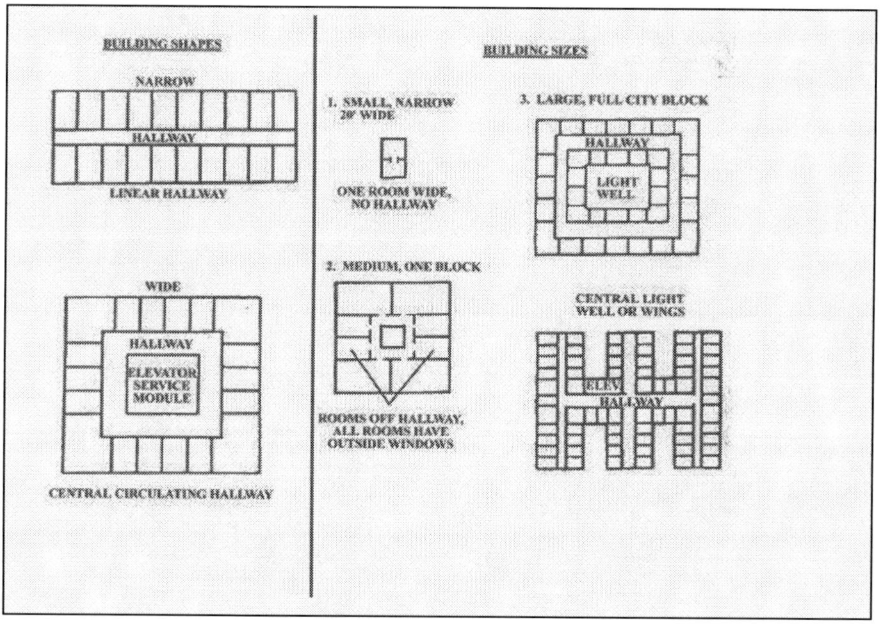

Figure I-17. Building Shapes and Sizes

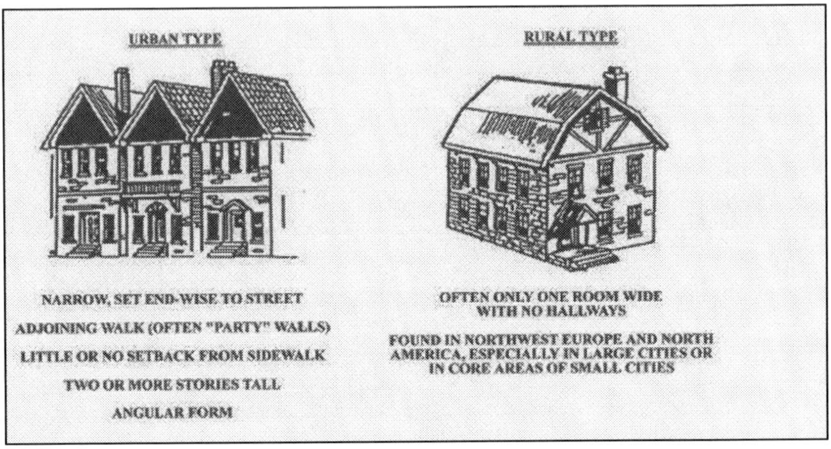

Figure I-18. Types of Housing

a. Another common type of building structure in cities with European influence is called the Hof-style apartment building (Figure I-19).

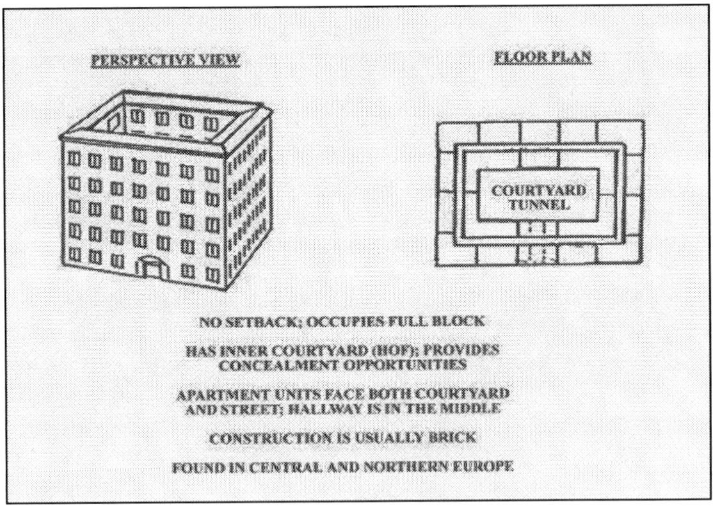

Figure I-19. Hof-Style Apartment Building

b. In the Middle East and tropical regions, the most common type of housing is the enclosed courtyard. Houses are added to one another with little regard to the street pattern. The result is a crooked, narrow maze that is harder to move through or fire in than dense European areas. (Figure I-20).

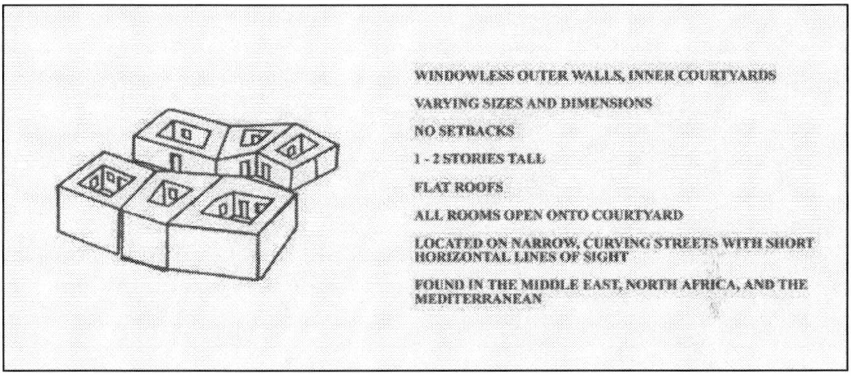

Figure I-20. Enclosed Courtyard

5. Characteristics of Buildings. Certain characteristics of both mass-construction and framed buildings can be helpful in analyzing a built-up area. Leaders can use Table I-1 to determine how to defend or attack a certain building given the unit's available weapons systems.

Type of Construction	Building Material	Height (Stories)	Average Wall Thickness (cm)
Mass	Stone	1 to 10	75
Mass	Brick	1 to 3	22
Mass	Brick	3 to 6	38
Mass	Concrete block	1 to 5	20
Mass	Concrete wall and slab	1 to 10	22 to 38
Mass	Concrete tilt-ups	1 to 3	18
Framed	Wood	1 to 5	3
Framed	Steel (heavy cladding)	3 to 50	30
Framed	Concrete/steel (light cladding)	3 to 100	2 to 8

Table I-1. Characteristics of Buildings

6. Distribution of Building Types. Certain types of buildings dominate certain parts of a city; this establishes patterns within a city. Analysis of the distribution and nature of these patterns has a direct bearing on military planning and weapon selection (Figure I-21).

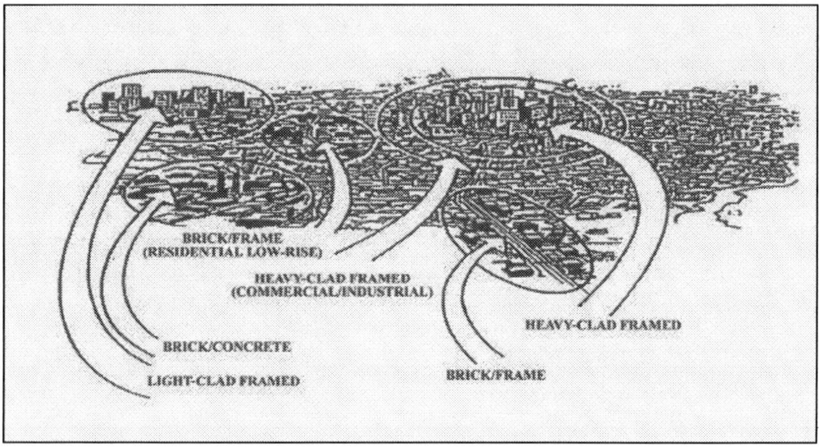

Figure I-21. Distribution of Building Types

a. Mass-construction buildings are the most common structures in built-up areas, forming about two-thirds of all building types. Brick structures account for nearly 60 percent of all buildings, especially in Europe.

b. Steel- and concrete-framed multistory buildings have an importance far beyond their one-third contribution to total ground floor area. They occupy core areas—a city's most valuable land—and as centers of economic and political power, they have a high potential military significance.

c. Open space accounts for about 15 percent of an average city's area. Many open spaces are grass-covered and are used for parks, athletic fields, and golf courses; some are broad, paved areas. The largest open spaces are associated with suburban housing developments where large tracts of land are recreation areas.

d. Streets serving areas consisting of mostly one type of building normally have a common pattern. In downtown areas, for example, high land values result in narrow streets. Street widths are grouped into three major classes: 7 to 15 meters, located in medieval sections of European cities; 15 to 25 meters, located in newer, planned sections of most cities; and 25 to 50 meters, located along broad boulevards or set far apart on large parcels of land. When a street is narrow, observing or firing into windows of a building across the street can be difficult because an observer is forced to look along the building rather than into windows.

When the street is wider, the observer has a better chance to look and fire into the window openings (Figure I-22).

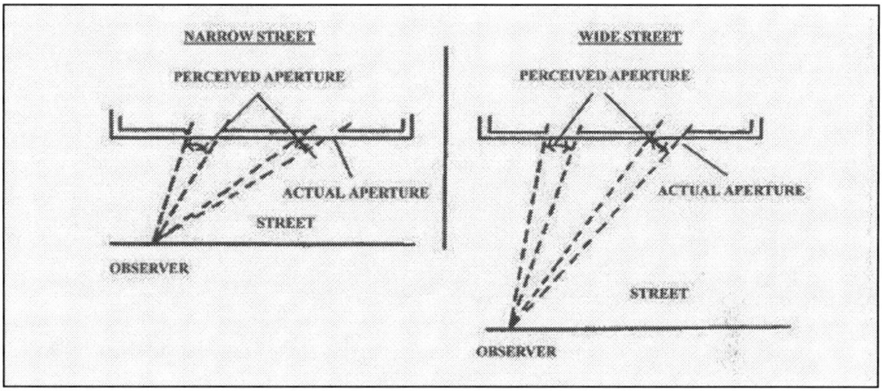

Figure I-22. LOS Distances and Angles of Obliquity

Appendix J

Lessons Learned from Russian Military Operations in Chechnya 1994-1996

1. Introduction. Many historical studies of urban combat have become dated. This is primarily the result of the impact that technology has had on military capabilities and evolving tactics. However, certain tenets of urban combat remain constant. The battles for the city of Grozny during the Russian intervention in the Republic of Chechnya represent a recent and critically important example of large scale operations in urban combat. Combat in Grozny was characterized by a large, technologically sophisticated military force (Russian) engaging and ultimately being defeated by a small, relatively primitive irregular force (Chechen). Grozny provides a number of fresh insights, and reinforcement of time honored tenets of urban warfare, across the scope of activities germane to modern urban combat.

2. Operations in Chechnya. Russia's war in Chechnya essentially began with the collapse of the Soviet Union, immediately after which pro-independence factions in Chechnya began agitating to free the Republic from Russian domination. This eventually led to civil war with pro-Russian factions within the Republic. Russia committed troops to Chechnya on 11 December 1994, for what many in the Russian government and military thought would be a quick campaign to restore Russian primacy. What followed was a bitter, costly, and protracted struggle that ended with the complete withdrawal of Russian military forces from Chechnya.

A primary focus of Russian operations in Chechnya was the capture of the capital city, Grozny. The initial attack on Grozny was conducted by a mechanized force, consisting of 6,000 troops mounted in tanks, BMP infantry fighting vehicles, and BTR armored personnel carriers. The Russians anticipated light opposition, but instead met a determined and heavy resistance from Chechens armed with a large number of antitank weapons. The attack was repulsed with significant Russian losses, both in personnel and armored vehicles (105 of 120 vehicles were lost). A second mechanized attack conducted on New Years Eve, 1994, was also repulsed with the loss of 140 of 200 tanks employed. The Russians captured Grozny some two months later, primarily through the use of excessive, overwhelming firepower, and at great cost to themselves and the local Chechen population.

The capture of Grozny marked the end of the first phase of Russian combat operations in Chechnya. Beginning in March 1995, the Russians became heavily engaged in anti-partisan operations as they tried to gain control of the country. Though the Russians controlled several major Chechen cities and a portion of the countryside by May 1995, they were never able to fully isolate the Republic. Chechen separatist fighters continued to receive a steady flow of arms and supplies from neighboring countries. Consequently, the Chechen fighters were able to maintain the initiative throughout this period, engaging Russian forces whenever and wherever they chose.

Even late in the campaign, Russian intelligence failed to detect Chechen fighters infiltrating Grozny, which enabled the Chechens to launch a major attack against the Russian-controlled city in early August 1996. The Chechens succeeded in capturing the city less than two weeks later.

.3

The recapture of Grozny was a significant loss for the Russians, precipitating a general cease-fire and the withdrawal of Russian forces from the Republic.

Besides the military aspects of combat in Chechnya, other issues such as ethnic, cultural, and religious divides impacted upon the nature of the conflict. The Russians failed to culturally orient their troops, which resulted in many serious cultural mistakes when dealing with the Chechen civilians. Once insulted or mistreated, the Chechen civilians became active participants in the struggle against Russia. The war eventually degenerated into one of abject ferocity and brutality on both sides. The Chechens are reported to have executed Russian prisoners, decapitated bodies, and boobytrapped the dead. Russian wounded and bodies were hung upside down outside of Chechen defensive positions in buildings, causing the Russians to have to shoot through their own people to engage the Chechens. The Russians, initially hesitant to cause civilian casualties, increasingly engaged Chechen units with a growing disregard for the safety of the local populace. As Russian discipline degraded, the incidences of Russian troops engaging in looting, arson, torture, even the summary execution of civilians, rose significantly. The result was the deaths of tens of thousands of civilians.

3. Lessons Learned. The intervention in Chechnya and the battles for Grozny generated many important lessons across the warfighting functions of combat in urbanized terrain. The following information is primarily from the Russian point of view, though some Chechen capabilities are included. The lessons learned are arrayed as warfighting functional considerations.

a. Command and Control.

(1) Russian military commanders lacked clear policy guidance. The Russian military's ability to attain and maintain steady, logical operational progress was hampered, if not negated, by a lack of clear governmental policy. The Russian government was split at the highest levels over the wisdom of committing forces to Chechnya, and the conduct of the intervention.

(2) Poor Russian command and control structure throughout all levels. There was no direct, unified chain of command for the operations in Chechnya. Operational directives often flowed from a number of governmental agencies directly to local unit commanders, often bypassing intermediate military commanders. This resulted in a lack of overall situational awareness, and exceptionally poor coordination among tactical forces.

(3) Russians were unable to contain/prohibit Chechen operations against Russia. When the Russian forces finally began to achieve progress in late May 1995, the Chechens initiated attacks against targets inside Russia. The result was a major propaganda victory for the Chechens.

(4) Excessive senior officer (flag level) interference contributed to operational confusion. There were eight major changes at the senior command level during the course of the intervention. Additionally, there were approximately one hundred general officers on the operational scene, about one general officer per 3,000 troops, which resulted in confusion as they got in each others way.

(5) Poor small unit leadership, particularly at the NCO level, was a primary cause of Russian tactical failures in Grozny. While the Russians recognize that urban combat is small unit intensive, they do not possess a professional NCO corps. Additionally, Russian forces lacked some 12,000 platoon leaders when the intervention began. This had a crippling effect on small unit operations.

(6) Russian inability to isolate Chechnya/Grozny allowed a constant flow of reinforcements and supplies into the country/city. Russian control of any given area was at best, porous. The Chechens continued to receive outside support throughout the campaign, as well as, a large supply of arms and equipment from poorly disciplined Russian troops inside Chechnya.

(7) Both sides used commercial off-the-shelf communication assets. The information war was as vitally important as any other aspect of the campaign. The Chechens had to make do with commercially available communication equipment. They fully exploited the use of cellular telephones, Motorola radios, improvised television stations, video cameras, and the internet. The Russians lacked sufficient military communication equipment, and had to supplement with commercial assets.

(8) Russian tactical communication was very difficult in Grozny. Urban structures in Grozny interfered with Russian military radios, severely hindering tactical communications. The Russians had to establish ground-based and aircraft-based relay stations to overcome the interference.

(9) Boundaries between units were tactical weak points. Units had to think of boundaries in both a horizontal and vertical context. There were many instances when the Chechens held the third floor and above, while the Russians held the first two floors and maybe the roof. This caused a lack of coordination, and mutual support. Additionally, due to poor command and control and coordination between units, boundaries became seams through which the Chechens could pass unopposed.

b. Intelligence.

(1) Lack of high quality intelligence made operations much more difficult and dangerous. Russian intelligence collection and dissemination capabilities were poor throughout the intervention. In March 1996, the Russian Minister of the Interior complained that poor reconnaissance and intelligence allowed Chechen fighters to move in and out of Grozny without opposition, which contributed to the heavy Russian losses.

(2) Directives to minimize civilian casualties ultimately put the Russians at a disadvantage. An advantage accrues to the side with less concern for the safety of the civilian population. Initially, the Russians attempted to limit civilian casualties. The Chechen

.3

fighters took advantage of this reticence, often using civilians, schools, and apartment buildings as a shield. While this may seem to be cold-blooded of the Chechen fighters, what must be understood is that the majority of the populace in Grozny were actually ethnic Russians, and that the ethnic Chechens had no remorse in using them in this manner.

(3) Russian concerns for the safety of the Chechen populace and property declined. As the Russian troops became increasingly frustrated in their attempts to separate Chechen fighters from the local populace, and their casualties began to mount, concerns for civilian safety waned, resulting in greater civilian casualties and further alienating the Chechen people.

(4) Rigorous communications security is essential, even against less sophisticated enemies.Much of the Russian tactical radio traffic was broadcast in the clear, which allowed the Chechens to intercept it. The Chechens were able to redirect Russian airstrikes onto Russian targets, and engage Russian artillery forward observers who broadcast their own position coordinates.

(5) The Russians made extensive use of RPVs for reconnaissance. The Chechen campaign saw the first Russian use of unmanned aerial vehicles for reconnaissance and intelligence gathering. There were primarily used by Russian airborne forces.

c. Maneuver.

(1) Inadequate training in the most basic maneuver and combat skills inhibited Russian operations. Overall, poor Russian combat performance could be traced to a lack of training in fundamental military skills, a situation that was then exacerbated by an additional lack of training in specific urban combat skills. Individual skills training was so inadequate that some units were entirely incapable of undertaking combat operations. The Russian's ability to conduct large scale operations had been seriously eroded by a complete cessation of exercises at the division level in the 2 years prior to the Chechen campaign. Even regimental and battalion level exercises had been curtailed by 75 percent or more. Additionally, The air forces were not any better trained or prepared for their operational role in Chechnya.

(2) Lack of cohesiveness among the Russian forces led to confusion and antagonism. The Russian ground forces consisted primarily of regular army units and Ministry of Interior Internal Troops, which were essentially a police force. With no real unified command and control, the two groups were usually conducting uncoordinated missions, which led to confusion and fratricide.

(3) The nature of cities channels combat operations along narrow lanes of activity. Combat conditions in Grozny were characterized by narrow fields of view, limited fields of fire, and constricted avenues of approach. For the Russians, this meant movement along approaches well known and heavily defended by the Chechens.

(4) Night fighting was the most difficult activity for Russian infantry. The Russian infantry, at all levels, was inadequately trained for night operations, and lacked night vision equipment.

(5) Armored vehicles cannot operate in cities without extensive dismounted infantry support. The Chechens formed antitank hunter-killer teams equipped with RPG-7 and RPG-18 antitank missiles. They engaged Russian tanks with volley fire from above, behind, and the sides. Armor columns not supported by dismounted infantry suffered great losses.

(6) Armored engineer vehicles are indispensable for removing obstacles. The Russians found that armored engineer vehicles were critical to restoring or improving mobility in Grozny. Eventually, the Russians formed special assault groups that included armored engineer vehicles in each group.

(7) The Chechens preferred small unit, hit-and-run ambush tactics. The Chechens relied on small units (15 to 20 troops) to conduct quick, sharp ambushes, from which they would disengage before the Russians could bring significant firepower to bear. Ambushes were often conducted in three tiers, with Chechens in the underground, on the ground floor, and on the roof, with each element having a different task in the ambush. They also used ambush tactics against Russian helicopters.

(8) Urban combat is extremely manpower intensive, and results in significant attrition of personnel and materiel for the attacker. The Russians discovered that a 5 to 1 manpower ratio was often not enough, due to the high casualty rates consistent with urban combat and the requirement to guard virtually every building taken.

(10) Tracked vehicles are preferable to wheeled vehicles in the urban environment. Tracked vehicles provided better mobility than wheeled vehicles due to their significantly greater ability to negotiate obstacles and rubble.

(11) Armored vehicles required additional protection (i.e., reactive armor, sandbags, wire mesh, etc.) to negate attacks from the sides, rear, and top. Russian tank design emphasized greater armor protection on the frontal arc, and less on other areas of the vehicle. This proved inadequate for urban combat, where attacks can come from any direction. The majority of lethal hits came from above, easily penetrating turrets and engine decks, and from the rear.

(12) The spatial qualities and perspectives of urban combat are more "vertical" than non-urban combat. The Chechens regularly fought from basements, rooftops, and upper-floor rooms, making them difficult to locate and engage. Russians soldiers trained in open areas, had to reorient to a confined, highly vertical environment.

.3

(13) Standard Russian military unit configurations were inappropriate for urban combat. Typical Russian army unit organizations, oriented to open spaces and mechanized operations, proved to be inadequate for urban combat. They eventually had to adopt specially configured assault teams that included, among other assets, significantly greater firepower and engineer assets. Essentially, they moved towards a better combined arms team.

(14) The Russians found their helicopters to be far too vulnerable to rooftop snipers and ambushes in the urban setting. While Russian doctrine had long specified the capture of a building from the top-down, it became apparent in Grozny that this tactic was too expensive in terms of helicopter losses. The Chechens made effective use of snipers and RPG-equipped ambush teams to damage or destroy helicopters supporting Russian ground forces.

d. Fires.

(1) Artillery was invaluable in the direct fire mode. The Russians found artillery, to include rocket launchers and mortars, to be very effective for direct fire against point targets , usually at ranges of 150 to 200 meters.

(2) Obscurants (smoke, white phosphorous, tear gas) are especially useful in urban combat. The Russians used significant amounts of smoke and white phosphorus to mask the movement of forces. Every fourth or fifth artillery or mortar round was either smoke or WP. WP also had a toxic effect, easily penetrating Chechen protective masks.

(3) The Chechens effectively used RPGs against helicopters. The Chechens made effect use of ambush teams equipped with RPGs to engage Russian helicopters, either by damaging or destroying the helicopter or simply prohibiting helicopterborne operations.

(4) Air defense guns (i.e., ZSU-23-4) were valuable in suppressing ground targets. The Russians found air defense guns effective against multi-story buildings because they had sufficient elevation to hit targets in the upper stories. Air defense guns became a primary asset in the Russian special assault groups.

(5) The Russians determined that overwhelming firepower can make-up for organizational and tactical deficiencies in the short run if collateral damage is disregarded. Whenever Russian military units proved incapable of effective urban operations, the Russians simply fell back on massive, overwhelming firepower to reduce a strong point. It was through the use of excessive, unrestrained firepower, and a complete lack of regard for collateral damage, that finally enabled the Russians to gain control of Grozny.

(6) The Chechens negated Russian supporting fires by "hugging" Russian units. Whenever possible, Chechen fighters adopted the tactic of staying as physically close to Russian units as possible, so that the Russians could not employ supporting arms without the risk of injuring their own troops.

(7) The Russians found "bunker busting" weapons to be invaluable. The Russians came to rely heavily on "bunker busting" weapons for engaging Chechens, primarily snipers, entrenched in buildings.

(8) Russian helicopters needed stand-off weapons, that would allow them to engage targets while outside the range of Chechen weapons. The Chechens employed 23mm cannons and heavy machine guns against Russian helicopters, causing the Russian's helicopter gunships to fire from ranges of 3,000 meters or more. This required the use of wire-guided munitions and large caliber (240mm) rockets.

(9) The Russians made extensive use of precision-guided weapons. When the weather allowed, the Russians were able to use precision guided weapons, such as laser-guided bombs and missiles. They had great effect against priority targets such as bridges, major road intersections, and buildings.

e. Logistics.

(1) Deficiencies in the Russian logistics system degraded combat effectiveness. Significant inadequacies in the Russian logistics apparatus effected virtually all combat forces and operations. Some soldiers actually entered combat in Grozny without weapons or ammunition. The Russian transportation system failed. Supply officers were unprepared for the high volume demands for ammunition, hand grenades, smoke grenades, demolition charges, and one-shot antitank weapons.

(2) Russian logistics personnel were ill-prepared in basic soldier skills. Russian infantry often had to be used to conduct logistics operations into Grozny, because the logistics personnel lack the basic skills to defend themselves against the Chechens.

(3) The Russian ground forces required special equipment not normally found in unit T/Es. The T/E for the typical Russian infantry unit did not include such items as ropes, grappling hooks, and ladders, all of which proved essential for urban combat.

(4) Armored vehicle recovery is extremely difficult and dangerous. The Russians found the recovery of vehicles to be extremely difficult due to the terrain (i.e., narrow streets, rubble and debris), and Chechen opposition.

(5) Some Russian equipment was modified in the field to counter Chechen tactics and equipment. The Russians had to adopt add-on armor for their vehicles. They fabricated metal screens and wire mesh guards, which helped to defeat RPG and Molotov cocktail attacks.

(6) Medical support was of great importance. Russian casualty evacuation was difficult due to the lack of vehicular mobility, and the vulnerability of helicopters to Chechen fire. Many casualties were psychological, due in great part to the intense, violent nature of close quarter combat. The Russians were ill-prepared for the level of mental health problems they encountered. Additionally, logistics units were often unable to provide fresh drinking water, which caused Russian troops to consume contaminated water that resulted in other health problems.

f. Force Protection.

(1) Operations took much longer than expected, and the cost in terms of casualties was much greater than anticipated. The Russians committed unprepared, poorly trained and poorly equipped forces to Chechnya, based on an unrealistic operational time frame. Ultimately, this cost the Russians significant casualties and the war.

(2) Fratricide was a serious and continuing problem throughout the campaign. The situation facing Russian combat forces was ripe for fratricide. Poorly trained units, operating in a confused and uncertain urban environment, often unable to tell friend from foe, and lacking quality leadership and inter-unit coordination, were often as dangerous to themselves as they were to the Chechens .

(3) Chechen use of adhoc air defense systems eliminated Russian air support, particularly helicopters, in Grozny. The Chechens were able to improvise effective anti-air defenses using a mixture of anti-air and ground (i.e., heavy machine guns, RPGs) weapons, and ambush tactics specifically adapted to the urban environment.

(4) The Chechens made extensive and effective use of snipers. Snipers fired from well inside rooms versus near window openings, as well as, from rooftops and basements. The Russians lacked an effective sniper and counter-sniper capability of their own.

(5) The Chechens improvised crude chemical weapons. The Chechens, lacking access to military chemical weapons, Used off-the-shelf materials and equipment to make chemical mines (i.e., chlorine gas), detonated remotely by radio.

(6) The Chechens made extensive and effective use of boobytraps. It seemed that the Chechens mined and boobytrapped 7-8everything. They had a good understanding of the average Russian soldier's actions and reactions, and boobytrapped accordingly. Additionally, the Russians found it difficult to maintain boobytrap awareness among their troops.

Appendix K

Glossary

I. Acronyms

AAA .... antiaircraft artillery
AAAV .... advanced assault amphibious vehicle
AAV .... assault amphibious vehicle
ACE .... aviation combat element
AO .... area of operations
APC .... armored personnel carrier
APDS-T .... armor-piercing, discarding sabot with tracer
APFSDS .... armor-piercing, fin-stabilized, discarding sabot
API .... armor-piercing incendiary
ARVN .... Army of the Republic of Vietnam
ATF .... amphibious task force
ATGM .... antitank guided missile
BDA .... battle damage assessment
BMNT .... .beginning of morning nautical twilight
CA .... civil affairs
CAG .... civil affairs group
CE .... command element
CI .... counterintelligence
COC .... combat operations center
CONAD .... consolidated administration center
CP .... command post
CSS .... combat service support
CSSE .... combat service support element
DOD .... Department of Defense
DS .... direct support
EOD .... explosive ordnance disposal
EPW .... enemy prisoner of war
EW .... electronic warfare
FAC(A) .... forward air controller (airborne)
FASCAM .... family of scatterable mines
FDC .... fire direction center
FEBA .... forward edge of the battle area
FLOT .... forward line of own troops
FM .... U.S. Army field manual
FM .... Frequency modulated
FMFM .... Fleet Marine Force manual
FMFRP .... Fleet Marine Force reference publication
FO .... forward observer
FPL .... final protective line
FRAGO .... fragmentary order
FSC .... fire support coordinator
GCE .... ground combat element
GPS .... global positioning system
GS .... general support
HE .... high explosive
HEAT .... high-explosive antitank
HEAT-MP .... HEAT-multipurpose
HEDP .... high-explosive dual-purpose
HEI-T .... high-explosive incendiary with tracer
HLZ .... helicopter landing zone
HMMWV .... high-mobility, multipurpose wheeled vehicle
HSS .... health services support
HUMINT .... human intelligence
IDF .... Israeli Defense Forces
IFV .... infantry fighting vehicle
IPB . . intelligence preparation of the battlefield/battlespace
ITOW .... improved TOW
JTF .... joint task force
KOCOA .... key terrain, observation and fields of fire, cover and concealment, obstacles, and avenues of approach
LAV .... light armored vehicle
LCAC .... landing craft air cushion
LD .... line of departure
LOS .... line of sight
LZ .... landing zone
MAGTF .... Marine air-ground task force
MCDP .... Marine Corps doctrinal publication
MCRP .... Marine Corps reference publication
MCWP .... Marine Corps warfighting publication
MEDEVAC .... medical evacuation
MEF .... Marine expeditionary force
MET .... meteorological
METT-T ... mission, enemy, terrain and weather, troops and support available-time available
MEU .... Marine expeditionary unit
MOOTW .... military operations other than war
MOPP .... mission-oriented protective posture
MOUT .... military operations on urbanized terrain
MP .... military police
MPF .... maritime prepositioning force
MULE .... modular universal laser equipment
NBC .... nuclear, biological, and chemical
NCA .... National Command Authorities
NCF .... naval construction force
NEF .... naval expeditionary force
NEO .... noncombatant evacuation operation
NGF .... naval gunfire
NSFS .... naval surface fire support
NVA .... North Vietnamese Army
NWP .... naval warfare publication

Appendix J

Glossary

I. Acronyms

OP .... observation post
OPLAN .... operation plan

OPORD
operation order
OPSEC
operations security
PDDA
power-driven decontamination apparatus
PGM
precision-guided munition
PLO
Palestine Liberation Organization
PLRS
position location reporting system
POL
petroleum, oils, and lubricants
PSYOP
psychological operations
ROE
rules of engagement
ROK
Republic of Korea
RP
red phosphorus
RSOP . . . . reconnaissance, selection, and occupation

of

position
SASR
special application sniper rifle
SMAW
shoulder-launched multipurpose assault

weapon
SOP
standing operating procedure
STANAG
NATO standardization agreement
STB
supertropical bleach
TOW
tube-launched, optically tracked, wire-guided

missile
UAV
unmanned aerial vehicle
UCMJ
Uniform Code of Military Justice
U.N.
United Nations
VC
Viet Cong
VHF
very high frequency

WP . . . . white phosphorus

II. Definitions

A

assault—1. The climax of an attack, closing with the enemy in hand-to-hand fighting. 2. To make a short, violent, but well-ordered attack against a local objective, such as a gun emplacement, a fort, or a machine gun nest. (Joint Pub 1-02)

assault—The culmination of an attack which closes with the enemy. (MCRP 5-2A)

attack—An offensive action characterized by movement supported by fire with the objective of defeating or destroying the enemy. (FMFRP 0-14)

B

battlespace—All aspects of air, surface, subsurface, land, space, and electromagnetic spectrum which encompass the area of influence and area of interest. (FMFRP 0-14)

block—1. A tactical task assigned to a unit that requires it to deny the enemy access to a given area or to prevent enemy advance in a given direction or an avenue of approach. It may be for a specified time. Units assigned this mission may have to retain terrain and accept decisive engagement. 2. An obstacle effect that integrates fire planning and obstacle effort to stop an attacker on a specific avenue of approach or to prevent an enemy from exiting an engagement area. (MCRP 5-2A)

breach—A tactical task where any means available are employed to break through or secure a passage through an enemy defense, obstacle, minefield, or fortification. (MCRP 5-2A)

built-up area—A concentration of structures, facilities, and populations, such as villages, cities, and towns, that forms the economic and cultural focus for the surrounding area. (MCRP 5-2A, modified)

bypass—A tactical task that involves maneuvering around an obstacle, position, or enemy force to maintain the momentum of advance. Bypassed obstacles and enemy forces are reported to higher headquarters. (MCRP 5-2A)

C

clear—A tactical task to remove all enemy forces and eliminate organized resistance in an assigned zone, area, or location by destroying, capturing, or forcing the withdrawal of enemy forces such that they cannot interfere with the friendly unit's ability to accomplish its mission. (MCRP 5-2A)

combat arms—Units who close with and destroy enemy forces or provide firepower and destructive capabilities on the battlefield. (MCRP 5-2A, modified)

combat operations center (COC)—The primary operational agency required to control the tactical operations of a command that employs ground and aviation combat, combat support, and combat service support elements or portions thereof. The combat operations center continually monitors, records, and supervises operations in the name of the commander and includes the necessary personnel and communications to do the same. (FMFRP 0-14)

combat service support (CSS)—The essential capabilities, functions, activities, and tasks necessary to sustain all elements of operating forces in theater at all levels of war. Within the national and theater logistic systems, it includes but is not limited to that support rendered by service forces in ensuring the aspects of supply, maintenance, transportation, health services, and other services required by aviation and ground combat troops to permit those units to accomplish their missions in combat. Combat service support encompasses those activities at all levels of war that produce sustainment to all operating forces on the battlefield. (Joint Pub 1-02)

combat support (CS)—Fire support and operational assistance provided to combat elements. (Joint Pub 1-02)

combat support (CS)—Units that provide critical combat functions in conjunction with combat arms units to secure victory (MCRP 5-2A, modified).

command post (CP)—A unit's or subunit's headquarters where the commander and the staff perform their activities. In combat, a unit's or subunit's headquarters is often divided into echelons; the echelon in which the unit or subunit commander is located or from which such commander operates is called a command post. (Joint Pub 1-02)

contain—A tactical task to restrict enemy movement. (MCRP 5-2A)

cover—1. Shelter or protection from enemy observation that reduces the effects of enemy direct and indirect fire. 2. A type of security operation that protects the force from surprise, develops the situation, and gives commanders time and space in which to respond to the enemy's actions. (MCRP 5-2A)

D

defend—A combat operation designed to defeat an attacker and prevent him from achieving his objectives. It employs all means and methods available to prevent, resist, or destroy an enemy attack. Forms of defensive operations are area and mobile. Choices of defensive operations are in-depth and forward. The defensive techniques are defend in sector, defend a battle position, and defend a strong point. (MCRP 5-2A)

destroy—1. A tactical task to physically render an enemy force combat-ineffective unless it is reconstituted. 2. To render a target so damaged that it cannot function as intended nor be restored to a usable condition without being entirely rebuilt. (MCRP 5-2A)

disrupt—A tactical task or obstacle effect that integrates fire planning and obstacle effort to break apart an enemy's formation and tempo, interrupt the enemy's timetable, or cause premature commitment of enemy forces, or the piecemealing of his attack. (MCRP 5-2A)

II. Definitions

F

fire support—Assistance to elements of the Marine air-ground task force engaged with the enemy rendered by other firing units, including (but not limited to) artillery, mortars, naval surface fire support, and offensive air support. (FMFRP 0-14)

fix—1. A tactical task in which actions are taken to prevent the enemy from moving any part of his forces either from a specific location or for a specific period of time by holding or surrounding them to prevent their withdrawal for use elsewhere. 2. A tactical obstacle effect that integrates fire planning and obstacle effort to slow an attacker within a specified area—normally an engagement area. (MCRP 5-2A)

H

helicopterborne operation—A military action in which combat forces and their equipment maneuver about the battlefield by helicopters or vertical-landed aircraft. Aviation activities are under the control of the aviation combat element commander who is assigned in direct or general support of one or more combat element(s). (FMFRP 0-14)

hold—1. To maintain or retain possession of by force, as a position or an area. 2. In an attack, to exert sufficient pressure to prevent movement or redisposition of enemy forces. (Joint Pub 1-02)

I

infiltration—The movement through or into an area or territory occupied by either friendly or enemy troops or organizations. The movement is made, either by small groups or by individuals, at extended or irregular intervals. When used in connection with the enemy, it infers that contact is avoided. (Joint Pub 1-02)

isolate—A tactical task given to a unit to seal off (both physically and psychologically) an enemy from his sources of support, to deny an enemy freedom of movement, and prevent an enemy unit from having contact with other enemy forces. An enemy must not be allowed sanctuary within his present position. (MCRP 5-2A)

L

linkup—An operation wherein two friendly ground forces join together in a hostile area. (FMFRP 0-14)

M

Marine air-ground task force (MAGTF)—A task organization of Marine forces (division, aircraft wing, and service support groups) under a single command and structured to accomplish a specific mission. (Joint Pub 1-02)

military operations on urbanized terrain (MOUT)—All military actions that are planned and conducted on a topographical complex and its adjacent natural terrain where manmade construction is the dominant feature. It includes combat in cities, which is that portion of MOUT involving house-to-house and street-by-street fighting in towns and cities (MCRP 5-2A, modified)

military operations other than war (MOOTW)—Operations that encompass the use of military capabilities across the range of military operations short of war. These military actions can be applied to complement any combination of the other instruments of national power and occur before, during, and after war. (Joint Pub 1-02)

N

neutralize—1. To render enemy personnel or material incapable of interfering with a particular operation. 2. To render safe mines, bombs, missiles, and boobytraps. (MCRP 5-2A)

O

occupy—A tactical task in which a force moves onto an objective, key terrain, or other man-made or natural terrain area without opposition, and controls that entire area. (MCRP 5-2A)

P

peace enforcement—Application of military force, or the threat of its use, normally pursuant to international authorization, to compel compliance with resolutions or sanctions designed to maintain or restore peace and order. (Joint Pub 1-02)

peacekeeping—Military operations undertaken with the consent of all major parties to a dispute, designed to monitor and facilitate implementation of an agreement (ceasefire, truce, or other such agreement) and support diplomatic efforts to reach a long-term political settlement. (Joint Pub 1-02)

protect—A tactical task to prevent observation of or engagement or interference with, a force or location. (MCRP 5-2A)

pursuit—An offensive operation designed to catch or cut off a hostile force attempting to escape, with the aim of destroying it. (Joint Pub 1-02)

R

reduce—1. A tactical task to gain control over an enemy position or objective. 2. A task to create lanes through or over an obstacle sufficient to allow the attacking force to accomplish its mission. (MCRP 5-2A)

retain—A tactical task to occupy and hold a terrain feature to ensure it is free of enemy occupation or use. (MCRP 5-2A)

rules of engagement (ROE)—Directives issued by competent military authority which delineate the circumstances and limitations under which United States forces will initiate and/or continue combat engagement with other forces encountered. (Joint Pub 1-02)

S

screen—A task to maintain surveillance, provide early warning to the main body, or

II. Definitions

impede, destroy, and harass enemy reconnaissance within its capability without becoming decisively engaged. (MCRP 5-2A)

search—A systematic reconnaissance of a defined area, so that all parts of the area have passed within visibility. (Joint Pub 1-02)

secure—A tactical task to gain possession of a position or terrain feature, with or without force, and to deploy in a manner which prevents its destruction or loss to enemy action. The attacking force may or may not have to physically occupy the area. (MCRP 5-2A)

seize—To clear a designated area and obtain control of it. (FMFRP 0-14)

suppression—A tactical task to employ direct or indirect fires, electronic attack, or smoke on enemy personnel, weapons, or equipment to prevent or degrade enemy fires and observation of the friendly forces. (FMFRP 0-14)

T

terrorism—The calculated use of violence or threat of violence to inculcate fear; intended to coerce or to intimidate governments or societies in the pursuit of goals that are generally political, religious, or ideological. (Joint Pub 1-02)

Appendix L

References and Related Publications

Field Manuals (FMs)

3-11	Flame, Riot Control Agent, and Herbicide Operations
5-34	Engineer Field Data
5-102	Countermobility
5-250	Explosives and Demolitions
7-98	Operations in a Low-Intensity Conflict
9-15	Explosive Ordnance Disposal Service and Unit Operations
20-32	Mine/Countermine Operations
21-75	Combat Skills of the Soldier
3-30	Grenades and Pyrotechnic Signals
23-34	TOW Weapon System
27-10	Law of Land Warfare
33-1	Psychological Operations
34-130	Intelligence Preparation of the Battlefield
41-5	Joint Manual for Civil Affairs
41-10	Civil Affairs Operations
90-8	Counterguerrilla Operations
100-20	Military Operations In Low Intensity Conflict

Technical Manual (TM)

9-2350-264-10-2	Operator's Manual for Operation Under Usual and Unusual Conditions for Tank, Combat, Full-Tracked: 120mm Gun, M1A1 General Abrams, with Change 4

Marine Corps Concept Paper

A Concept for Future Military Operations on Urbanized Terrain

Marine Corps Doctrinal Publications (MCDPs)

1	Warfighting
1-3	Tactics

Marine Corps Reference Publications (MCRPs)

5-2A	Operational Terms and Graphics

Marine Corps Warfighting Publications (MCWPs)

3-15.3	Scout Sniping (under development)
3-37.1	Chemical Operations
3-37.4	NBC Reconnaissance

Fleet Marine Force Manuals (FMFMs)

0-9	Field Firing for the M16A2
0-25	The Law of Land Warfare
1-3	Tactics
2-6	MAGTF Rear Area Security
2-11	Antiarmor Operations
3-5	Employment of Military Police in Combat
3-30	Communications
6	Ground Combat Operations
6-7	Scouting and Patrolling for Infantry Units
6-8	Supporting Arms Observer, Spotter, and Controller
6-9	Marine Artillery Support
6-21	Tactical Fundamentals of Helicopterborne Operations
7-14	Combating Terrorism
7-34	MAGTF Civil Affairs
7-34-1	Civil Affairs Guidance for the Small-Unit Leader
7-40	Helicopter Insertion/Extraction
8-2	Counterinsurgency Operations
11-9	NBC Protection
11-10	NBC Decontamination
11-11	Treatment of Chemical Agent Casualties and Conventional Military Chemical Injuries
11-12	Chemical Hazard Assessment Guide
11-13	NBC Handbook
11-17	Chemical and Biological Contamination Avoidance
11-18	Nuclear Contamination Avoidance
11-20	NBC Reconnaissance
13-7	MAGTF Breaching Operations

Fleet Marine Force Reference Publications (FMFRPs)

0-6	Marine Troop Leader's Guide
0-14	Marine Corps Supplement to DOD Dictionary of Military and Associated Terms
4-26	Enemy Prisoners of War and Civilian Internees
7-14A	The Individual's Guide for Understanding and Surviving Terrorism
7-37	Vehicle Bomb Search
12-15	Small Wars Manual

12-18	Mao Tse-Tung on Guerrilla Warfare
12-25	The Guerrilla and How to Fight Him
13-5	Engineer Field Data

Joint Publications

1-02	DOD Dictionary of Military and Associated Terms
3-07	Joint Doctrine for Military Operations Other Than War
3-11	Joint Doctrine for Nuclear, Biological, and Chemical (NBC) Defense
3-15	Joint Doctrine for Barriers, Obstacles, and Mine Warfare
3-53	Doctrine for Joint Psychological Operations
4-06	JTTP for Mortuary Affairs in Joint Operations

Standard NATO Agreements (STANAGs)

2003	Patrol Reports by Army Forces
2014	Operations Orders
2029	Method of Describing Ground Locations, Areas, and Boundaries
2070	Emergency War Burial Procedures
2088	Battlefield Illumination
2101	Establishing Liaison
2868	Land Force Tactical Doctrine (ATP-35(B))

Miscellaneous

Fratricide: Reducing Self-Inflicted Losses, Newsletter No. 92-4, Center for Army Lessons Learned (CALL), U.S. Army Combined Arms Command, Fort Leavenworth, Kansas, April 1992.

HR, 25, Annex to Hague Convention No. IV, *Embodying Regulations Respecting the Laws and Customs of War on Land*, 18 October 1907.

Modern Experience in City Combat, U.S. Army Human Engineering Laboratory, Aberdeen Proving Ground, Maryland, 1987.

Made in United States
Orlando, FL
05 November 2026

90229825R10324